Advanced Membrane Materials for CO_2 Capture and Separation

Advanced Membrane Materials for CO_2 Capture and Separation

Guest Editors

Naiying Du
Haiqing Lin

MDPI

Basel • Beijing • Wuhan • Barcelona • Belgrade • Novi Sad • Cluj • Manchester

Guest Editors

Naiying Du
Clean Energy Innovation Research Center
National Research Council Canada
Ottawa
Canada

Haiqing Lin
Department of Chemical and Biological Engineering
University at Buffalo, State University of New York
Buffalo
USA

Editorial Office
MDPI AG
Grosspeteranlage 5
4052 Basel, Switzerland

This is a reprint of the Special Issue, published open access by the journal *Membranes* (ISSN 2077-0375), freely accessible at: https://www.mdpi.com/journal/membranes/special_issues/Y25HYIDGS9.

For citation purposes, cite each article independently as indicated on the article page online and as indicated below:

Lastname, A.A.; Lastname, B.B. Article Title. *Journal Name* **Year**, *Volume Number*, Page Range.

ISBN 978-3-7258-6808-7 (Hbk)
ISBN 978-3-7258-6809-4 (PDF)
https://doi.org/10.3390/books978-3-7258-6809-4

Contents

About the Editors

Naiying Du

Naiying Du is a Senior Research Officer at the National Research Council Canada (NRC), where she leads research in polymer chemistry and functional materials for energy, environmental, and sustainability applications. Her scientific work focuses on the rational design, synthesis, and characterization of advanced polymeric and hybrid materials, with a particular emphasis on understanding structure–property relationships, materials processing, and surface/interface engineering to achieve tailored transport, mechanical, and barrier properties. She has spearheaded projects on high-performance membranes and coatings for gas separation and carbon capture, polymer-based materials for electrochemical energy conversion and storage, and sustainable materials that support low-carbon and circular-economy technologies. Her research has enabled the development of novel polymer platforms with enhanced durability, selectivity, and efficiency, bridging fundamental polymer chemistry with applied energy and environmental solutions. Over the course of her career, she has authored numerous peer-reviewed publications, contributed to advanced materials patents, and mentored a generation of researchers in the field of functional and sustainable polymer materials, establishing herself as a leading scientist in the integration of chemistry, materials science, and sustainable technologies.

Haiqing Lin

Haiqing Lin is a Professor of Chemical and Biological Engineering and the Director of Graduate Studies at the University at Buffalo (SUNY). His research focuses on advanced membrane materials for gas separations, including CO_2 capture from flue gas and syngas, and water purification, with an emphasis on structure–property relationships, transport phenomena, and antifouling strategies. Lin's work spans the design, synthesis, characterization, and engineering of membranes that enable energy-efficient separations for clean energy and environmental applications. He has made significant contributions in developing high-performance membranes for CO_2/H_2 and CO_2/N_2 separations, advancing performance beyond commercial standards, and exploring membranes with enhanced selectivity and antifouling properties for water purification applications. Over his career, Lin has published nearly 160 peer-reviewed papers, authored multiple book chapters, co-invented 10 patents, and mentored numerous graduate and undergraduate researchers. His contributions to separation science have been recognized with awards including the NSF CAREER Award and the AIChE Institute Award for Excellence in Industrial Gases Technology. Lin received his Ph.D. in Chemical Engineering from the University of Texas, Austin, in 2005; worked in industry, where he developed commercial membrane technologies; and joined the University of Buffalo's faculty in 2013, where he continues to lead innovative research in membrane science and engineering.

membranes

Review

Membrane Separation Technology in Direct Air Capture

Pavlo Ignatusha [1,2], Haiqing Lin [3,*], Noe Kapuscinsky [1,4], Ludmila Scoles [1], Weiguo Ma [1], Bussaraporn Patarachao [1] and Naiying Du [1,*]

1 Energy, Mining and Environment Research Center, National Research Council of Canada, Ottawa, ON K1A 0R6, Canada; noe.kapuscinsky@nrc-cnrc.gc.ca (N.K.); ludmila.scoles@nrc-cnrc.gc.ca (L.S.)
2 Department of Biochemistry, Microbiology and Immunology, University of Ottawa, Ottawa, ON K1N 6N5, Canada
3 Department of Chemical and Biological Engineering, University at Buffalo, The State University of New York, Buffalo, NY 14260, USA
4 Department of Chemical and Biological Engineering, University of Ottawa, Ottawa, ON K1N 6N5, Canada
* Correspondence: haiqingl@buffalo.edu (H.L.); naiying.du@nrc-cnrc.gc.ca (N.D.)

Abstract: Direct air capture (DAC) is an emerging negative CO_2 emission technology that aims to introduce a feasible method for CO_2 capture from the atmosphere. Unlike carbon capture from point sources, which deals with flue gas at high CO_2 concentrations, carbon capture directly from the atmosphere has proved difficult due to the low CO_2 concentration in ambient air. Current DAC technologies mainly consider sorbent-based systems; however, membrane technology can be considered a promising DAC approach since it provides several advantages, e.g., lower energy and operational costs, less environmental footprint, and more potential for small-scale ubiquitous installations. Several recent advancements in validating the feasibility of highly permeable gas separation membrane fabrication and system design show that membrane-based direct air capture (m-DAC) could be a complementary approach to sorbent-based DAC, e.g., as part of a hybrid system design that incorporates other DAC technologies (e.g., solvent or sorbent-based DAC). In this article, the ongoing research and DAC application attempts via membrane separation have been reviewed. The reported membrane materials that could potentially be used for m-DAC are summarized. In addition, the future direction of m-DAC development is discussed, which could provide perspective and encourage new researchers' further work in the field of m-DAC.

Keywords: direct air capture; membrane; carbon dioxide; high permeance

1. Introduction

The terms climate change and climate action have been on the tip of everyone's tongues for the last few years. Although carbon capture is a promising start, with the carbon budget clock ticking away, it is clear that simply lowering emissions, e.g., capturing CO_2 from point sources, will not make enough of an impact. It is not long until the carbon budget is depleted, and every additional ton of CO_2 will need to be managed by the use of negative emission technologies [1,2]. In fact, in order to meet the Paris Climate Agreement goals of preventing a 1.5–2 °C temperature increase over preindustrial levels, 10 $GtCO_2/yr$ will need to be removed from the atmosphere by the midcentury, increasing to 20 $GtCO_2/yr$ by the end of the century [3]. Failing to prevent the increase in temperature has a very real social cost. For example, wheat, rice, maize, and soybean represent over 67% of human caloric intake; only a 1 °C increase in temperature will reduce the global production of wheat by 6%, rice by 3%, maize by 7.4%, and soybean by 3.1% [4]. Additionally, changes in weather patterns, acidification of oceans, and melting of polar ice caps place much of the planet's biodiversity at risk. Therefore, the development of cost-effective negative emission technologies becomes essential to remediate climate change. As one of the negative emission technologies, direct air capture (DAC) describes a process by which CO_2 is removed directly from the atmosphere rather than from higher concentration point sources. This proves to be quite challenging as the concentration of CO_2 in the atmosphere is only ~400 ppm [5]. Nonetheless, it is important to develop methods to capture low-concentration atmospheric CO_2 since capturing all emissions from point sources would fail to accomplish even an 80% emission reduction by 2050 [6], while direct air capture would be able to target the CO_2 from the billions of small point sources which account for 1/3 to 1/2 of society's CO_2 emissions [7].

Currently, sorbent-based DAC technologies are under development at a pilot scale, including solid sorbents and liquid solvents [8,9]. These two technologies rely on absorption/desorption technologies, which require high energy inputs and large location-dependent installations. In addition, these processes often involve the use of chemicals and, therefore, introduce added environmental and safety risks [5,9–12]. Liquid solvent DAC utilizes contactors where the gas encounters a basic solution. The resulting compounds need intense heating to release captured CO_2 [8,11,13]. Because of the use of strong bases, the sorbent liquids are usually no more than 30% concentrated, which greatly decreases their binding potential with CO_2 [14]. The use of strong chemicals, e.g., KOH, also risks a negative environmental impact in the form of spills. In addition, liquid solvents use 1–7 tons of water to capture 1 ton of CO_2 [9]. Recent research on liquid solvent-based DAC focuses on lowering the consumption of energy and water, e.g., IPDA (3-(aminomethyl)-3,5,5-trimethylcyclohexylamine) liquid to solid carbamic acid conversion for CO_2 which can capture low (400 ppm) CO_2 with >99% removal at a lower desorption temperature of 333 K [15]. Solid sorbent methods involve pushing air into a specially designed sorbent until it is saturated. The sorbent is then heated and/or vacuumed to desorb CO_2 [14]. Solid sorbents have been made of several different materials like metal-organic frameworks (MOFs), mixed metal oxides, poly(ethylenimine) etc. [16]. Metal-organic frameworks are formed through the linkage of organic and inorganic constituents, which form highly structured and microporous materials with high free volumes. The performance of solid sorbent DAC depends on the conditions of the air being processed, including temperature and humidity. High temperatures have been shown to increase energy requirements, leading to a loss of efficiency and an increase in cost [17]. These factors lead to location dependence for DAC installation for both technologies. The source of energy used to power these sorbent-based DAC plants must also be considered when evaluating their level of negative emissions and may limit the location even more [9]. The KOH liquid absorption method reportedly requires 8.81 GJ of natural gas or 5.25 GJ of gas and 366 kWh of electricity for every ton of CO_2 captured [11]. Solid sorbent energy requirements are around 6 GJ of thermal energy and 1.5 GJ of electricity per ton of CO_2 [14]. These energy demands arise primarily due to the desorption steps. Although sorbent-based DAC technologies being developed to a plant scale is a great start and is paving a path for DAC, the impact on the environment should not be underestimated, and other environmentally friendly approaches should be explored.

The idea of m-DAC was proposed twenty years ago [18]; however, only recently has a more detailed study been reported that proved that membrane processes could be considered as a new DAC approach [5]. It is a new and rather exciting area of research that shows promise for lower-cost direct air carbon capture and can lower the risk of environmental impact associated with sorption technologies [5]. Theoretically, considering the process only requires energy to blow air through the membranes, advances in membrane materials should drastically decrease the cost of operation, especially given that thermodynamic energy requirements are 20–30 times lower than that of the best DAC methods currently in use [19,20]. However, presently, m-DAC will not be a competitive option to sorbents-based DAC unless major breakthroughs are made in increasing membrane selectivity for the CO_2:N_2 gas pair and CO_2 permeance. Currently, it is widely accepted that m-DAC could play an active role in hybrid system designs that incorporate other DAC technologies. In the past few decades, large amounts of funding have been allocated worldwide for the R&D of membrane-based CO_2 capture from point sources, and a vast number of published reports related to high permeable polymeric materials for CO_2 capture membranes could shed light on the selection of DAC membrane materials. Rather than introducing the m-DAC hybrid system, this review gives a perspective of the most recent research on membrane-based direct air capture systems and potential polymeric membrane materials for DAC. Although there is currently very little literature on m-DAC, given its novelty, this review intends to illustrate the need for membrane DAC research and encourage researchers in the field to explore further.

2. Feasibility of m-DAC

Although membrane separation is a promising technology for capturing CO_2 and, to date, a few pilot plants for point source CO_2 capture have been operated; the pervasive belief is that m-DAC was implausible because the concentration of CO_2 in the air is only about 400 ppm which leads to an insufficient driving force for CO_2 permeation through the membranes [21].

Recent achievements in polymer membrane materials with ultrahigh gas permeance and selectivity have exposed the silver lining that membrane separation could potentially be applied in DAC processes. Fujikawa et al. explored the potential of m-DAC based on process simulation with consideration of the state-of-the-art CO_2 separation membrane performance. The results of the four-stage separation could encourage researchers to explore more realistic membrane performances and process conditions for m-DAC. In their incredibly comprehensive modeling study, they modeled a 4-stage separation process after highly permeable membranes (CO_2 Permeance of 40,000 GPU (Gas Permeation Unit (GPU) = 3.35×10^{-10} $\frac{mol}{m^2 \cdot s \cdot Pa}$) and CO_2/N_2 selectivity of 70) and showed that a CO_2 concentration of ~30% was achievable at the final stage of a four-stage system with only a membrane area of ~3.2 m^2/kg CO_2/day (as a flat sheet) and 16 kWh/kg-CO_2/day. Their model system assumed a feed pressure of 101.3 kPa and a permeate pressure of 5 kPa and maintained a pre-industrial retentate CO_2 concentration of ~300 ppm [12].

3. Considerations for m-DAC

Direct air capture by membranes depends on several factors, which include membrane properties as well as process parameters.

3.1. Permeance

Gas permeance is a measure of the gas transport ability of a material at a given thickness; therefore, it depends on the permeability and thickness of the membrane. Separation by membranes typically follows the Solubility-Diffusion model, which is based on the solubility and diffusion rate of gases going through the membrane [22,23]. Some membranes are capable of facilitated transport mechanisms where carriers react reversibly with CO_2 selectively, while other gasses only cross by the Solubility-Diffusion mechanism at a lower rate. As a result, these membranes offer high selectivity and gas permeance [24].

$$P = \frac{K}{l} \tag{1}$$

$$K = S \times D \tag{2}$$

Membranes with a high Solubility-Diffusion relationship and low thickness are candidates for increased permeance based on Equations (1) and (2) where P is gas permeance, K is permeability (Barrer = 3.35×10^{-10} $\frac{mol \cdot m}{m^2 \cdot s \cdot Pa}$), l is membrane thickness, S is solubility and D is diffusion rate. For power plant post-combustion CO_2 capture, Merkel et al. concluded that improved gas permeance in membranes is more critical for reducing capture cost than enhanced selectivity [20]. While selectivity is also extremely crucial for DAC, only membranes with high permeance can be considered as an attractive option for DAC. Thinning of membranes is a large area of research for increasing permeance. The thickness of DAC membranes should be reduced to the nanometer level in order to increase membrane permeance significantly because the required membrane area is inversely proportional to permeance. However, it should be noted that studies have shown permeability has a tendency to decrease for many materials in the submicron range, which can lead to a decrease in permeance [25,26].

3.2. Selectivity

Membrane gas selectivity (CO_2/N_2) is of great interest for m-DAC design, given that nitrogen is the largest component of the earth's atmosphere and the CO_2 concentration is only about 400 ppm. Ideal gas selectivity is defined as the ratio of CO_2 to N_2 permeability, as shown below.

$$\alpha_{CO_2/N_2} = \frac{K_{CO_2}}{K_{N_2}} \tag{3}$$

where α is the selectivity, K_{CO_2} is CO_2 permeability and K_{N_2} is N_2 permeability in single gas tests [27]. The determination of selectivity becomes more complicated when it comes to mixed gas testing. Low selectivity leads to a higher migration of unwanted gases and a lower concentration of CO_2 in the permeate. As a rule of thumb, permeance should be maximized to decrease membrane area [12], but selectivity should be optimized as membranes with a selectivity that is too high require more membrane area and show little benefit in CO_2 purity. Merkel et al. compared two membranes for point source CO_2 capture, the "best case" membrane and membrane B. The "best case" membrane had a CO_2 permeance of 1000 GPU and a CO_2/N_2 selectivity of 50, while membrane B had a CO_2 permeance of 1000 GPU and a CO_2/N_2 selectivity of 200. They found that the best-case membrane produced a permeate with 46% CO_2 at 2.1 MM m^2, while membrane B yielded 55% CO_2 with an area nearly three times larger at 5.7 MM m^2 [20]. The concentration of CO_2 in the permeate increases greatly with selectivity until a selectivity of about 30 when it begins to plateau. The selectivity also decreases the energy required for vacuuming, which similarly plateaus. Fujikawa et al. showed that a membrane with a CO_2 permeance of 10,000 GPU and a CO_2/N_2 selectivity of 50 could reach 66.7% CO_2 with only 4.63 m^2/kg CO_2/day and ~12.3–16 kWh/kg-CO_2/day at a pressure ratio of 50 using a feed pressure of 110 kPa and a permeate pressure of 2 kPa. The membrane area required increases with selectivity almost linearly while the benefits in CO_2 concentration, energy requirement, and CO_2 emission related to the energy production decrease with each 10-step selectivity increase [12].

Although less important than the previous two parameters, CO_2/O_2 selectivity should also be considered depending on downstream applications of captured CO_2 as processes like the reduction of CO_2 to CO and CH_4 are typically hindered due to the oxidative pressures of O_2 [12].

3.3. Pressure Ratio

The pressure ratio is feed pressure divided by permeate pressure.

$$\varphi = \frac{p_f}{p_p} \tag{4}$$

$$X_p \leq X_f \times \varphi \tag{5}$$

where p_f and p_p are feed and permeate pressures respectively. Equation (5), where X_p is the mol fraction in the permeate and X_f is the mol fraction in the feed, shows that the mole fraction of CO_2 on the permeate side is limited by the pressure ratio independently of other parameters, notably selectivity [12,23]. For flux across the membrane in the desired direction, the partial permeate CO_2 pressure cannot exceed the partial feed CO_2 pressure in order to maintain a favorable cross-membrane pressure gradient. Because the mole fraction in the permeate is limited proportionally by the pressure ratio, a higher pressure ratio leads to greater permeate CO_2 concentrations. The largest effect of pressure ratio on CO_2 permeate concentration is observed under $\varphi = 30$, after which the effect begins to plateau [12]. Vacuuming processes have been shown to be more energy efficient than feed compression systems for gas separation [23].

Fujikawa et al. studied the effect of pressure ratio on m-DAC via process simulation with the same CO_2 retentate concentration (~300 ppm) and the same membrane (permeance of 40,000 GPU and selectivity of 70) (Table 1); however, the pressure ratio is different. When the pressure at the permeate side is 4 kPa (φ = 25), the final CO_2 concentration can exceed 40%. Membrane area and CO_2 emissions related to the energy production decrease from 3.19 to 2.6 $m^2/kg\text{-}CO_2/day$ and 0.6 to 0.54 $kgCO_2^{emitted}/kgCO_2^{captured}$, respectively, when the pressure ratio is raised from 20 to 25. It is important to note that at each stage, the energy and membrane area required falls dramatically when compared to the first stage. This illustrates the importance of process parameters in addition to material properties like permeance and selectivity.

Table 1. The m-DAC separation outcomes with different pressure ratios [12]. Data for the Pressure ratio of 20 is summated while the pressure ratio of 25 is not (each stage).

	Pressure Ratio = 20 Permeate Pressure = 5 kPa					Pressure Ratio = 25 Permeate Pressure = 4 kPa				
Number of Separation Stages	1	2	3	4	Total	1	2	3	4	Total
CO_2 concentration in permeate (%)	0.6	2.9	10.8	29.8	-	0.7	3.9	15.5	42.4	-
Membrane area ($m^2/kg\text{-}CO_2/day$)	2.57	(0.47)	(0.12)	(0.03)	3.19	2.15	0.35	0.08	0.02	2.6
Energy required for vacuuming ($kWh\text{-}CO_2/day$)	12.7	(2.4)	(0.7)	(0.2)	16.0	11.6	1.9	0.5	0.2	14.2
CO_2 emission related to the energy production ($kgCO_2^{emitted}/kgCO_2^{captured}$)	0.48	(0.09)	(0.02)	(0.01)	0.6	0.44	0.07	0.02	0.01	0.54

3.4. Stage Cut

Stage cut is defined as permeate flow divided by feed flow.

$$\varnothing = \frac{f_p}{f_f} \tag{6}$$

where $\varnothing$ is Stage cut and f_p and f_f are permeate and feed flows, respectively [12,28]. A high feed flow leads to a lower change in feed gas concentrations, which makes the difference in driving forces for unwanted gases more negligible when compared to CO_2. As a result, the purity of the permeate is higher, but the recovery % is lower as less CO_2 crosses at a low-stage cut. Decreasing the flow rate of the feed increases the amount of CO_2 that crosses but, in doing so, increases the driving force for other gases. Therefore, a higher stage cut leads to a lower purity but a higher recovery % of CO_2. Membranes with higher areas lead to greater stage cuts, lower purity, and a higher % CO_2 recovery [12]. All of these parameters should be fined-tuned for m-DAC systems depending on the final goal, whether that be total recovery or a high purity for downstream applications.

In addition, these parameters may vary at each stage of a multi-step separation system. Energy requirements for installation, operation as well as manufacture of membranes must also be considered. Due to the infancy of m-DAC technology, there are very few parametric and gas module separation studies available. With that being said, several models that take these parameters into account have shown the scope of the possibility of m-DAC. High selectivity membrane material (HPM)with a CO_2 permeance of 2500 GPU and CO_2/N_2 selectivity of 680 has been shown to be capable of a capture purity of nearly 20% in a single stage at a pressure ratio around 0.02, and commercial Polaris membranes (CO_2 permeance of 2000 GPU, CO_2/N_2 selectivity of 30) can reach 50% with two stages [29]. HPM showed an exponential increase in CO_2 capture purity when decreasing the pressure ratio from 0.02, while no significant effect was observed for the Polaris membrane due to a negligible effect from concentration polarization [29]. For m-DAC, HPM and Polaris membranes required 3000 and 18,000 kWh/ton at one stage, respectively, to achieve 20 and 2.5% purity [29]. The study outlined the importance of a low-stage cut for improving the recovery ratio [29].

Although m-DAC CO_2 concentrations are not commonly tested, Lee et al. showed their ionic liquid + graphene oxide PIL-IL-GO facilitated transport membrane could produce a purity of 32% at 410 ppm CO_2 using only a single stage at 1 bar (1 bar = 100 kPa) of feed pressure [30]. Under these conditions, the membrane displayed a permeance of 3090 GPU and a CO_2/N_2 selectivity of 1180 [30]. Unfortunately, the stage cut and pressure ratio were not reported. Proposed multistage membrane separation models have shown improved CO_2 separation with each stage and notably outline that the first stage has the highest energy, area, and emission values, after which each sequential stage shows an exponential decay [12].

Although very few papers have reported membrane separation under m-DAC conditions, the former examples indicate real promise and should encourage more widespread testing of membranes with atmospheric CO_2 capture in mind.

4. Potential Membrane Materials

According to the concern above, the thickness of polymeric selective layers for DAC needs to be less than 1 μm. Hence, a typical membrane for DAC should consist of a thin selective layer deposited on an intermediate gutter layer on a porous substrate support, as shown in Figure 1. The gutter layer serves to protect from the penetration of the selective layer material into the porous support [25]. Highly permeable polydimethylsiloxane (PDMS) is frequently used as the gutter layer, and polyacrylonitrile (PAN) is used as the substrate [25,31–33]. Although each layer of material introduces transport resistance, the PDMS and PAN are so porous that their effect is negligible. Selective layers are manufactured by several different methods, including spin coating, dip coating, roll coating, blading, solution casting, dry-jet wet spinning (hollow fiber), and spray coating, with spin coating being the most common method for ultrathin membranes [32,34–37].

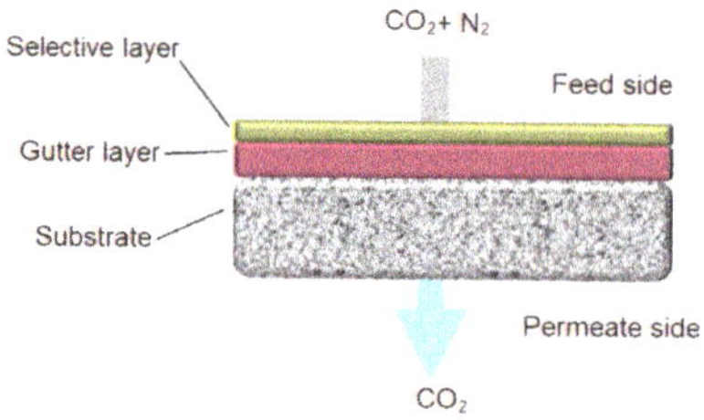

Figure 1. Sample diagram of typical gas separation membrane layers.

The selective layer is the most vital element of composite membranes and is modified with permeance and selectivity in mind. Fujikawa et al. recommend membranes with a CO_2 permeance of >10,000 GPU and CO_2/N_2 selectivity of >30 for m-DAC. However, most current membranes fail to reach these specifications [12]. In the last decade, numerous promising polymeric membranes have stood out. Materials such as polymers of intrinsic microporosity (PIMs), polymers with ethylene oxide/ethylene glycol groups, mixed matrix membranes (MMMs), and facilitated transport membranes might meet the standards set by Fujikawa if their thickness falls below 1 μm as the permeance could be increased. Inorganic membranes have been explored in the past but often suffer drawbacks due to price and scalability [26].

In the last two decades, researchers have tried different approaches to increase permeability and selectivity, as well as to reduce the aging of membrane materials, for instance, introducing more rigid units into the polymer structure, post-modifying polymer structure with different functional groups, and adding functionalized nanofillers to membranes creating MMMs [38]. In this paper, we summarize the potential polymeric membrane materials for DAC. These materials with high permeability/permeance and/or high CO_2/N_2 selectivity near or above the 2008 Robeson upper bound are mainly applied to CO_2 capture from point sources (Table 2). However, they could still be good starting candidates for further m-DAC development [39].

Table 2. Selection of recent promising membrane materials showing CO_2 permeabilities $\geq$ 1000 Barrer or with CO_2/N_2 Selectivity's $\geq$ 30 at temperatures $\leq$ 35 °C. Inorganic membranes have not been included. Some additions are not mutually exclusive and could belong to several categories. Data is from pure gas permeation measurements. A more complete summary can be found in Table S1.

Polymeric Membranes				T (°C)	Pressure (kPa)	CO_2 Permeability (Barrer)	Gas Selectivity (CO_2/N_2)	Ref.
Type		Polymers						
Polymer membranes	Copolymers	KAUST-PI-1		35	200	2389	33	[40]
		PIM-BTrip (160 μm)	Aged 490 days	25	100	6060	31.0	[41]
			Aged 120 days	25	100	6040	30.2	[41]
		SFX-PIM-33 (Aged 130 days)	(Aged 130 days)	25	200	1848	30.8	[42]
		BPM-50		35	350	4883	43	[43]
		VAP7		30	100	1370	32	[44]
		PTCNSi(OMe)$_3$		20–22	100	2000	35.7	[45]
		Polaris™ gen1		-	-	1000 (Commercially Available)	50	[46]
		PolyActive™/85		-	-	1480	55	[47]
	Copolymers with post modification	TZ-PIM-1		25	440	~3000	~30	[48]
		AO-PIM-1 + Methanol		35	200	1153	35	[49]
		MTZ100-PIM *		25	350	1391	22.2	[50]
		Thioamide-PIM-1 + Ethanol		25	100	1120	30.3	[51]
		cPIM-1		25	200	3739 ± 32	34.9	[52]
Mixed matrix membranes		6FDA-durene/Si-5		25	200	3785	31	[53]
		PIM-MFI3		25	100	2530	30	[54]
		Pebax-2533/ZIF	35 wt%	25	200	1287	32.3	[55]
		SPEEK/MIL-101 (Cr) 40 wt% *		30	100	30	40	[56]
		SPEEK/S-MIL-101 (Cr) 40 wt% *		30	100	35	41	[56]
		UiO-66-CN@sPIM-1 *		25	140	16,121.3	27	[57]
		PDMS-SAPO-34 (PM-30 wt%)		25	2000	5753	31	[58]
		PIM-1/GO		30	400	6169	123	[59]
		CNT-ZIF-8-PDMS		25	100	8705	45.6	[60]
		PAO-PIM-1/NH2-UiO-66	7 wt%	35	100	3825	30.0	[61]
		PEO/HPNs	0.5 wt%	35	100	~1400	~41	[62]
			1 wt%	35	100	~1900	~44	[62]
Facilitated transport membranes		Pebax [C4MIM][Gly] 20 wt%		25	100	~1100	~110	[63]
		C(30)-P(1:1)		25	200	~1650	~55	[64]
		Pebax-PEI-MCM-41-20		25	100	1521	102	[65]
		15 wt% ([Cu(6)]2+@13X)/6FDA-Durene		35	200	~1034	38.3	[66]
		Pebax 1657/MWNTs-NH2/GTA (P10CN1G25)		35	700	1408	~40	[67]
		Pebax 1657/SG 20 wt%		25	200	~1200	~55	[68]
		CA/PM-4 (1:3 wt%)		35	300	3000	59	[69]
		PIM-Py-Cl 15 wt%		25	200	4959.8	42	[70]
		PIM-Py-Ac 15 wt%		25	200	6204.8	62	[70]
		PIM-Py-BF4 15 wt%		25	200	5584.3	46	[70]

The membranes marked with a * do not meet the table's criteria through pure gas testing but do meet the criteria under mixed gas conditions or humidified conditions which are covered later.

4.1. Copolymers

Various copolymers with high CO_2 permeability and CO_2/N_2 selectivity have been explored. Their structural designs focus on adding rigid structures to reduce interchain packing and create a microporous structure, in addition to frequently introducing ethylene oxide and ethylene glycol groups to enhance CO_2 affinity.

Aging is the major drawback of these microporous polymers [38,71]. Aging refers to the thermodynamic stabilization of the membrane polymer packing, which leads to a collapse of pores and a loss of free volume, resulting in a decrease in permeability. This phenomenon is observed after short periods of time and at accelerated rates with thinner membranes [25,34,72,73]. Membranes also suffer from plasticization, where CO_2 sorption at high pressures increases polymer chain movement. This leads to an increased permeability for all gasses due to an increase in free volume and, consequently, a decrease in selectivity [74].

Introducing rigid and bulky units into the polymer chain can efficiently increase CO_2 permeabilities with or without aging while still maintaining moderate selectivity due to their special chain rigidity. PIMs are a good example of polymers that successfully increase permeability by introducing rigid structures into the polymer chain. PIMs are usually composed of two components: (i) a structural unit that possesses concavities and introduces a site of contortion into the polymer chain and (ii) a linking group (e.g., derived from dibenzodioxin or imide formation) that fuses the structural units together during polymerization and leads to inefficient packing and high free volume [25,73,75]. Some of these polymers offer high permeability as well as cheap facile manufacture. The main building blocks of PIMs are (i) spirobisindane (SBI); (ii) phenazine; (iii) ethanoanthracene (EA); (iv) triptycene (Tridp); (v) benzotriptycene (BTrip); (vi) spirobifluorene (SBF); (vii) Tröger's base (TB); and (viii) tetraphenylethylene etc. [76]. Polyimides of intrinsic microporosity (PIM-PIs) can also be prepared when structural PIM motifs containing rigid and bulky contortion sites are introduced into polyimide backbones. The structural design can be manipulated in many different ways depending on the location of the contortion site, for example, either in dianhydride or diamine monomers for PIMs-PIs. The method in which PIMs are synthesized has been shown to have an effect on chain packing, membrane structure, performance, and aging.

PIMs containing Trip and BTrip (Figure 2a) units exhibit excellent CO_2 permeabilities (up to ~22,000 Barrer) [41]. Their highly rigid structure leads to observed ultra-microporosity, which facilitates the transport of smaller gas molecules whilst increasing the activation energy for larger molecules like N_2. Experimentally, PIM-BTrip has been shown to have a high diffusivity, leading to greater permeability whilst maintaining selectivity due to the selective molecular sieving based on kinetic diameter [41]. KAUST-PI-1 is another example of a co-polymer that boasts high permeability due to its rigid backbone structure, which effectively leads to the formation of ultra-porous structures (Figure 2b). It was also observed that the choice of bridgehead group had a significant effect on the permeability of KAUST-PI-1. The bulky isopropyl bridgeheads and methyl-substituted diamines increased intrachain rigidity and improved CO_2 permeability 4-fold [40]. Co-polymer designs with bulky groups in mind, like SFX-PIMs (Figure 2c), have shown CO_2 permeabilities comparable to the PIM-1 standard but with improved CO_2/N_2 selectivity around 30 [42]. PIM-bpy-x gas separation performance was enhanced by incorporating bulky structures into the polymer through the polycondensation of a tetraphenyl bipyrimidine monomer (Figure 2d). PIM-bpy-x exhibited excellent gas separation performance (4234 barrer)—a 21% improvement in CO_2 permeability. The increase in CO_2 permeability is due to the affinity of the N-rich bipyrimidine units for CO_2 [77].

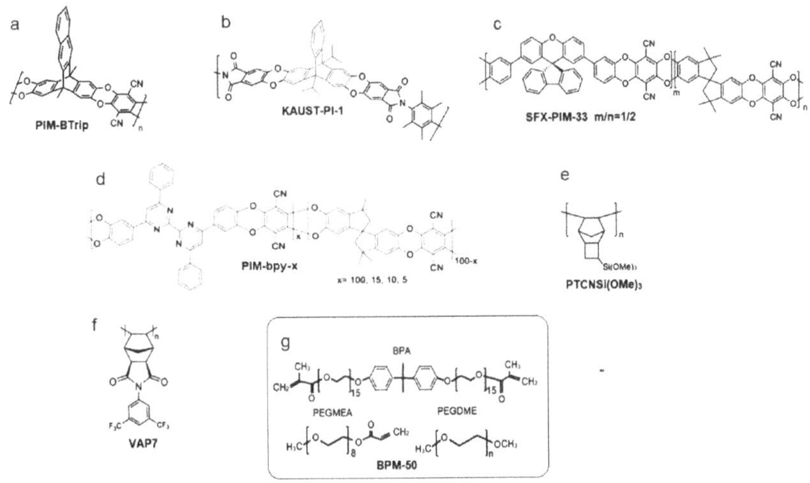

Figure 2. Chemical structures of high permeable copolymers, with CO_2 permeabilities ≥ 1000 Barrer or with CO_2/N_2 Selectivity's ≥ 30 at temperatures ≤ 35 °C. (**a**) PIM-BTrip [41], (**b**) KAUST-PI-1 [40], (**c**) SFX-PIM-33 (m/n = 1/2) [42], (**d**) PIM-bpy-x [77], (**e**) PTCNSi(OMe)3 [45], (**f**) VAP7 [44], (**g**) BPM-50 [43].

High permeable polymers synthesized from different modified monomers have also shown the improvement of gas separation properties. Grazia Bezzu et al. conducted a combined simulation and experimental study to investigate the effect on polymer microporosity and gas permeability of PIM-SBFs by placing simple substituents such as methyl, t-butyl, and fused benzo groups onto spirobifluorene monomers. It was shown that methyl or t-butyl substituents both cause a large increase in gas permeabilities, with four methyl groups enhancing the concentration of ultramicropores (<0.7 nm), which contribute to selective gas transport. The t-butyl substituents lower selectivity by generating a greater concentration of larger, less selective micropores (>1.0 nm) due to their size [78]. Alentiev et al. wonderfully demonstrated the variability that can occur with subtle modifications of side groups on the monomers (Figure 2e). Following this strategy of monomer substituent modification, new high molecular weight metathesis and addition polynorbornenes with $(AlkO)_3$Si-groups of different lengths (Alk = Me, Et, n-Pr, n-Bu) were synthesized. These polymers, with similar scaffolding, consist of two parts—glassy (rigid polymer main chains) and rubbery (flexible side chains), which allowed dramatic tuning of polymer properties by the modification of polymer main chain structures and the length of trialkoxysilyl side groups. PTCNSi(OMe)$_3$, for example, showed a selectivity of 35.7 while increasing the length of substituents with PTCNSI(OEt)$_3$ led to a selectivity drop down to 21.3 [45]. The drop in selectivity was likely due to the longer chains creating larger pores. Nazarov et al. performed a similar study exploring several modified monomers that formed vinyl-addition polymers. VAP7 was the most promising for DAC application (Figure 2f). VAP7′s performance can be attributed to the trifluoromethyl groups, which hinder inter-chain interactions, leading to an increase in free volume due to a decrease in favorable chain packing [44].

Ethylene oxide and ethylene glycol groups have been exploited in membrane design due to their high affinity for CO_2 by Lewis base interactions, which often lead to a higher selectivity for CO_2 [79,80]. Membranes are not made solely of polyethylene glycol due to the high crystallinity induced by strong hydrogen bonding, which is not attractive for gas permeability. Both Polaris and Polyactive (shown in Table 2) are commercial copolymers containing PEO groups [80,81]. Both of these PEO-containing copolymers show CO_2/N_2 selectivity over 30 and respectable CO_2 permeance [46,47]. Pebax 1657 is also a promising PEO-containing commercial polymer that is frequently used in MMM applications. BPM-50 (Figure 2g), an oxygen-containing copolymer that was formed through the UV polymerization of PEGMEA, BPA, and PEGDME, boasts an impressive CO_2 permeability of 4883 Barrer and CO_2/N_2 selectivity of 43. In addition, the method of preparation for BPM-50 is clean, rapid, and solvent-free [43].

Another effective approach to significantly improving the permeability and selectivity is post-modifying copolymers. The functionalization of promising copolymers allows for further enhancements to gas separation performance (Table 2). Taking TZPIM as an example (Figure 3a), the tetrazole structures from the modification of nitrile groups in PIM-1 raise the affinity for CO_2, hence increasing the single gas selectivity of CO_2/N_2 to 30. These tetrazole groups also serve to increase rigidity through hydrogen bonding, which in turn increases plasticization resistance from CO_2 [48]. Further modifying TZPIM to MTZ100-PIM (Figure 3b) increases solvent processability over TZPIM-1, which is of critical importance for large-scale membrane fabrication [50]. MTZ100-PIM showed a significant change in performance between pure and mixed gas permeation testing. Under pure gas testing, MTZ100-PIM showed a CO_2 permeability of 1391 barrer and a CO_2/N_2 selectivity of 22.2. Conversely, mixed gas testing with a 2:8 CO_2:N_2 per volume mixture yielded more impressive results with a CO_2 permeability 2057 barrer and a CO_2/N_2 selectivity of 41.6. This behavior is attributed to a suppression of N_2 permeability as condensable CO_2 occupies sorption sites far more than the non-condensable N_2, reducing the rate of passage for N_2. AO-PIM-1 is an excellent example of the common observations that come with functionalization (Figure 3c). The oxygen and nitrogen groups lead to more interchain interactions, which causes a net decrease in microporosity and the formation of more narrow pores. This, in turn, results in a decrease in CO_2 permeability and an increase in CO_2/N_2 selectivity to 35 due to the electronegative groups interacting favorably with CO_2 [49]. Thioamide-PIM (Figure 3d) continues this trend with lower CO_2 permeability but higher CO_2/N_2 selectivity (30.3) when compared to PIM-1 [51]. Yu et al. recently tested di-substituted PIM-1 (D-cPIM-1) and branched PIM-1 (B-cPIM-1) containing carboxylic acid groups (Figure 3e). They found that the 70% hydrolyzed di-substituted PIM-1 showed a CO_2 permeance of 7700 GPU and CO_2/N_2 selectivity of 56. More importantly, although the branched hydrolyzed PIM showed a lower CO_2 permeance of 3200 GPU, it showed no decrease in permeance after 60 days, indicating the branched structure may be vital for aging resistance [71]. These hydrolyzed PIMs have also been investigated for increasing CO_2 permeability due to quadrupole interactions, as further illustrated by cPIM-1 [52].

PIM and several polymers with similar structures are often treated with ethanol or methanol to remove residual solvent from casting and increase the diffusion coefficient, thus increasing permeability [82,83]. UV treatment has also been shown to increase performance due to chain scission and oxidation of the polymers [84].

Although the CO_2/N_2 gas separation performance of some polymers synthesized via thermal or UV treatment meets our summary criteria showing increased separation properties, e.g., carbon molecular sieves, TOX-PIM-1 [85], PIM-300 [86], and PIM-1-UV/Ozone [84], the high fabrication temperature, brittleness of thin sheets and poor solubility after treatment could limit their feasibility in m-DAC.

4.2. MMMs

The introduction of filler materials into polymeric membranes integrates their advantageous properties and provides a feasible approach to fabricating membranes for DAC [87–89]. These fillers offer a wide range of functionality due to their varying structures and often have low cost. Most added fillers help with molecular sieving due to their innate structure or, based on their dispersion/interaction with the polymeric material, can help create free volume and improve the gas-membrane affinity to provide effective solutions for overcoming the performance trade-off effect [57]. These structures can range from 0D to 3D with varying complexities and molecular interactions with gas molecules [90].

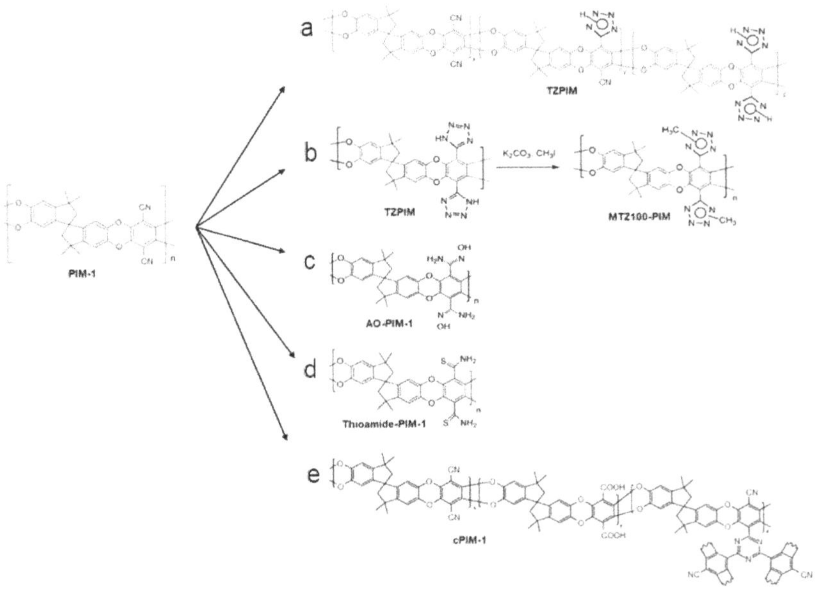

Figure 3. Chemical structures of copolymers with post modification (CO_2 permeabilities $\geq$ 1000 Barrer or with CO_2/N_2 selectivity's $\geq$ 30 at temperatures $\leq$ 35 °C). (**a**) TZPIM [48], (**b**) MTZ100-PIM [50], (**c**) AO-PIM-1 [49], (**d**) Thioamide-PIM-1 [51], (**e**) cPIM-1 [71].

Amongst the crowd of MMM fillers, MOFs stand out as they consistently have positive effects on membrane performance depending on their loading [57,91–93]. For example, UiO-66-CN covalently linked with s-PIM boasted a CO_2 permeability of 16,121 Barrer and CO_2/N_2 single gas selectivity of 27.0. In addition, when tested under 1:1 CO_2:N_2 by volume, the selectivity increased to 53.5 (Figure 4a). This is explained by the high porosity and rich sorption sites in the 3-D MOF domain, where CO_2 is competitively adsorbed over N_2 [57]. UiO66-NH_2 dopped with Ag^+ leads to great increases in gas separation characteristics when tested under 1:9 CO_2:N_2 by volume as Ag^+ raises the hydrophilicity and facilitates CO_2 transport while also leading to an increase in free volume, which characteristically increases permeability to >15,000 Barrer with selectivity 30 [94]. Wang et al. reported a hybrid membrane containing UiO66-NH_2 in PAO-PIM (Figure 4b). The amidoxime and amine groups tend to form hydrogen bonds, creating a hydrogen bond network between the two phases, which raised the CO_2 permeability from 2902 to 3825 Barrer due to an increase in the diffusivity coefficient while still maintaining a respectable selectivity [61].

3D MOF fillers are not the only inorganic fillers being explored for mixed matrix membranes. The addition of silica nanoparticles is an efficient strategy for boosting the initial gas permeability and suppressing physical aging. By incorporating 0.05 wt% sulfonic acid-functionalized silica nanosheets (S-SN) into the PIM-1 polymer, the resulting membrane showed a 40% higher CO_2 permeability and 22% enhancement in CO_2/N_2 selectivity. 150-day-aged freestanding PIM/SN1 and 28-days-aged TFN PIM/S-SN0.05 showed 70% and 5 times greater CO_2 permeability than the pure thick PIM-1 and TFC PIM-1 membranes, respectively [35] (Figure 4c). Similarly, Nafisi et al. studied the effect of silica nanoparticle loading in both Pebax and 6FDA-durene matrices. They found that the integration of silica nanoparticles raised the permeability in both cases, more dramatically with 6FDA durene (Figure 4d). They observed an increase in Fractional Free Volume (FFV) with filler loading; FFV increases as the density of the membrane decreases. The more drastic increase in permeability observed with 6FDA durene is hypothesized to be due to its glassy nature; therefore, the presence of nanoparticles has a greater effect on the FFV [53].

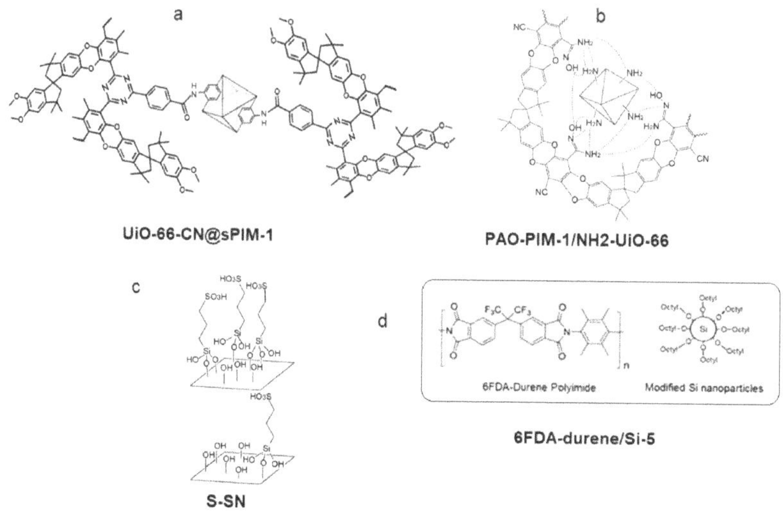

Figure 4. Examples of high permeable MMMs with MOF and Silica nanoparticles (CO_2 permeabilities $\geq$ 1000 Barrer or with CO_2/N_2 Selectivity's $\geq$ 30 at temperatures $\leq$ 35 °C). (**a**) UiO-66-CN@sPIM-1 [57], (**b**) PAO-PIM-1/NH_2-UiO-66 [61], (**c**) S-SN [95], (**d**) 6FDA-durene/Si-5 [53].

Other inorganic fillers, such as synthetic zeolite—SAPO-34, can also largely increase the CO_2 permeability of MMMs [95]. Similarly, SAPO-34 raises the free volume of polymer membranes, leading to an increase in permeability and raising selectivity when selectively sieving molecules based on kinetic diameter [58]. It was reported that a PDMS-SAPO-34 (PM-30 wt%) membrane exhibited a CO_2 permeability of 5753 Barrer with an ideal CO_2/N_2 selectivity of about 31 at 2000 kPa (20 bar) and 25 °C, which surpassed the 2008 Robeson upper bound for CO_2/N_2 separation.

Also, a large number of novel inorganic fillers could not only help tune the free volume of the polymer matrix but also create the coefficient path. Carbon nanotubes and graphene oxide (GO) nanosheets are examples of such fillers. Efficient CO_2 transport pathways were constructed by the homogeneous dispersion of carbon nanotubes (CNTs) and GO within MMMs, leading to both enhanced permeability and selectivity. Li et al. showed that mixing both carbon nanotubes and GO fillers into Matrimid membranes displayed a synergistic relationship [96]. The extraordinarily smooth walls of CNTs acted as a highway for high permeability, whereas the graphene oxide nanosheets acted as a selective barrier to create a more tortuous diffusion pathway for N_2, meanwhile providing a more selective path for CO_2 through interactions with the hydroxyl and carboxyl groups on the GO surface (Figure 5). The membrane combining 5 wt% of CNTs and 5 wt% of GO (Matrimid®-CNTs/GO-5/5) displayed optimum performance with a CO_2 permeability of 38.07 Barrer, a CO_2/N_2 selectivity of 81.00. Compared to pure Matrimid® membranes, the CO_2 permeability and CO_2/N_2 selectivity of the Matrimid®-CNTs/GO-5/5 membranes were increased by 331% and 147%, respectively. Although the reported 38.07 Barrer is still much lower than the expected permeability of CO_2 for m-DAC, the method inspires future research on designing high-performance membranes for DAC. More effort needs to be put into choosing the proper polymer matrix. Furthermore, to achieve high performance, ideal polymer–filler interface morphology and homogenous filler dispersion have to be realized. These challenges can be addressed by appropriate screening or modification of fillers to balance the polymer–filler, and filler–filler interactions [97].

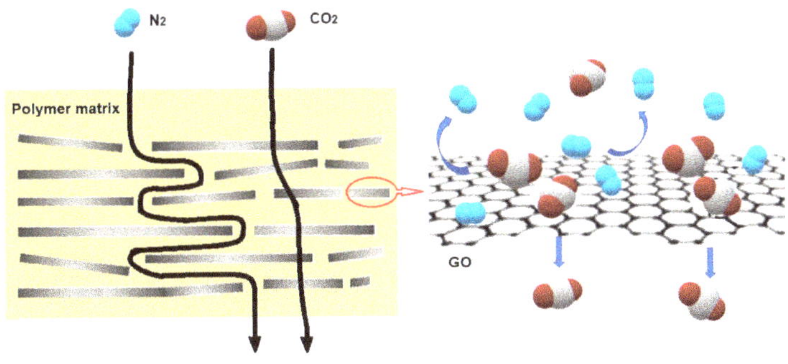

Figure 5. Schematic of the mechanism for gas travel through MMMs with GO fillers.

Besides inorganic fillers, organic hollow nanoparticles have great promise for creating new highly permeable membranes. Ding et al. fabricated hollow polyamide nanoparticles (HPN) with a dense shell via interfacial polymerization in a surfactant-free microemulsion. The hollow nanoparticles combined with liquid acrylate monomers to form mixed matrix membranes via UV-induced photo-polymerization. The nanoparticles with an average diameter of 42 nm dispersed uniformly in the membranes. The CO_2 permeability and CO_2/N_2 permselectivity of the MMMs both increased as the nanofiller loading increased. The gas permeation and separation performance exceeded the Robeson upper bound line with a maximum CO_2 permeability of 1898 Barrer and a maximum CO_2/N_2 permselectivity of 43.9 at 1 wt% nanofiller loading. The improvement mainly arose from the increase in the CO_2 solubility, the N_2 diffusivity, and the CO_2/N_2 solubility selectivity [62].

In addition, several novel molecular-scale fabrication methods render highly efficient polymeric composites and could be potentially applied for m-DAC. Instead of using traditional substrates and gutter layers, Ashtiani et al. developed membranes with carbon nanotube support structures and a ZIF-8 gutter layer (zeolitic imidazolate frameworks-ZIF). The surface was spray-coated with PDMS, which helped fill any gaps or interfacial defects. Due to the ZIF-8′s metal sites, CO_2 was effectively permeated through feeble electrostatic fields and the quadrupole moments of CO_2 [60]. The in-situ micelle-induced blending approach is another significant step toward realizing high-performance CO_2 separation membranes. Seong et al. reported highly CO_2-permeable membranes obtained by blending in situ self-assembled micellar-structured poly(poly(ethylene glycol) methyl ether acrylate) (PPEGMEA) with poly(ether-block-amide) (Pebax) (Figure 6). The Pebax/PPEGMEA (30/70 *w*/*w*) blend membrane comprised highly CO_2-philic PPEGMEA micelles, which appreciably increased the d-spacing of the Pebax matrix. Consequently, the blend membrane containing a high molecular weight of PPEGMEA exhibited an unprecedented CO_2 permeability enhancement of 1054% compared to the pristine Pebax membrane while maintaining good CO_2 selectivity relative to N_2 because of the enriched polyethylene glycol moieties. This excellent separation performance was maintained for 100 h up to 10 atm, validating the good long-term separation performance of the membrane [98]. Furthermore, the membrane exhibited good mechanical strength and plasticization tolerance. Similarly, membranes formed through the radical polymerization of PEGMEA, PEGDA, and PEGDME presented CO_2 permeability of 2980 Barrer and CO_2/N_2 selectivity of 45.7 with stability over 500 h [99]. Gel mixed matrix composites provide a class of membranes with high CO_2 permeability and stability performance. The fillers incorporate alongside small molecule liquid substances to increase free volume [100]. Chen et al. prepared EM400/MIL-101(Cr)–NH_2 MMMs; the addition of MIL-101(Cr)–NH_2 improved the mechanical properties of the membrane and increased the selectivity of CO_2/N_2. The EM400/MIL-101(Cr)–NH_2 MMMs were further modified with tripropionin. The tripropionin greatly improved the CO_2/N_2 separation performance of the g-MMMs by increasing the FFV and the CO_2 solubility coefficients at the same time. The CO_2 permeability was increased from 213 to 1182 Barrer when compared to the EM400/MIL-101(Cr)-NH_2 MMMs without TPP. This kind of gel-MMM also showed stability over 168 h of aging testing [100]. An alternative study tested SPEEK/MIL-101 (Cr) and SPEEK/S-MIL-101 (Cr) membranes at 40% loading under humidified conditions. Under these conditions, both membranes showed permeability improvements from 30 to 1623 Barrer and 35 to 2064 Barrer, respectively. Although the selectivity for SPEEK/MIL-101 (Cr) membranes did not change from 40, the selectivity did improve from 41 to 53 in the case of SPEEK/S-MIL-101 (Cr). The proportionality between permeability and water content was attributed to an increase in pore swelling, while the increase in selectivity can be explained by the decreased level of transport resistance for CO_2 with water as compared to N_2 [56].

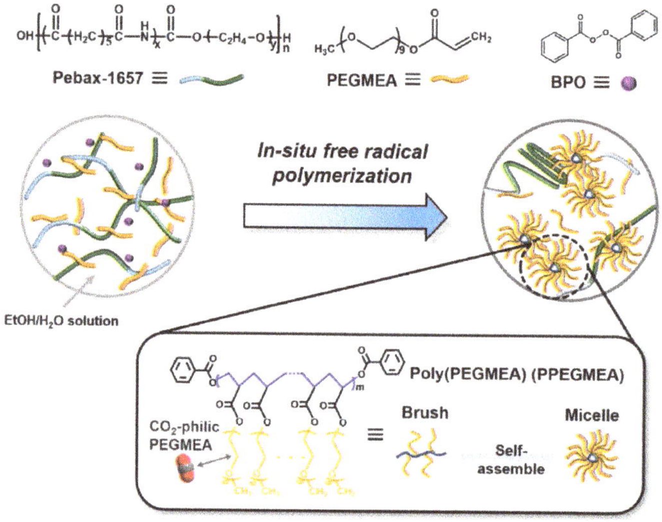

Figure 6. Schematic of Pebax/PPEGMEA membranes obtained via the free radical polymerization of PEGMEA in Pebax in the presence of BPO. Reprinted with permission from Ref. [99]. Copyright 2022 Elsevier.

4.3. Facilitated Transport Membranes

Finally, facilitated transport membranes are unique in their exceptionally high selectivity for CO_2 compared to typical solubility/diffusion membranes due to reversible reactions between the carriers in the membrane and CO_2. The carriers can be divided into fixed or mobile carriers (Figure 7). These reversible reactions require water to occur, setting a dependence on the humidity of the feed gas [101]. When considering m-DAC, the need for humidity can lead to tighter requirements for installation limited to humid areas or introduce higher operating costs and water requirements for humidifying the feed gas. With that being said, PIL-IL-GO membranes have been tested under m-DAC conditions of 410 ppm CO_2 and showed CO_2 permeances between 3000 and 4000 GPU with very high CO_2/N_2 selectivity (>1000) [30]. There are various carrier structures, but most are based on amines. For example, P(DADMACA-co-VAm) membranes containing primary amino groups, carbonate groups, and quaternary ammonium groups show synergistic increases in membrane separation performance over their homogeneous counterparts (PVAm/PDADMACA) with a CO_2 permeance of 1842 GPU and a CO_2/N_2 selectivity of 160 [63]. In addition, PVA membranes have been modified with both SiO_2 nanofillers and [bmim][Tf_2N] ionic liquid to produce membranes with a CO_2 permeance of 3016 GPU and a CO_2/N_2 selectivity of 62 [102]. The use of ionic liquids may not be the best practice as they can be high-cost and harmful to the environment and living organisms, therefore eliminating one of the advantages of m-DAC over the other sorption technologies [103].

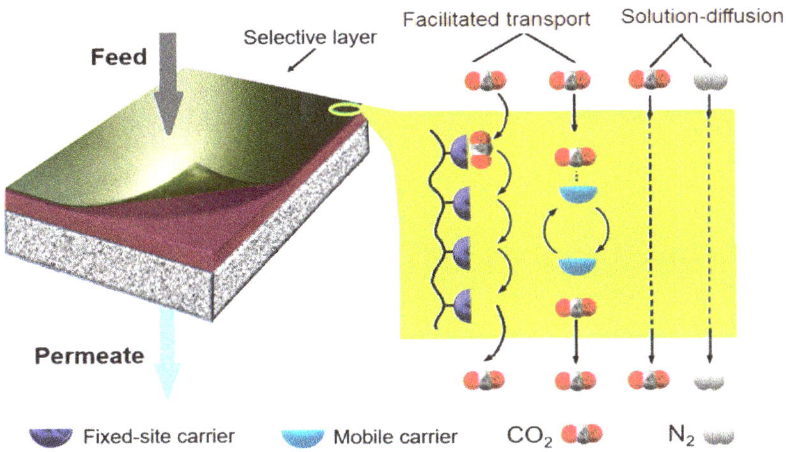

Figure 7. CO_2/N_2 separation through the facilitated transport membrane [12,104].

5. M-DAC Application

Once CO_2 is captured by DAC, it can be stored or recycled for use in several different downstream applications, as summarized in Figure 8. One important parameter for any CO_2 utilization after capture is the percent purity of CO_2. The most common method of CO_2 storage is geological sequestration; generally, geological storage of CO_2 requires CO_2 gas at a high purity (more than 98%) [5,105]. Currently, a purity over 90% is not possible to achieve by m-DAC efficiently in 1 stage; therefore, geological storage is an unlikely application of m-DAC without several stages or pairing with other CO_2 capture approaches [5,12,16]. A simulation-based study using a membrane with a CO_2 permeance of 1850 GPU and CO_2/N_2 selectivity of 80 for hybrid membrane cryogenic (HMC) capture showed a 9% reduction in capture cost at a capture ratio of 85% compared to conventional monoethanolamide (MEA) carbon capture [106,107]. Similarly, another HMC process using a membrane with a CO_2 permeance of 1000 GPU and CO_2/N_2 selectivity of 100 was shown to reduce the energy requirement compared to MEA capture from 4.409 GJ_{th}/t_{CO2} to 3.25 GJ_{th}/t_{CO2} from a stream of 15% CO_2 and a pressure ratio of 11.11 achieving a capture ratio of 85% and a purity over 89% [107,108]. Unfortunately, there is a lack of research on hybrid separation systems using membranes alongside other DAC methods, which have the possibility for efficient cost reductions. A recent engineering parametric study, which investigated the impact of material performance, process design, and operating conditions on membrane-based DAC, shed light on the future of this technique. The study concluded that with existing commercial membrane materials, the maximal CO_2 output concentration from a single-stage separation would be ~2%. A similar two-stage process could increase this to ~50%, noting that O_2 and water would be co-permeated. High-performance materials (not commercialized) could achieve ~12% in a single stage and up to ~99% in two stages [16]. For these high-performance materials in a two-stage process, costs and energy fall in the affordable range, e.g., 103 to 104$ per tCO_2 with a first-order optimized energy requirement of 101 GJ per tCO_2 [16]. Currently, multi-stage membrane modules are still needed if high-purity outputs are desired since the performance might worsen due to pressure drop and concentration polarization [16]. Hence, it could be an efficient approach to produce high-concentration CO_2 via a hybrid system, which combines m-DAC and other CO_2 capture systems, although more research is needed in this area.

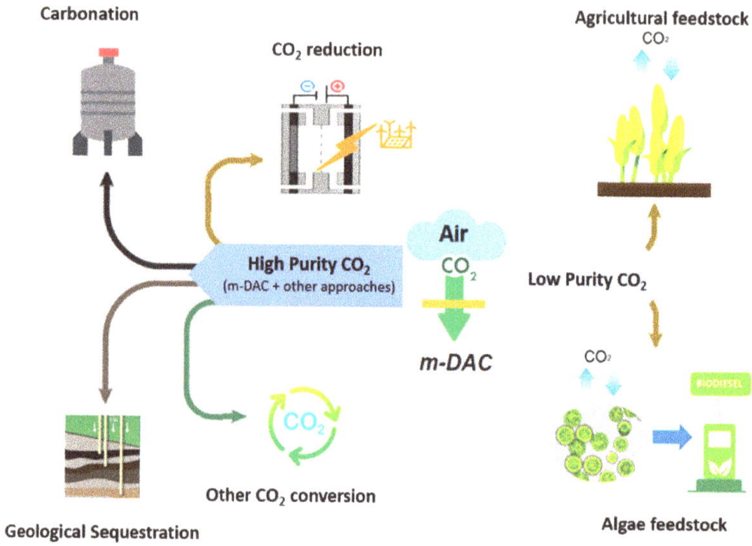

Figure 8. Examples of post-capture utilization pathways for CO_2 depending on purity.

With m-DAC, the contaminants impacting CO_2 purity are gases like O_2 and N_2, not NO_x and SO_x, which are picked up by flue gas capture. It seems only agricultural applications can use dilute CO_2 [5]. Biological fixation through photosynthetic organisms is the planet's natural method for CO_2 recycling; photosynthetic microorganisms offer fast reproduction, adaptability, and efficient conversion of CO_2 into sugars and typically require low CO_2 (<~40%) concentrations [16,109] (Table S2). With currently commercially available membranes, the 2% CO_2 concentration could be of interest for the intensification of greenhouses or algae ponds. Captured high-purity CO_2 can also be used to enhance oil recovery. It can be used for welding, dry ice, soda, and feedstock for greenhouses and other agricultural installations. Furthermore, CO_2 can be used for conversion to several commercial chemicals such as ammonia for fertilizer, plastics which reduce the use of petrochemical products, formic acid, synthetic fuels by methanation, and methanol, to name a few [109]. In that event, multistage membrane units or hybrid processes could be of interest [29]. Although most electrochemical reduction of CO_2 uses pure CO_2, recent studies have shown that CO_2 reduction to CO can be achieved with low CO_2 purity, indicating another realistic use for m-DAC CO_2 [110]. The CO produced can be used to produce useful hydrocarbons for downstream applications. For all the aforementioned uses, the source of energy (renewable or not) for conversion and the processes' emissions should be considered to determine the overall environmental impact and the extent of negative emissions.

Currently, since m-DAC is rather expensive with current membranes, the most attractive selling point is the modularity. Fujikawa et al. showed through process simulation that a 4-stage separation module using current highly permeable membranes can fit into 0.01 m^3 and capture 1 kg-CO_2 per day [12]. Dittmeyer et al. proposed DAC-integrated AC units that could convert captured CO_2 into hydrocarbon fuels. Their paper provided perspective examples of towns fueling their cars using the fuels generated by negative emission technologies. This approach could add an incentive for the public to get involved in DAC technology [25]. Small-scale m-DAC installations have the potential to be installed in office buildings and schools where the CO_2 concentration reaches up to 1000 ppm [34]. Areas with elevated CO_2 concentrations compared to the atmosphere would increase the capture efficiency of m-DAC. Moreover, the location independence of m-DAC allows for installation near CO_2 storage and recycling sites. This process can avoid the construction of pipelines needed to transport the CO_2 recovered from location-dependent capture sites to storage sites [7]. This also avoids the energy required to compress the gas for transport. Using the previously mentioned utilization of CO_2 as an agricultural feedstock as an example, m-DAC modules could become a standard installation in every greenhouse.

6. Perspective

The m-DAC process has been highlighted as a promising complementary technology to sorbent-based DAC. Currently, the major barriers facing m-DAC's full-scale implementation include membrane material performance, membrane fabrication, and system process development [12]:

(1) Membrane materials

The performance of membrane materials is significantly affected by the trade-off of permeance/selectivity as well as plasticization and physical aging. More research needs to be conducted on m-DAC materials to meet the criteria of CO_2 permeance > 10,000 GPU and CO_2/N_2 selectivity > 30. Manipulation of the physical (mainly free volume) and chemical (mainly reactivity with CO_2) membrane environment to improve performance via novel chemistries and preparative methods should continue to be an active area of research [86]. Besides employing novel chemistries to prepare polymers via improving the chain rigidity by introducing more inflexible units, the development of polymer nanocomposites through different approaches, e.g., "in-situ micelle-induced blending" and "gel mixed matrix compositing". Additionally, using proper nanomaterials, such as selected functionalized MOF, silica, and GO, looks promising in terms of CO_2/N_2 separation performance, cost, synthesis, and feasibility. Notably, some inorganic/organic fillers could not only help tune the free volume of the polymer matrix but also increase CO_2 solubility or create a coefficient path [98].

In addition, exploiting facilitated transport mechanisms can also provide effective solutions for overcoming the performance trade-off effect [57]. Low-cost and high-performance membranes for DAC are required by industry. Hence the exploration of cheaper and affordable materials should always be pursued.

(2) Membrane fabrication

To meet the performance expectation of DAC and be employed in large-scale applications, membrane materials should be capable of forming thin membranes with selective layers less than 1 μm to enable high gas permeance and be packaged into large spiral wound or hollow fiber membrane modules to maximize the surface area to volume ratio [46]. In this sense, the nanofillers for mixed matrix membranes need to be carefully chosen and appropriately dispersed. Polymers that are not solvent-processable or prepared at high temperatures should attract less attention. It seems that commercially available Pebax (PEO-based polymers) and PVAm-based facilitated transport membranes are good candidates for selective coating layers [64,79].

(3) System and process

Improvements to the process durability of m-DAC systems against the typical, high TRL (Technology Readiness Level) separation operations, such as adsorption, absorption, and cryogenic distillation, should be made. Compared to CO_2 separation from point sources, rather than impurities such as SO_x, NO_x, H_2S, light hydrocarbons, or aromatics, the main concern of feed gases for m-DAC is water [111]. As a condensable gas, it could lead to an increase in free volume, causing a significant decline in performance. In most facilitated transport membranes where amine groups are used as the carriers, the oxidation of carriers by oxidizing gases (mainly O_2) resulted in the deterioration of membrane performance [111]. The effective pre-treatment of feed gases and exploration of anti-oxidized carriers could help to solve the problem.

Because of the demand for CO_2 at high concentrations, more research needs to focus on multistep and hybrid separation systems [28]. Also, the testing of these systems should be conducted physically with the support of computer modeling. More investigation is needed on process cost analysis since DAC conditions are different from those operating in typical CO_2 capture, and process cost reductions make the process feasible. Advancements in membrane technology and m-DAC system tunability should be conducted with future downstream applications in mind, which may require a focus on CO_2/O_2 selectivity and permeate purity [9,12,28]. At the same time, more attention should be brought to using lower purity CO_2 in different utilization pathways with a focus on mixtures of $O_2/N_2/CO_2$ typical of m-DAC [111].

Although most membrane materials do not currently meet the performance criteria for m-DAC, these improved membranes should be further investigated in conjunction with current DAC methods to potentially create more efficient hybrid capture systems. For example, m-DAC can be used as a first separation stage before solid sorbents do the heavy lifting of selective capture. Other innovative ideas, such as membranes used as contactors for liquid solvents or a support structure for solid sorbent monoliths in DAC adsorbers, will inspire researchers to leverage the advantages of the polymeric membrane in DAC processes.

In addition, as suggested by Castro-Muñoz et al., membrane performance should be measured in GPU for permeance rather than Barrer for permeability for standardization across experiments with variability in membrane thickness [26].

Ultimately, the focus of new research should be on increasing membrane permeance, optimizing selectivity, increasing scalability, decreasing cost, and use of hazardous materials, which are commonly associated with DAC methods currently in use. Future m-DAC systems should provide life cycle analyses to finitely determine the extent of negative emissions and operational cost to better illustrate the feasibility of using membranes for direct air capture.

7. Conclusions

Increasing CO_2 levels in the atmosphere and 2050 Net Zero Emission requirements have aroused many research efforts in exploring efficient negative CO_2 emission technologies to capture CO_2 from the atmosphere. In the past decades, the intrinsic advantages of membrane technology have promoted its application in CO_2 capture from point sources. Previous efforts in membrane technology have shown the possibility of membrane processes being considered as a new approach to DAC. Due to its great potential for unprecedented improvements in capture cost, installation applications, and environmental impact compared to the current sorbent-based DAC, it is believed that m-DAC will become an important addition to the portfolio of methods aimed at realizing ubiquitous CO_2 capture and could change the outlook on CO_2 capture in the near future. To promote the practical applications of m-DAC, challenges should be overcome through the intersection and application of chemistry, materials science, and engineering. The research on membrane materials is fundamental as the efficiency of m-DAC systems is heavily reliant on the permeance and selectivity of the membranes.

In this mini-review, the required properties of membranes for DAC, process parameters for modeling/system design, potential materials for membranes, and applications of m-DAC have been outlined. Particularly, we summarize the advances up to 2023 in high permeability polymer-based membrane materials for CO_2 separations which could be potential DAC membrane materials, including polymeric membranes, mixed matrix membranes, and CO_2-facilitated transport membranes. A number of representative examples of recent advances are highlighted, followed by a brief perspective on the direction of future research and development. This work sheds light on the progress of m-DAC and encourages further research at the intersection of disciplines, such as polymer chemistry, inorganic chemistry, nanoscience and technology, and chemical engineering, drawing an inspiring picture of membranes that are more robust, have lower cost and higher CO_2 separation performance.

Supplementary Materials: The following supporting information can be downloaded at: https://www.mdpi.com/article/10.3390/membranes14020030/s1, Table S1: Summary of recent promising membrane materials showing permeabilities ≥ 1000 Barrer or permeances > ~1000 GPU or greater with CO_2/N_2 Selectivity's ≥ 20 at temperatures ≤ 35 °C; Table S2: CO_2 purity requirements for downstream applications. References [112–176] are cited in the Supplementary Materials.

Author Contributions: Writing—original draft preparation, P.I. and N.D.; writing—review and editing, H.L., N.K., L.S., W.M. and B.P. All authors have read and agreed to the published version of the manuscript.

Funding: This research was funded by Carbon Capture Utilization and Storage (CCUS) program from Natural Resources Canada's Office of Energy Research and Development (OERD) and Clean Production Program A1-022694 from National Research Council Canada.

Institutional Review Board Statement: Not applicable.

Data Availability Statement: Data openly available in a public repository that issues datasets with and without DOIs.

Conflicts of Interest: The authors declare no conflict of interest.

References

1. Minx, J.C.; Lamb, W.F.; Callghan, M.W.; Fuss, S.; Hilaire, J.; Creutzig, F.; Amann, T.; Beringer, T.; De Oliveira Garcia, W.; Hartmann, J. Negative emissions-Part1: Research landscape and synthesis. *Environ. Res. Lett.* **2018**, *13*, 063001. [CrossRef]
2. IPCC. 2023: Summary for Policymakers. In *Climate Change 2023: Synthesis Report. A Report of the Intergovernmental Panel on Climate Change. Contribution of Working Groups I, II and III to the Sixth Assessment Report of the Intergovernmental Panel on Climate Change*; Core Writing Team, Lee, H., Romero, J., Eds.; IPCC: Geneva, Switzerland, 2023; 36p, *in press*.
3. National Academies of Sciences, Engineering, and Medicine. *Negative Emissions Technologies and Reliable Sequestration: A Research Agenda*; The National Academies Press: Washington, DC, USA, 2019. [CrossRef]
4. Zhao, C.; Liu, B.; Piao, S.; Wang, X.; Lobell, D.B.; Huang, Y.; Huang, M.; Yao, Y.; Bassu, S.; Ciais, P.; et al. Temperature increase reduces global yields of major crops in four independent estimates. *Proc. Natl. Acad. Sci. USA* **2017**, *114*, 9326–9331. [CrossRef] [PubMed]
5. Fujikawa, S.; Selyanchyn, R. Direct air capture by membranes. *MRS Bull.* **2022**, *47*, 416–423. [CrossRef]
6. Chiwaye, N.; Majozi, T.; Daramola, M.O. On optimisation of N_2 and CO_2-selective hybrid membrane process systems for post-combustion CO_2 capture from coal-fired power plants. *J. Membr. Sci.* **2021**, *638*, 119691. [CrossRef]
7. Lackner, K.S.; Brennan, S.; Matter, J.M.; van der Zwaan, B. The urgency of the development of CO_2 capture from ambient air. *Proc. Natl. Acad. Sci. USA* **2012**, *109*, 13156–13162. [CrossRef] [PubMed]
8. Shayegh, S.; Bosetti, V.; Tavoni, M. Future Prospects of Direct Air Capture Technologies: Insights from an Expert Elicitation Survey. *Front. Clim.* **2021**, *3*, 630893. [CrossRef]
9. Ozkan, M. Direct air capture of CO_2: A response to meet the global climate targets. *MRS. Energy Sustain.* **2021**, *8*, 51–56. [CrossRef]
10. Beuttler, C.; Charles, L.; Wurzbacher, J. The Role of Direct Air Capture in Mitigation of Anthropogenic Greenhouse Gas Emissions. *Front. Clim.* **2019**, *1*, 10. [CrossRef]
11. Keith, D.W.; Holmes, G.; St. Angelo, D.; Heidel, K. A Process for Capturing CO_2 from the Atmosphere. *Joule* **2018**, *2*, 1573–1594. [CrossRef]
12. Fujikawa, S.; Selyanchyn, R.; Kunitake, T. A new strategy for membrane-based direct air capture. *Polym. J.* **2021**, *53*, 111–119. [CrossRef]
13. Socolow, R.; Desmond, M.; Aines, R.; Blackstock, J.; Bolland, O.; Kaarsberg, T.; Lewis, N.; Mazzotti, M.; Pfeffer, A.; Sawyer, K.; et al. *Direct Air Capture of CO_2 with Chemicals: A Technology Assessment for the APS Panel on Public Affairs*; American Physical Society: College Park, MD, USA, 2011.
14. McQueen, N.; Gomes, K.V.; McCormick, C.; Blumanthal, K.; Pisciotta, M.; Wilcox, J. A review of direct air capture (DAC): Scaling up commercial technologies and innovating for the future. *Prog. Energy* **2021**, *3*, 032001. [CrossRef]
15. Kikkawa, S.; Anamoto, K.; Fujiki, Y.; Hirayama, J.; Kato, G.; Miura, H.; Shishido, T.; Yamazoe, S. Direct Air Capture of CO_2 Using a Liquid Amine-Solid Carbamic Acid Phase-Separation System Using Diamine Bearing an Aminocyclohexyl Group. *ACS Environ. Au* **2022**, *2*, 354–362. [CrossRef] [PubMed]
16. Erans, M.; Sanz-Pérez, E.S.; Hanak, D.P.; Clulow, Z.; Reiner, D.M.; Mutch, G.A. Direct air capture: Process technology, Techno-economic and socio political challanges. *Energy Environ. Sci.* **2022**, *15*, 1360–1405. [CrossRef]
17. Wiegner, J.F.; Grimm, A.; Weimann, L.; Gazzani, M. Optimal Design and Operation of Solid Sorbent Direct Air Capture Processes at Varying Ambient Conditions. *Ind. Eng. Res.* **2022**, *61*, 12649–12667. [CrossRef]
18. Lackner, K.S. A Guide to CO_2 Sequestration. *Science* **2003**, *300*, 1677–1678. [CrossRef] [PubMed]
19. Osterloh, F.E. The Low Concentration of CO_2 in the Atmosphere Is an Obstacle to a Sustainable Artificial Photosynthesis Fuel Cycle Based on Carbon. *ACS Energy Lett.* **2016**, *1*, 1060–1061. [CrossRef]

20. Merkel, T.C.; Lin, H.; Wei, X.; Baker, R. Power plant post-combustion carbon dioxide capture: An opportunity for membranes. *J. Membr. Sci.* **2010**, *359*, 126–139. [CrossRef]
21. Keith, D.W.; Heidel, K.; Cherry, R. *Geoengineering Climate Change. Environment Necessity or Pandora's Box? Capturing CO_2 from Atmosphere: Rationale and Process Design Considerations*; Brian, L., Ed.; Cambridge University Press: Cambridge, UK, 2010; Chapter 6; pp. 107–126.
22. Selyanchyn, R.; Fujikawa, S. Membrane thinning for efficient CO_2 capture. *Sci. Technol. Adv. Mater.* **2017**, *18*, 816–827. [CrossRef]
23. Xie, K.; Fu, Q.; Qiao, G.G.; Webley, P.A. Recent progress on fabrication of polymeric thin film gas separation membranes for CO_2 capture. *J. Membr. Sci.* **2019**, *572*, 38–60. [CrossRef]
24. Chen, Y.; Zhao, L.; Wang, B.; Dutta, P.; Ho, W.S.W. Amine-containing polymer/zeolite Y composite membranes for CO_2/N_2 separation. *J. Membr. Sci.* **2016**, *497*, 21–28. [CrossRef]
25. Dittmeyer, R.; Klumpp, M.; Kant, P.; Ozin, G. Crowd oil not crude oil. *Nat. Commun.* **2019**, *10*, 1818. [CrossRef] [PubMed]
26. Fujikawa, S.; Ariyoshi, M.; Selyanchyn, R.; Kunitake, T. Ultra-fast, Selective CO_2 Permeation by Free-standing Siloxane Nanomembranes. *Chem. Lett.* **2019**, *48*, 1351–1354. [CrossRef]
27. Setiawan, W.K.; Chiang, K. Amine-functionalized biogenic silica incorporation effect on poly (ether-block-amide) membrane CO_2/N_2 separation performance. *J. Membr. Sci.* **2023**, *680*, 121732. [CrossRef]
28. Castro-Muñoz, R.; Ahmad, M.Z.; Malankowska, M.; Coronas, J. A new relevant membrane application: CO_2 direct air capture (DAC). *J. Chem. Eng.* **2022**, *446*, 137047. [CrossRef]
29. Castel, C.; Bounaceur, R.; Favre, E. Membrane Processes for Direct Carbon Dioxide Capture From Air: Possibilities and Limitations. *Front. Chem. Eng.* **2021**, *3*, 668867. [CrossRef]
30. Lee, Y.Y.; Gurkan, B. Graphene oxide reinforced facilitated transport membrane with poly(ionic liquid) and ionic liquid carriers for CO_2/N_2 separation. *J. Membr. Sci.* **2021**, *638*, 119652. [CrossRef]
31. Yave, W.; Car, A.; Wind, J.; Peinemann, K. Nanometric thin film membranes manufactured on square meter scale: Ultra-thin films for CO_2 capture. *Nanotechnology* **2010**, *21*, 395301. [CrossRef]
32. Selyanchyn, C.; Selyanchyn, R.; Fujikawa, S. Critical Role of the Molecular Interface in Double-Layered Pebax-1657/PDMS Nanomembranes for Highly Efficient CO_2/N_2 Gas Separation. *ACS Appl. Mater. Interfaces* **2020**, *12*, 33196–33209. [CrossRef]
33. Yoo, M.J.; Kim, K.H.; Lee, J.H.; Kim, T.W.; Chung, C.W.; Cho, Y.H.; Park, H.B. Ultrathin gutter layer for high-performance thin-film composite membranes for CO_2 separation. *J. Membr. Sci.* **2018**, *566*, 336–345. [CrossRef]
34. Seppänen, O.A.; Fisk, W.J.; Mendell, W.J. Association of Ventilation Rates and CO_2 Concentrations with Health and Other Responses in Commercial and Institutional Buildings. *Indoor Air* **1999**, *9*, 226–252. [CrossRef]
35. Mohsenpour, S.; Guo, Z.; Almansour, F.; Holmes, S.M.; Budd, P.M.; Gorgojo, P. Porous silica nanosheets in PIM-1 membranes for CO_2 separation. *J. Membr. Sci.* **2022**, *661*, 120889. [CrossRef]
36. Vendamme, R.; Onoue, S.; Nakao, A.; Kunitake, T. Robust free-standing membranes of organic/inorganic interpenetrating networks. *Nat. Mater.* **2006**, *5*, 494–501. [CrossRef] [PubMed]
37. Wanatabe, H.; Kunitake, T. A Large, Freestanding, 20 nm Thick Nanomembrane Based on an Epoxy Resin. *Adv. Mater.* **2007**, *19*, 909–912. [CrossRef]
38. He, S.; Zhu, B.; Li, S.; Zhang, Y.; Jiang, X.; Lau, C.H.; Shao, L. Recent progress in PIM-1 based membranes for sustainable CO_2 separations: Polymer structure manipulation and mixed matrix membrane design. *Sep. Purif. Technol.* **2022**, *284*, 120277. [CrossRef]
39. Robeson, L.M. The upper bound revisited. *J. Membr. Sci.* **2008**, *320*, 390–400. [CrossRef]
40. Swaidan, R.; Al-Saeedi, M.; Ghanem, B.; Litwiller, E.; Pinnau, I. Rational Design of Intrinsically Ultramicroporous Polyimides Containing Bridgehead-Substituted Triptycene for Highly Selective and Permeable Gas Separation Membranes. *Macromolecules* **2014**, *47*, 5104–5114. [CrossRef]
41. Comesaña-Gándara, B.; Chen, J.; Bezzu, B.G.; Carta, M.; Rose, I.; Ferrari, M.; Esposito, E.; Fuoco, A.; Jansen, J.C.; Mckeown, N.B. Redefining the Robeson upper bounds for CO_2/CH_4 and CO_2/N_2 separations using a series of ultrapermeable benzotriptycene-based polymers of intrinsic microporosity. *Energy Environ. Sci.* **2019**, *12*, 2733–2740. [CrossRef]
42. Shrimant, B.; Kharul, U.K.; Wadgaonkar, P.P. Spiro[fluorene-9,9′-xanthene]-containing copolymers of intrinsic microporosity: Synthesis, characterization and gas permeation properties. *React. Funct. Polym.* **2018**, *133*, 153–160. [CrossRef]
43. Sun, W.; Yin, M.; Zhang, W.; Li, S.; Wang, N.; An, Q. Green Techniques for Rapid Fabrication of Unprecedentedly High-Performance PEO Membranes for CO_2 Capture. *ACS Sustain. Chem. Eng.* **2021**, *9*, 10167–10175. [CrossRef]
44. Nazarov, I.V.; Khrychikova, A.P.; Medentseva, E.I.; Bermesheva, E.V.; Borisov, I.L.; Yushkin, A.A.; Volkov, A.V.; Wozniak, A.I.; Petukhov, D.I.; Topchiy, M.A.; et al. CO_2-selective vinyl-addition polymers from nadimides: Synthesis and performance for membrane gas separation. *J. Membr. Sci.* **2023**, *677*, 121624. [CrossRef]
45. Alentiev, D.A.; Egorova, E.S.; Bermeshev, M.V.; Starannikova, L.E.; Topchiy, M.A.; Asachenko, A.F.; Gribanov, P.S.; Nechaev, M.S.; Yampolskii, Y.P.; Finkelshtein, E.S. Janus tricyclononene polymers bearing tri(n-alkoxy)silyl side groups for membrane gas separation. *J. Mater. Chem. A* **2018**, *6*, 19393–19408. [CrossRef]
46. White, L.S.; Amo, K.D.; Wu, T.; Markel, T.C. Extended field trials of Polaris sweep modules for carbon capture. *J. Membr. Sci.* **2017**, *542*, 217–225. [CrossRef]
47. Brinkmann, T.; Lillepärg, J.; Notzke, H.; Pohlmann, J.; Shishatskiy, S.; Wind, J.; Wolff, T. Development of CO_2 selective poly(ethylene oxide)-based membranes: From laboratory to pilot plant scale. *Engineering* **2017**, *3*, 485–493. [CrossRef]

48. Du, N.; Park, H.B.; Robertson, G.P.; Dal-Cin, M.M.; Visser, T.; Scoles, L.; Guiver, M.D. Polymer nanosieve membranes for CO_2-capture applications. *Nat. Mater.* **2011**, *10*, 372–375. [CrossRef] [PubMed]
49. Swaidan, R.; Ghanem, B.S.; Litwiller, E.; Pinnau, I. Pure- and mixed-gas CO_2/CH_4 separation properties of PIM-1 and an amidoxime-functionalized PIM-1. *J. Membr. Sci.* **2014**, *457*, 95–102. [CrossRef]
50. Du, N.; Robertson, G.P.; Dal-Cin, M.M.; Scoles, L.; Guiver, M.D. Polymers of intrinsic microporosity (PIMs) substituted with methyl tetrazole. *Polymer* **2012**, *53*, 4367–4372. [CrossRef]
51. Manson, C.R.; Maynard-Atem, L.; Al-Harbi, N.M.; Budd, P.M.; Bernardo, P.; Bazzarelli, F.; Clarizia, G.; Jansen, J.C. Polymer of intrinsic Microporosity Incorporating Thioamide Functionality: Preparation and Gas Transport Properties. *Macromolecules* **2011**, *44*, 6471–6479. [CrossRef]
52. Han, W.; Zhang, C.; Zhao, M.; Yang, F.; Yang, Y.; Weng, Y. Post-modification of PIM-1 and simultaneously in situ synthesis of porous polymer networks into PIM-1 matrix to enhance CO_2 separation performance. *J. Membr. Sci.* **2021**, *636*, 119544. [CrossRef]
53. Nafisi, V.; Hägg, M. Development of Nanocomposite Membranes Containing Modified Si Nanoparticles in PEBAX-2533 as a Block Co-polymer and 6FDA-Durene Diamine as a Glassy Polymer. *ACS Appl. Mater. Interfaces* **2014**, *6*, 15643–15652. [CrossRef]
54. Mason, C.R.; Buonomenna, M.G.; Golemme, G.; Budd, P.M.; Galiano, F.; Figoli, A.; Friess, K.; Hynek, V. New organophilic mixed matrix membranes derived from a polymer of intrinsic microporosity and silicalite-1. *Polymer* **2013**, *54*, 2222–2230. [CrossRef]
55. Nafisi, V.; Hägg, M. Development of dual layer of ZIF-8/PEBAX-2533 mixed matrix membrane for CO_2 capture. *J. Membr. Sci.* **2014**, *459*, 244–255. [CrossRef]
56. Xin, Q.; Liu, T.; Li, Z.; Wang, S.; Li, Y.; Li, Z.; Ouyang, J.; Jiang, Z.; Wu, H. Mixed matrix membranes composed of sulfonated poly(ether ether ketone) and a sulfonated metal–organic framework for gas separation. *J. Membr. Sci.* **2015**, *488*, 67–78. [CrossRef]
57. Yu, G.; Zou, X.; Sun, L.; Liu, B.; Wang, Z.; Zhang, P.; Zhu, G. Constructing Connected Paths between UiO-66 and PIM-1 to improve Membrane CO_2 Separation with Crystal-Like Gas Selectivity. *Adv. Mater.* **2019**, *31*, e1806853. [CrossRef] [PubMed]
58. Haider, B.; Dilshad, M.R.; Akram, M.S.; Islam, A.; Kaspereit, M. Novel Polydimethylsiloxane membranes impregnated with SAPO-34 zeolite particles for gas separation. *Chem. Pap.* **2021**, *75*, 6417–6431. [CrossRef]
59. Chen, M.; Soyekwo, F.; Zhang, Q.; Hu, C.; Zhu, A.; Liu, Q. Graphene oxide nanosheets to improve permeability and selectivity of PIM-1 membrane for carbon dioxide separation. *J. Ind. Eng. Chem.* **2018**, *63*, 296–302. [CrossRef]
60. Ashtiani, S.; Sofer, Z.; Průša, F.; Friess, K. Molecular-level fabrication of highly selective composite ZIF-8-CNT-PDMS membranes for effective CO_2/N_2, CO_2/H_2 and olefin/paraffin separations. *Sep. Purif. Technol.* **2021**, *274*, 119003. [CrossRef]
61. Wang, Z.; Ren, H.; Zhang, S.; Zhang, F.; Jin, J. Polymers of intrinsic microporosity/metal-organic framework hybrid membranes with improved interfacial interaction for high-performance CO_2 separation. *J. Mater. Chem. A* **2017**, *5*, 10968–10977. [CrossRef]
62. Ding, X.; Tan, F.; Zhao, H.; Hua, M.; Wang, M.; Xin, Q.; Zhang, Y. Enhancing gas permeation and separation performance of polymeric membrane by incorporating hollow polyamide nanoparticles with dense shell. *J. Membr. Sci.* **2019**, *570–571*, 53–60. [CrossRef]
63. Li, X.; Ding, S.; Zhang, J.; Wei, Z. Optimizing microstructure of polymer composite membranes by tailoring different ionic liquids to accelerate CO_2 transport. *Int. J. Greenh. Gas Control* **2020**, *101*, 103136. [CrossRef]
64. Liu, Y.; Yu, S.; Wu, H.; Li, Y.; Wang, S.; Tian, Z.; Jiang, Z. High permeability hydrogel membranes of chitosan/poly ether-block-amide blends for CO_2 separation. *J. Membr. Sci.* **2014**, *469*, 198–208. [CrossRef]
65. Wu, H.; Li, X.; Li, Y.; Wang, S.; Guo, R.; Jiang, Z.; Wu, C.; Xin, Q.; Lu, X. Facilitated transport mixed matrix membranes incorporated with amine functionalized MCM-41 for enhanced gas separation properties. *J. Membr. Sci.* **2014**, *465*, 78–90. [CrossRef]
66. Mashhadikhan, S.; Moghadassi, A.; Amooghin, A.E.; Sanaeepur, H. Interlocking a synthesized polymer and bifunctional filler containing the same polymer's monomer for conformable hybrid membrane systems. *J. Mater. Chem. A* **2020**, *8*, 3942–3955. [CrossRef]
67. Zhao, D.; Ren, J.; Wang, Y.; Qiu, Y.; Li, H.; Hua, K.; Li, X.; Ji, J.; Deng, M. High CO_2 separation performance of Pebax®/CNTs/GTA mixed matrix membranes. *J. Membr. Sci.* **2017**, *521*, 104–113. [CrossRef]
68. Zhang, H.; Tian, H.; Zhang, J.; Guo, R.; Li, X. Facilitated transport membranes with an amino acid salt for highly efficient CO_2 separation. *Int. J. Greenh. Gas Control* **2018**, *78*, 85–93. [CrossRef]
69. Akbarzadeh, E.; Shockravi, A.; Vatanpour, V. High performance compatible thiazole-based polymeric blend cellulose acetate membrane as selective CO_2 absorbent and molecular sieve. *Carbohydr. Polym.* **2021**, *252*, 117215. [CrossRef] [PubMed]
70. Wang, C.; Guo, F.; Li, H.; Xu, J.; Hu, J.; Liu, H.; Wang, M. A porous ionic polymer bionic carrier in a mixed matrix membrane for facilitating selective CO_2 permeability. *J. Membr. Sci.* **2020**, *598*, 117215. [CrossRef]
71. Yu, M.; Foster, A.B.; Alshurafa, M.; Luque-Alled, J.M.; Gorgojo, P.; Kentish, S.E.; Scholes, C.A.; Budd, P.M. CO_2 separation using thin film composite membranes of acid-hydrolyzed PIM-1. *J. Membr. Sci.* **2023**, *679*, 121697. [CrossRef]
72. Swaidan, R.; Ghanem, B.; Litwiller, E.; Pinnau, I. Physical Aging, Plasticization and Their Effects on Gas Permeation in "Rigid" Polymers of Intrinsic Microporosity. *Macromolecules* **2015**, *48*, 6553–6561. [CrossRef]
73. Ahmad, M.Z.; Castro-Muñoz, R.; Budd, P.M. Boosting gas separation performance and suppressing the physical aging of polymers of intrinsic microporosity (PIM-1) by nanomaterial blending. *Nanoscale* **2020**, *12*, 23333–23370. [CrossRef]
74. Norahim, N.; Yaisanga, P.; Faungnawakij, K.; Charinpanitkul, T.; Klaysom, C. Recent Membrane Developments for CO_2 Separation and Capture. *Chem. Eng. Technol.* **2018**, *41*, 211–223. [CrossRef]

75. Budd, P.M.; Msayib, K.J.; Tattershall, C.E.; Ghanem, B.S.; Reynolds, K.J.; Mckeown, N.B.; Fritsch, D. Gas separation membranes from polymers of intrinsic microporosity. *J. Membr. Sci.* **2005**, *251*, 263–269. [CrossRef]
76. Wang, Y.; Ma, X.; Ghanem, B.S.; Alghunaimi, F.; Pinnau, I.; Han, Y. Polymers of intrinsic microporosity for energy-intensive membrane-based gas separations. *Mater. Today Nano* **2018**, *3*, 69–95. [CrossRef]
77. Han, X.; Zhang, J.; Yue, C.; Pang, J.; Zhang, H.; Jiang, Z. Novel copolymers with intrinsic microporosity containing tetraphenyl-bipyrimidine for enhanced gas separation. *J. Ind. Eng. Chem.* **2020**, *91*, 102–109. [CrossRef]
78. Bezzu, C.G.; Carta, M.; Ferrari, M.; Jansen, J.C.; Monteleone, M.; Elisa, E.; Fuoco, A.; Hart, K.; Liyana-Arachchi, T.P.; Colina, C.M.; et al. The synthesis, chain-packing simulation and long-term gas permeability of highly selective spirobifluorene-based polymers of intrinsic microporosity. *J. Mater. Chem A* **2018**, *6*, 10507–10514. [CrossRef]
79. Zhu, B.; Jiang, X.; He, S.; Xiaobin, Y.; Long, J.; Zhang, Y.; Shao, L. Rational design of poly(ethylene oxide) based membranes for sustainable CO_2 capture. *J. Mater. Chem. A* **2020**, *8*, 24233–24252. [CrossRef]
80. Bandehali, S.; Moghadassi, A.; Parvizian, F.; Hosseini, S.M.; Matsuura, T.; Joudaki, E. Advances in high carbon dioxide separation performance of poly (ethylene oxide)-based membranes. *J. Energy Chem.* **2020**, *46*, 30–52. [CrossRef]
81. Han, Y.; Ho, W.S.W. Recent advances in polymeric membranes for CO_2 capture. *Chin. J. Chem. Eng.* **2018**, *26*, 2238–2254. [CrossRef]
82. Rogan, Y.; Starannikova, L.; Ryzhikh, V.; Yampolskii, V.; Yampolskii, Y.; Bernardo, P.; Bazzarelli, F.; Jansen, J.C.; McKeown, N.B. Synthesis and gas permeation properties of novel spirobisindane-based polyimides of intrinsic microporosity. *Polym. Chem.* **2013**, *4*, 1820–3813. [CrossRef]
83. Ghanem, B.S.; McKeown, N.B.; Budd, P.M.; Al-Harbi, N.M.; Fritsch, D.; Heinrich, K.; Starannikova, L.; Tokarev, A.; Yampolskii, Y. Synthesis, Characterization, and Gas Permeation Properties of a Novel Group of Polymers with Intrinsic Microporosity: PIM-Polyimides. *Macromolecules* **2009**, *42*, 7881–7888. [CrossRef]
84. Song, Q.; Cao, S.; Zavala-Rivera, P.; Lu, L.P.; Li, W.; Ji, Y.; Al-Muhtaseb, S.A.; Cheetham, A.K.; Sivaniah, E. Photo-oxidative enhancement of polymeric molecular sieve membranes. *Nat. Commun.* **2013**, *4*, 1918. [CrossRef]
85. Song, Q.; Cao, S.; Pritchard, R.H.; Ghalei, B.; Al-Muhtaseb, S.A.; Terentjev, E.M.; Cheetham, A.K.; Sivaniah, E. Controlled thermal oxidative crosslinking of polymers of intrinsic microporosity towards tunable molecular sieve membranes. *Nat. Commun.* **2014**, *5*, 4813. [CrossRef] [PubMed]
86. Li, F.Y.; Xiao, Y.; Chung, T.; Kawi, S. High-Performance Thermally Self-Cross-Linked Polymer of Intrinsic Microporosity (PIM-1) Membranes for Energy Development. *Macromolecules* **2012**, *45*, 1427–1437. [CrossRef]
87. Luo, S.; Zhang, Q.; Zhu, L.; Lin, H.; Kazanowska, B.A.; Doherty, C.M.; Hill, A.J.; Gao, P.; Guo, R. Highly Selective and Permeable Microporous Polymer Membranes for Hydrogen Purification and CO_2 Removal from Natural Gas. *Chem. Matter.* **2018**, *30*, 5322–5332. [CrossRef]
88. Ma, X.; Salinas, O.; Litwiller, E.; Pinnau, I. Pristine and thermally-rearranged gas separation membranes for novel o-hydroxyl-functionalized spirobifluorene-based polyimides. *Polym. Chem.* **2014**, *5*, 6913–6922. [CrossRef]
89. Bandehali, S.; Amooghin, A.E.; Sanaeepur, H.; Ahmadi, R.; Fuoco, A.; Jansen, J.C.; Shirazian, S. Polymers of intrinsic microporosity and thermally rearranged polymer membranes for highly efficient gas separation. *Sep. Purif. Technol.* **2021**, *278*, 119513. [CrossRef]
90. Janakiram, S.; Ahmadi, M.; Dai, Z.; Ansaloni, L.; Deng, L. Performance of Nanocomposite Membranes Containing 0D to 2D Nanofillers for CO_2 Separation: A Review. *Membranes* **2018**, *8*, 24. [CrossRef]
91. Liu, M.; Nothling, M.D.; Webly, P.A.; Jin, J.; Fu, Q.; Qiao, G.G. High-throughput CO_2 capture using PIM-1@MOF based thin film composite membranes. *J. Chem. Eng.* **2020**, *396*, 125328. [CrossRef]
92. Muldoon, P.F.; Venna, S.R.; Gidley, D.W.; Baker, J.S.; Zhu, L.; Tong, Z.; Xiang, F.; Hopkinson, D.P.; Yi, S.; Sekizkardes, A.K.; et al. Mixed Matrix Membranes from a Microporous Polymer Blend and Nanosized Metal-Organic Frameworks with Exceptional CO_2/N_2 Separation Performance. *ACS Mater. Lett.* **2020**, *2*, 821–828. [CrossRef]
93. Fan, S.; Wang, J.; Liao, L.; Feng, J.; Li, B.; Zhang, S. Enhanced selectivity in thin film composite membrane for CO_2 capture through improvement to support layer. *J. Chem. Eng.* **2023**, *468*, 143645. [CrossRef]
94. Lin, Z.; Yuan, Z.; Wang, K.; He, X. Synergistic tuning mixed matrix membranes by Ag^+-doping in UiO-66-NH_2/polymers of intrinsic microporosity for remarkable CO_2/N_2 separation. *J. Membr. Sci.* **2023**, *681*, 121775. [CrossRef]
95. Messaoud, S.B.; Takagaki, A.; Sugawara, T.; Kikuchi, R.; Oyama, S.T. Mixed matrix membranes using SAPO-34/polyetherimide for carbon dioxide/methane separation. *Sep. Purif. Technol.* **2015**, *148*, 38–48. [CrossRef]
96. Li, X.; Ma, L.; Zhang, H.; Wang, S.; Jiang, Z.; Guo, R.; Wu, H.; Cao, X.; Yang, J.; Wang, B. Synergistic effect of combining carbon nanotubes and graphene oxide in mixed matrix membranes for efficient CO_2 separation. *J. Membr. Sci.* **2015**, *479*, 1–10. [CrossRef]
97. Wang, S.; Li, X.; Wu, H.; Tian, Z.; Xin, Q.; He, G.; Peng, D.; Chen, S.; Yin, Y.; Jiang, Z.; et al. Advances in high permeability polymer-based membrane materials for CO_2 separations. *Energy Environ. Sci.* **2016**, *9*, 1863–1890. [CrossRef]
98. Seong, M.K.; Yu, H.J.; Ha, S.Y.; Chang, W.S.; Kim, H.; Lee, J.S. Poly(poly(ethylene glycol) methyl ether acrylate) micelles for highly CO_2 permeable membranes. *J. Membr. Sci.* **2022**, *662*, 120917. [CrossRef]
99. Jiang, X.; Li, S.; Shao, L. Pushing CO_2-philic membrane performance to the limit by designing semi-interpenetrating networks (SIPN) for sustainable CO_2 separations. *Energy Environ. Sci.* **2017**, *10*, 1339–1344. [CrossRef]
100. Chen, S.; Zhao, D.; Feng, Y.; Liu, H.; Li, S.; Qiu, Y.; Ren, J. The preparation and characterization of gel-mixed matrix membranes (g-MMMs) with high CO_2 permeability and stability performance. *J. Membr. Sci.* **2022**, *652*, 120471. [CrossRef]

101. Guo, H.; Wei, J.; Deng, J.; Yi, S.; Wang, B.; Deng, L.; Jiang, X.; Dai, Z. Facilitated transport membranes for CO_2/CH_4 separation—State of the art. *Adv. Membr.* **2022**, *2*, 100040. [CrossRef]
102. Hong, C.; Leo, C.P.; Ahmad, N.N.R.; Ahmad, A.L.; Mohammad, A.W. Polyvinyl alcohol membrane incorporated with amine-modified silica nanoparticles and ionic liquid for facilitated transport of CO_2. *Int. J. Greenh. Gas Control* **2022**, *120*, 103774. [CrossRef]
103. Costa, S.P.F.; Azevedo, A.M.O.; Pinto, P.C.A.G.; Saraiva, M.L.M.F.S. Environmental Impact of Ionic Liquids: Recent Advances in (Eco)toxicology and (Bio)degradability. *Eur. J. Chem.* **2017**, *10*, 2321–2347. [CrossRef]
104. Han, Y.; Ho, W.S.W. Recent advances in polymeric facilitated transport membranes for carbon dioxide separation and hydrogen purification. *J. Polym. Sci.* **2020**, *58*, 2435–2449. [CrossRef]
105. IEAGHG. *Effecs of Impurities on Geological Storage of CO_2*; IEAGHG: Cheltenham, UK, 2011.
106. Anantharaman, R.; Berstad, D.; Roussanaly, S. Techno-economic Performance of a Hybrid Membrane—Liquefaction Process for Post-combustion CO_2 Capture. *Energy Procedia* **2014**, *61*, 1244–1247. [CrossRef]
107. Sreenath, S.; Sam, A.A. Hybrid membrane-cryogenic CO_2 capture technologies: A mini-review. *Front. Energy Res.* **2023**, *11*, 1167024. [CrossRef]
108. Belaissaoul, B.; Le Moullec, Y.; Willson, D.; Favre, E. Hybrid membrane cryogenic process for post-combustion CO_2 capture. *J. Membr. Sci.* **2012**, *415–416*, 424–434. [CrossRef]
109. Zhang, Z.; Pan, S.; Li, H.; Cai, J.; Olabi, A.G.; Anthony, E.J.; Manovic, V. Recent advances in carbon dioxide utilization. *Renew. Sustain. Energy Rev.* **2020**, *25*, 109799. [CrossRef]
110. Kim, B.; Ma, S.; Jhong, H.M.; Kenis, P.J. Influence of dilute feed and pH on electrochemical reduction of CO_2 to CO on Ag in a continuous flow electrolyzer. *Electrochim. Acta* **2015**, *166*, 271–276. [CrossRef]
111. Tsuji, T.; Sorai, M.; Shiga, M.; Fujikawa, S.; Kunitake, T. Geological storage of CO_2-N_2-O_2 mixtures produced by membrane-based direct air capture (DAC). *Greenh. Gases Sci. Technol.* **2021**, *11*, 610–618. [CrossRef]
112. Longo, M.; De Santo, M.P.; Esposito, E.; Fuoco, A.; Monteleone, M.; Giorno, L.; Comesaña-Gándara, B.; Chen, J.; Bezzu, C.G.; Carta, M.; et al. Correlating Gas Permeability and Young's Modulus during the Physical Aging of Polymers of Intrinsic Microporosity Using Atomic Force Microscopy. *Ind. Eng. Chem. Res.* **2020**, *12*, 5381–5391. [CrossRef]
113. Carta, M.; Bernardo, P.; Clarizia, G.; Jansen, J.C.; McKeown, N.B. Gas Permeability of Hexaphenylbenzene Based Polymers of Intrinsic Microporosity. *Macromol.* **2014**, *47*, 8320–8327. [CrossRef]
114. Santiago-Garcia, J.L.; Álvarez, C.; Sánchez, F.; de la Campa, J.G. Gas transport properties of new aromatic polyimides based on 3,8-diphenylpyrene-1,2,6,7-tetracarboxylic dianhydride. *J. Membr. Sci.* **2015**, *476*, 442–448. [CrossRef]
115. Ma, X.; Pinnau, I. A novel intrinsically microporous ladder polymer and copolymers derived from 1,1′,2,2′-tetrahydroxytetraphenylethylene for membrane-based gas separation. *Polym. Chem.* **2016**, *7*, 1244–1248. [CrossRef]
116. Wang, Z.; Wang, D.; Jin, J. Microporous Polyimides with Rationally Designed Chain Structure Achieving High Performance for Gas Separation. *Macromol.* **2014**, *47*, 7477–7483. [CrossRef]
117. Ma, X.; Zhu, Z.; Shi, W.; Ji, W.; Li, J.; Wang, Y.; Pinnau, I. Unprecedented gas separation performance of difluoro-functionalized triptycene-based ladd PIM membrane at low temperature. *J. Mater. Chem. A* **2021**, *9*, 5404–5414. [CrossRef]
118. Ghanem, B.S.; Swaidan, R.; Ma, X.; Litwiller, E.; Pinnau, I. Energy-Efficient Hydrogen Separation by AB-Type Ladder-Polymer Molecular Sieves. *Adv. Mater.* **2014**, *26*, 6696–6700. [CrossRef] [PubMed]
119. Ghanem, B.S.; Alghunaimi, F.; Wang, Y.; Genduso, G.; Pinnau, I. Synthesis of Highly Gas-Permeable Polyimides of Intrinsic Microporosity Derived from 1,3,6,8-Tetramethyl-2,7-diaminotriptycene. *ACS Omega* **2018**, *3*, 11874–11882. [CrossRef] [PubMed]
120. Hu, X.; Lee, W.H.; Bae, J.Y.; Zhao, J.; Kim, J.S.; Wang, Z.; Yan, J.; Lee, Y.M. Highly permeable polyimides incorporating Tröger's base (TB) units for gas separation membranes. *J. Membr. Sci.* **2020**, *615*, 118533. [CrossRef]
121. Ma, X.; Ghanem, B.; Salines, O.; Litwiller, E.; Pinnau, I. Synthesis and Effect of Physical Aging on Gas Transport Properties of a Microporous Polyimide Derived from a Novel Spirobifluorene-Based Dianhydride. *ACS Macro Lett.* **2015**, *4*, 231–235. [CrossRef] [PubMed]
122. Ma, X.; Abdulhamid, M.A.; Pinnau, I. Design and Synthesis of Polyimides Based on Carbocylic Pseudo-Tröger's Base-Derived Dianhydrides for Membrane Gas Separation Applications. *Macromolecules* **2017**, *50*, 5850–5857. [CrossRef]
123. Esposito, E.; Mazzei, I.; Monteleone, M.; Fuoco, A.; Carta, M.; McKeown, N.B.; Mapass-Evans, R.; Jansen, J.C. Highly Permeable Matrimid®/PIM-EA(H2)-TB Blend Membrane for Gas Separation. *Polymers* **2019**, *11*, 46. [CrossRef]
124. Carta, M.; Croad, M.; Malpass-Evans, R.; Jansen, J.C.; Bernardo, P.; Clarizia, G.; Friess, K.; Lanč, M.; McKeown, N.B. Triptycene Induced Enhancement of Membrane Gas Selectivity for Microporous Tröger's Base Polymers. *Adv. Mater.* **2014**, *26*, 3526–3531. [CrossRef]
125. Fritsch, D.; Bengtson, G.; Carta, M.; McKeown, N.B. Synthesis and Gas Permeation Properties of Spirobischromane-based Polymers of Intrinsic Microporosity. *Macromol. Chem. Phys.* **2011**, *212*, 1137–1146. [CrossRef]
126. Yuan, K.; Liu, C.; Zhang, S.; Jiang, L.; Liu, C.; Yu, G.; Wang, J.; Jian, X. Phthalazinone-based copolymers with intrinsic microporosity (PHPIMs) and their separation performance. *J. Membr. Sci.* **2017**, *541*, 403–412. [CrossRef]
127. Jiang, X.; Goh, K.; Wang, R. Air plasma assisted spray coating of Pebax-1657 thin-film composite membranes for post-combustion CO_2 capture. *J. Membr. Sci.* **2022**, *658*, 120741. [CrossRef]
128. Zhang, G.; Tran, N.T.; Huang, L.; Deng, E.; Blevins, A.; Guo, W.; Ding, Y.; Lin, H. Thin-film composite membranes based on hyperbranched poly(ethylene oxide) for CO_2/N_2 separation. *J. Membr. Sci.* **2022**, *644*, 120184. [CrossRef]

129. Li, S.; Wang, Z.; Zhang, C.; Wang, M.; Yuan, F.; Wang, J.; Wang, S. Interfacially polymerized thin film composite membranes containing ethylene oxide groups for CO_2 separation. *J. Membr. Sci.* **2013**, *436*, 121–131. [CrossRef]
130. Chua, M.L.; Xiao, Y.C.; Chung, T. Effects of thermally labile saccharide units on the gas separation performance of highly permeable polyimide membranes. *J. Membr. Sci.* **2012**, *415-416*, 375–382. [CrossRef]
131. Do, Y.S.; Lee, W.H.; Seong, J.G.; Kim, J.S.; Wang, H.H.; Doherty, C.M.; Hill, A.J.; Lee, Y.M. Thermally rearranged (TR) bismaleimide-based network polymers for gas separation membranes. *Chem. Commun.* **2016**, *52*, 13556–13559. [CrossRef] [PubMed]
132. Kim, S.; Jo, H.J.; Lee, Y.M. Sorption and transport of small gas molecules in thermally rearranged (TR) polybenzoxazole membranes based on 2,2-bis(3-amino-4-hydroxyphenyl)-hexafluoropropane (bisAPAF) and 4,4′-hexafluoroisopropylidene diphthalic anhydride (6FDA). *J. Membr. Sci.* **2013**, *441*, 1–8. [CrossRef]
133. Han, S.H.; Lee, J.E.; Lee, K.; Park, H.B.; Lee, Y.M. Highly gas permeable and microporous polybenzimidazole membrane by thermal rearrangement. *J. Membr. Sci.* **2010**, *357*, 143–151. [CrossRef]
134. Choi, J.I.; Jung, C.H.; Han, S.H.; Park, H.B.; Lee, Y.M. Thermally rearranged (TR) poly(benzoxazole-co-pyrrolone) membranes tuned for high gas permeability and selectivity. *J. Membr. Sci.* **2010**, *349*, 358–368. [CrossRef]
135. Du, N.; Dal-Cin, M.M.; Robertson, G.P.; Guiver, M.D. Decarboxylation-Induced Cross-Linking of Polymers of Intrinsic Microporosity (PIMs) for Membrane Gas Separation. *Macromolecules* **2012**, *45*, 5134–5139. [CrossRef]
136. Li, F.Y.; Xiao, Y.; Ong, Y.K.; Chung, T. UV-Rearranged PIM-1 Polymeric Membranes for Advanced Hydrogen Purification and Production. *Adv. Energy Mater.* **2012**, *2*, 1456–1466. [CrossRef]
137. Sun, Y.; Zhang, J.; Li, H.; Fan, F.; Zhao, Q.; He, G.; Ma, C. Ester-crosslinked polymers of intrinsic microporosity membranes with enhanced plasticization resistance for CO_2 separation. *Sep. Purif. Technol.* **2023**, *314*, 123623. [CrossRef]
138. Hazazi, K.; Ma, X.; Wang, Y.; Ogieglo, W.; Alhazmi, A.; Han, Y.; Pinnau, I. Ultra-selective carbon molecular sieve membranes for natural gas separations based on a carbon-rich intrinsically microporous polyimide precursor. *J. Membr. Sci.* **2019**, *585*, 1–9. [CrossRef]
139. Chuah, C.Y.; Lee, J.; Bao, Y.; Song, J.; Bae, T. High-performance porous carbon-zeolite mixed-matrix membranes for CO_2/N_2 separation. *J. Membr. Sci.* **2021**, *622*, 119031. [CrossRef]
140. Chi, W.S.; Kim, S.J.; Lee, S.; Bae, Y.; Kim, J.H. Enhanced Performance of Mixed-Matrix Membranes through a Graft Copolymer-Directed Interface and Interaction Tuning Approach. *ChemSusChem* **2014**, *8*, 650–658. [CrossRef] [PubMed]
141. Min, H.J.; Kang, M.; Bae, Y.; Blom, R.; Grande, C.A.; Kim, J.H. Thin-film composite mixed-matrix membrane with irregular micron-sized UTSA-16 for outstanding gas separation performance. *J. Membr. Sci.* **2023**, *669*, 121295. [CrossRef]
142. Luo, W.; Niu, Z.; Mu, P.; Li, J. MXene/poly(ethylene glycol) mixed matrix membranes with excellent permeance for highly efficient separation of CO_2/N_2 and CO_2/CH_4. *Colloids Surf. A Physico. Chem. Eng. Asp.* **2022**, *640*, 128481. [CrossRef]
143. Pu, Y.; Yang, Z.; Wee, V.; Wu, Z.; Jiang, Z.; Zhao, D. Amino-functionalized NUS-8 nanosheets as fillers in PIM-1 mixed matrix membranes for CO_2 separations. *J. Membr. Sci.* **2022**, *641*, 119912. [CrossRef]
144. Khdhayyer, M.R.; Esposito, E.; Fuoco, A.; Monteleone, M.; Giorno, L.; Jansen, J.C.; Attfield, M.P.; Budd, P.M. Mixed matrix membranes based on UiO-66 MOFs in the polymer of intrinsic microporosity PIM-1. *Sep. Purif. Technol.* **2017**, *173*, 304–313. [CrossRef]
145. Lee, C.S.; Kang, M.; Kim, K.C.; Kim, J.H. In-situ formation of asymmetric thin-film, mixed-matrix membranes with ZIF-8 in dual-functional imidazole-based comb copolymer for high-performance CO_2 capture. *J. Membr. Sci.* **2022**, *642*, 119913. [CrossRef]
146. Wu, X.; Tian, Z.; Wang, S.; Peng, D.; Yang, L.; Wu, Y.; Xin, Q.; Wi, H.; Jiang, Z. Mixed matrix membranes comprising polymers of intrinsic microporosity and covalent organic framework for gas separation. *J. Membr. Sci.* **2017**, *528*, 273–283. [CrossRef]
147. Hao, L.; Liao, K.; Chung, T. Photo-oxidative PIM-1 based mixed matrix membranes with superior gas separation performance. *J. Mater. Chem. A* **2015**, *3*, 17273–17281. [CrossRef]
148. Kinoshita, Y.; Wakimoto, K.; Gibbons, A.H.; Isfahani, A.P.; Kusuda, H.; Sivaniah, E.; Ghalei, B. Enhanced PIM-1 membrane gas separation selectivity through efficient dispersion of functionalized POSS fillers. *J. Membr. Sci.* **2017**, *539*, 178–186. [CrossRef]
149. Husna, A.; Hossain, I.; Choi, O.; Lee, S.; Kim, T. Efficient CO_2 Separation Using a PIM-PI-Functionalized UiO-66 MOF Incorporated Mixed Membrane in a PIM-PI-1 Polymer. *Macromol. Mater. Eng.* **2021**, *306*, 2100298. [CrossRef]
150. Sun, H.; Gao, W.; Zhang, Y.; Cao, X.; Bao, S.; Li, P.; Kang, Z.; Niu, J. Bis(phenyl)fluorene-based polymer of intrinsic microporosity/functionalized multi-walled carbon nanotubes mixed matrix membranes for enhanced CO_2 separation performance. *React. Funct. Polym.* **2020**, *147*, 104465. [CrossRef]
151. Sánchez-Laínez, J.; Pardillos-Ruiz, A.; Carta, M.; Malpass-Evans, R.; McKeown, N.B.; Téllez, C.; Coronas, J. Polymer engineering by blending PIM-1 and 6FDA-DAM for ZIF-8 containing mixed matrix membranes applied to CO_2 separations. *Sep. Purif. Technol.* **2019**, *224*, 456–462. [CrossRef]
152. Ji, Y.; Zhang, M.; Guan, K.; Zhao, J.; Liu, G.; Jin, W. High-Performance CO_2 Capture through Polymer-Based Ultrathin Membranes. *Adv. Funct. Mater.* **2019**, *29*, 1900735. [CrossRef]
153. Khdhayyer, M.; Bushell, A.F.; Budd, P.M.; Attfield, M.P.; Jiang, D.; Burrows, A.D.; Esposito, E.; Bernardo, P.; Monteleone, M.; Fuoco, A.; et al. Mixed matrix membranes based on MIL-101 metal–organic frameworks in polymer of intrinsic microporosity PIM-1. *Sep. Purif. Technol.* **2019**, *212*, 545–554. [CrossRef]
154. Kim, N.U.; Park, B.J.; Lee, J.H.; Kim, J.H. High-performance ultrathin mixed-matrix membranes based on an adhesive PGMA-co-POEM comb-like copolymer for CO_2 capture. *J. Mater. Chem. A* **2019**, *7*, 14723–14731. [CrossRef]

155. Deng, G.; Wang, Y.; Luo, J.; Zong, X.; Zhang, C.; Song, X. Synthesis and gas transport properties of hyperbranched network polyimides derived from Tris(4-aminophenyl) benzene. *Polymer* **2020**, *203*, 122776. [CrossRef]
156. Scofield, J.M.P.; Gurr, P.A.; Kim, J.; Fu, Q.; Kentish, S.E.; Qiao, G.G. Development of novel fluorinated additives for high performance CO_2 separation thin-film composite membranes. *J. Membr. Sci.* **2016**, *499*, 191–200. [CrossRef]
157. Hao, L.; Li, P.; Chung, T. PIM-1 as an organic filler to enhance the gas separation performance of Ultem polyetherimide. *J. Membr. Sci.* **2014**, *453*, 614–623. [CrossRef]
158. Yong, W.F.; Li, F.Y.; Xiao, Y.C.; Li, P.; Pramoda, K.P.; Tong, Y.W.; Chung, T.S. Molecular engineering of PIM-1/Matrimid blend membranes for gas separation. *J. Membr. Sci.* **2012**, *407-408*, 47–57. [CrossRef]
159. Yechan, L.; Chuah, C.Y.; Lee, J.; Bae, T. Effective functionalization of porous polymer fillers to enhance CO_2/N_2 separation performance of mixed-matrix membranes. *J. Membr. Sci.* **2022**, *647*, 120309. [CrossRef]
160. Park, C.; Kong, C.; Kim, E.; Lee, C.; Kim, K.; Lee, J.; Moon, S. High-flux CO_2 separation using thin-film composite polyether block amide membranes fabricated by transient-filler treatment. *J. Chem. Eng.* **2023**, *455*, 140883. [CrossRef]
161. Jeong, I.; Hossain, I.; Husna, A.; Kim, T. Development of CO_2-Philic Blended Membranes Using PIM-PI and PIM-PEG/PPG. *Macromol. Mater. Eng.* **2022**, *308*, 2200596. [CrossRef]
162. Sekizkardes, A.K.; Budhathoki, S.; Zhu, L.; Kusuma, V.; Tong, Z.; McNally, J.S.; Steckel, J.A.; Yi, S.; Hopkinson, D. Molecular design and fabrication of PIM-1/polyphosphazene blend membranes with high performance for CO_2/N_2 separation. *J. Membr. Sci.* **2021**, *640*, 119764. [CrossRef]
163. Yong, W.F.; Li, F.Y.; Chung, T.S.; Tong, Y.W. Molecular interaction, gas transport properties and plasticization behavior of cPIM-1/Torlon blend membranes. *J. Membr. Sci.* **2014**, *462*, 119–130. [CrossRef]
164. Liu, J.; Pan, Y.; Xu, J.; Wang, Z.; Zhu, H.; Liu, G.; Zhong, J.; Jin, W. Introducing amphipathic copolymer into intermediate layer to fabricate ultra-thin Pebax composite membrane for efficient CO_2 capture. *J. Membr. Sci.* **2023**, *667*, 121183. [CrossRef]
165. Han, H.; Scofield, J.M.P.; Gurr, P.A.; Webley, P.A.; Qiao, G.G. Ultrathin membrane with robust and superior CO_2 permeance by precision control of multilayer structures. *J. Chem. Eng.* **2023**, *462*, 142087. [CrossRef]
166. Li, S.; Wang, Z.; Yu, X.; Wang, J.; Wang, S. High-Performance Membranes with Multi-permselectivity for CO_2 Separation. *Adv. Mater.* **2012**, *24*, 3196–3200. [CrossRef] [PubMed]
167. Yuan, Y.; Qiao, Z.; Xu, J.; Wang, J.; Zhao, S.; Cao, X.; Wang, Z.; Guiver, M.D. Mixed matrix membranes for CO_2 separations by incorporating microporous polymer framework fillers with amine-rich nanochannels. *J. Membr. Sci.* **2021**, *620*, 118923. [CrossRef]
168. Gao, Y.; Quiao, Z.; Zhao, S.; Wang, Z.; Wang, J. In situ synthesis of polymer grafted ZIFs and application in mixed matrix membrane for CO_2 separation. *J. Mater. Chem. A* **2018**, *6*, 3151–3161. [CrossRef]
169. Li, N.; Wang, Z.; Wang, J. Biomimetic hydroxypropyl-β-cyclodextrin (Hβ-CD)/polyamide (PA) membranes for CO_2 separation. *J. Membr. Sci.* **2023**, *668*, 121211. [CrossRef]
170. Li, N.; Wang, Z.; Wang, J. Water-swollen carboxymethyl chitosan (CMC)/polyamide (PA) membranes with octopus-branched nanostructures for CO_2 capture. *J. Membr. Sci.* **2022**, *642*, 119946. [CrossRef]
171. Shen, Q.; Cong, S.; Zhu, J.; Zhang, Y.; He, R.; Yi, S.; Zhang, Y. Novel pyrazole-based MOF synergistic polymer of intrinsic microporosity membranes for high-efficient CO_2 capture. *J. Membr. Sci.* **2022**, *664*, 121107. [CrossRef]
172. Blunt, M.; Fayers, F.J.; Orr, F.M., Jr. Carbon dioxide in enhanced oil recovery. *Energy Convers. Manag.* **1993**, *34*, 1197–1204. [CrossRef]
173. Esposito, E.; Dellamuzia, L.; Moretti, U.; Fuoco, A.; Giorno, L.; Jansen, J.C. Simultaneous production of biomethane and food grade CO_2 from biogas: An industrial case study. *Energy Environ. Sci.* **2019**, *12*, 281–289. [CrossRef]
174. Carbon Dioxide (CO_2) Purity Grade Chart | CO2Meter.com. Available online: https://www.co2meter.com/en-jp/blogs/news/tagged/carbon-dioxide-co2?page=3 (accessed on 20 September 2022).
175. Iglina, T.; Iglin, P.; Pashchenko, D. Industrial CO_2 Capture by Algae: A Review and Recent Advances. *Sustainability* **2022**, *14*, 3801. [CrossRef]
176. Greenhouse Carbon Dioxide Supplementation. Available online: https://extension.okstate.edu/fact-sheets/greenhouse-carbon-dioxide-supplementation.html (accessed on 30 September 2023).

membranes

MDPI

Review

Biocatalytic Membranes for Carbon Capture and Utilization

Jialong Shen and Sonja Salmon *

Department of Textile Engineering, Chemistry and Science, Wilson College of Textiles, North Carolina State University, Raleigh, NC 27695-8301, USA; jshen3@ncsu.edu
* Correspondence: sisalmon@ncsu.edu

Abstract: Innovative carbon capture technologies that capture CO_2 from large point sources and directly from air are urgently needed to combat the climate crisis. Likewise, corresponding technologies are needed to convert this captured CO_2 into valuable chemical feedstocks and products that replace current fossil-based materials to close the loop in creating viable pathways for a renewable economy. Biocatalytic membranes that combine high reaction rates and enzyme selectivity with modularity, scalability, and membrane compactness show promise for both CO_2 capture and utilization. This review presents a systematic examination of technologies under development for CO_2 capture and utilization that employ both enzymes and membranes. CO_2 capture membranes are categorized by their mode of action as CO_2 separation membranes, including mixed matrix membranes (MMM) and liquid membranes (LM), or as CO_2 gas–liquid membrane contactors (GLMC). Because they selectively catalyze molecular reactions involving CO_2, the two main classes of enzymes used for enhancing membrane function are carbonic anhydrase (CA) and formate dehydrogenase (FDH). Small organic molecules designed to mimic CA enzyme active sites are also being developed. CO_2 conversion membranes are described according to membrane functionality, the location of enzymes relative to the membrane, which includes different immobilization strategies, and regeneration methods for cofactors. Parameters crucial for the performance of these hybrid systems are discussed with tabulated examples. Progress and challenges are discussed, and perspectives on future research directions are provided.

Keywords: biocatalyst; carbonic anhydrase; CO_2 capture; CO_2 reduction; enzyme; formate dehydrogenase; immobilization; membrane

1. Introduction

Carbon dioxide (CO_2) interacts with atmospheric, oceanic, and terrestrial biospheres [1] and is a part of feedback mechanisms that are responsible for natural glacial cycles [2]. Within the past century, anthropogenic CO_2 emissions have skyrocketed, causing an abrupt increase in atmospheric CO_2 levels [3] which is speeding up global warming and risks irreversible climate changes [4]. Carbon capture technologies that reduce the emissions from large point sources and capture legacy emissions directly from air are necessary to combat this crisis [5]. Converting captured CO_2 into valuable chemical feedstocks and products that replace current fossil-based materials is increasingly important for its double benefit and has incentivized the growth of an emerging market for carbon utilization [6].

Already, there are four types of industrially used CO_2 capture technologies. They include absorption in chemical or physical solvents, adsorption on solid absorbents, cryogenic processes, and membrane-based separations [7]. Each have their own advantages and limitations depending on the specific implementation conditions, such as temperature, pressure, and CO_2 concentrations. It is therefore common to combine more than one technology in efforts to reduce the overall cost of the capture process. For example, a conventional membrane separation process can be assisted by a cryogenic unit to improve CO_2 purity to a level that is not attainable by the membrane alone [8]. In another hybrid example, the porous structure of a gas–liquid membrane contactor (GLMC) is used to increase the contact surface areas of a gas stream with conventional chemical absorption solvents to achieve higher CO_2 capture efficiency with a smaller footprint [9]. The development of more sophisticated hybrid systems that combine the relatively newer concepts of metal organic frameworks (MOFs), ionic liquid (IL), and enzyme-based CO_2 capture and reactor systems, together with conventional CO_2 capture technologies, offer new possibilities for carbon capture and utilization applications [10]. By comparing their features, this review examines the range of hybrid technologies that employ both enzymes and membranes, and provides critical analyses of the promises and challenges within the field of biocatalytic membranes for carbon capture and utilization (CCU). This review provides a concise account of both membrane and biocatalysis concepts relevant for CCU. The analyses and discussions are intended to stimulate inspiration for novel ideas, and especially, to encourage collaboration among researchers from different academic backgrounds.

1.1. *Enzymes for CO_2 Capture and Utilization*

1.1.1. Carbonic Anhydrases

Among the many chemical reactions catalyzed by enzymes in biological systems, the hydration and dehydration reactions of CO_2 and bicarbonate (Equation (1)) related to cell respiration and blood pH homeostasis are among the most critical. Carbonic anhydrases (CAs) are a ubiquitous class of metal-containing enzymes found in all domains of life. At ideal conditions, CAs can catalyze the conversion of up to the order of one million molecules of substrate per molecule of enzyme per second [11]. Interest in using CA for engineering purposes was initially limited to CO_2 separation from life supporting closed spaces [12], such as submarines and spacecraft, or in biomedical devices, such as artificial lungs [13]. Now interest has been rekindled for its potential industrial-scale use [14] in mitigating the negative effects of CO_2 on climate change.

$$CO_2 + H_2O \leftrightarrow H^+ + HCO_3^- \quad (1)$$

In recent years, CA has been evaluated both in a dissolved form [15] or immobilized on packing materials [16,17] as a promoter in conventional CO_2 chemical absorption processes for enhancing CO_2 capture from power plant emissions using benign low regeneration energy solvents, such as aqueous solvents containing potassium carbonate, K_2CO_3. These inorganic salt-based solvents enhanced by CA offer the special advantage of minimizing pore wetting for novel gas–liquid membrane contactor applications, which is discussed in detail below. With the goal of finding lower cost synthetic alternatives, various CA mimic enzymes such as Zinc-based [18] and Cobalt-based [19] catalysts have been devised. Because water (moisture) is essential for CA enzymes to provide an enhancement effect, membrane developments that utilize CA and its mimics focus on "wet" membranes as the catalytic reaction cannot occur in dry membranes [20].

Like many other enzymes, CAs generally have low tolerance for elevated temperature environments, such as those encountered in the CO_2 stripper of reactive absorption processes. However, some CAs are naturally thermostable [14], and non-natural variants have been made by protein engineering to create ultra-thermostable CAs that can tolerate temperatures of up to 107 °C with pH > 10.0 in amine solvents [21]. Furthermore, enzyme immobilization can circumvent CA instability by retaining CAs in the lower temperature absorber column, thus preventing them from being exposed to the high temperature environment, and can also stabilize CAs against denaturing in harsh solvents for improved longevity [22].

1.1.2. Formate Dehydrogenases

Due to its vast quantity, captured CO_2 needs to be stored permanently, or ideally turned into valuable materials that displace CO_2 emissions from traditional processes. Such processes include the cement and chemical industries, which account for 7% of global CO_2 emissions and 7% of all oil extracted, respectively [23]. Formate dehydrogenase (FDH) catalyzes the reversible reduction of CO_2 to formic acid (Equation (2)), a commodity chemical with an existing market of more than one million tons/year [24]. Theoretically, the market size would grow much larger if efforts to use formate for liquid hydrogen fuel storage [25] or as a carbon source for microbial growth and biosynthesis of higher carbon chemicals [26,27] can be realized.

$$CO_2 + H^+ + 2e^- \leftrightarrow HCOO^- \quad (2)$$

The two major categories of FDH are metal-independent and metal-dependent. Metal-independent FDHs have limited CO_2 reduction activity [28], preferring to catalyze the formate oxidation reaction. CO_2 reduction requires a proton and two reducing electrons, which are supplied by enzyme cofactors, such as the reduced form of nicotinamide adenine dinucleotide (NADH). To convert each mole of CO_2, an equal molar amount of NADH is converted to the oxidized NAD^+ form. This means that NADH must be continuously supplied or regenerated to operate enzymatic membrane reactors continuously. Efficient regeneration of natural cofactor or artificial electron donors and carriers is, therefore, the bottleneck in enzyme efficiency that has spurred many innovative strategies reflected in the recent literature in this field. For example, Song et al. [29] covalently linked copper nanoparticles (CuNPs) with FDH for the regeneration of the enzyme cofactor NADH that is also tethered on the FDH through a flexible polyethylene glycol (PEG) swing arm. In addition, the rhodium bipyridine complex has been widely used to modify electrode surfaces for cofactor regeneration [30].

1.1.3. Enzyme Cascade with Other Oxidoreductases

Enzymatically produced formic acid has many downstream pathways for utilization. Enzymatic conversion to formaldehyde and methanol, both of which are among the top 10 petrochemicals produced in the world, is of interest in the search for sustainable low-temperature processes [28]. Redox reactions of formic acid catalyzed by formaldehyde dehydrogenase (FaldDH) and then alcohol dehydrogenase (ADH) produce formaldehyde and methanol in sequence. Such an enzyme cascade reaction that incorporates relevant enzymes in the same system in close proximity to each other could have additive rate enhancement effects due to raised local substrate concentration and reduced product inhibition [28]. However, there are several common limitations on current oxidoreductase for CO_2 reduction systems that need to be addressed. These include low enzyme activity in the carbon reduction direction, low efficiency of cofactor regeneration, low efficiency of electron transfer in cofactor-free systems, and CO_2 solubility and mass transfer limitations [31]. Research activities in protein engineering, enzyme immobilization [32], and reactor design and integration with complementary systems [33] have all contributed to overcoming these limitations [31]. Details of recent advances in performance and understanding of cascade systems involving membranes are discussed below.

1.1.4. Enzyme Immobilization

Enzymes are commonly immobilized on solid carriers—membranes being one of the most versatile—to improve their stability, facilitate their reuse, and reduce overall processing and chemical conversion costs [34]. Membranes, as enzyme carriers, can either be fabricated before (pre-existing) or during (formed in situ) the enzyme immobilization process.

On pre-existing (often commercially available) membrane carriers, enzymes are immobilized through various mechanisms including adsorption, covalent bonding, and affinity binding [32]. Sun et al. [35] used water plasma for treating a polyvinylidene fluoride (PVDF) flat sheet membrane, followed by silanization to introduce amine and epoxide groups on the membrane surface for subsequent covalent attachment of enzymes. Another versatile surface coating reagent, dopamine (DA), can impart amine functionalities on almost any type of material surface. Sun et al. co-deposited polyethylenimine (PEI) with DA on PVDF and polyethylene (PE) membranes and covalently attached CA enzyme through glutaraldehyde (GA) crosslinking with high activity recovery and excellent reusability for enhancing CO_2 mineralization through $CaCO_3$ precipitation [36]. In a technique that is especially suitable for membrane carriers, enzymes can be immobilized by direct and deliberate membrane fouling. Luo et al. [37] used a simple pressure driven filtration system to immobilize FDH, FaldDH, and ADH into flat sheet membrane pores for biocatalytic production of methanol from CO_2 without the need for chemical reagents.

Enzymes can also be immobilized in matrices by encapsulation or entrapment mechanisms during the phase inversion process of membrane fabrication. Ren et al. [38] encapsulated CA in zeolitic imidazolate framework 8 (ZIF-8), a type of MOF, and embedded the biocatalytic MOF in a poly(vinyl alcohol)/chitosan (PVA/CS) composite membrane. In a $CaCO_3$ precipitation test, the membrane structures formed during the immobilization process assisted biocatalyst recovery and reuse and exhibited a 1.63-fold improvement, compared with non-membrane MOF containing CA. Wen et al. [39] created CA nanoflowers, prepared by co-crystallization with Cu^{2+} and Zn^{2+} metal ions, and then embedded these in a CS/PVA hydrogel membrane. The amount of $CaCO_3$ produced by the biocatalytic composite membrane was nine-fold and two-fold higher compared with the free CA or dispersed nanoflowers alone, respectively. Enzyme immobilization by entrapment is the method most commonly used to fabricate mixed matrix membranes (MMM) with CA, or with CA mimics, for facilitated CO_2 transport separation applications [19].

While enzyme immobilization is its own extensive discipline, the introduction above should suffice for purposes of the current review topic. Interested readers are referred to classical protocol compilations for more detailed discussions about specific techniques for enzyme immobilization [40,41]. In addition, as awareness of biomimetic CO_2 mitigation solutions increases, the field of CA immobilization has matured. These advances are found in recent review articles on CA immobilization for CO_2 capture technologies [42], its industrial implementation [43], and its use in reactive absorption processes [44]. Reviews on enzyme immobilization for biocatalytic membranes in general are also available [32].

1.1.5. Comparisons of Biocatalysts with Electrocatalysts for CO_2 Reduction Reaction

When the reducing electrons are supplied by an external power source, the biocatalyst is effectively functioning as an electrocatalyst. However, there are several characteristics that make biocatalysts different in term of advantages and disadvantages. Enzymes have three-dimensional active sites that are able to have very specific interactions with a particular substrate from a mixture of similar compounds [45]. This high selectivity is one of the most sought-after features of biocatalysts. However, the drawback of high substrate selectivity is that the CO_2 reduction reaction can only proceed one step at a time by matching each specific substrate with a specific type of enzyme. This means a cascade of multiple enzyme systems is needed for complete oxidation or reduction of substrate to product. An example of the intricate reaction cascades utilized by nature is the Krebs's cycle, which is essential to the energy generation in cells [46]. Secondly, enzymes are complex folded long-chain protein molecules that are, for the most part, insulators against electron transfers. Therefore, effectively "wiring" enzymes to electrodes becomes especially challenging [47]. This is not the case for inorganic electrocatalysts used for CO_2 reduction reactions, including metal alloys, metal oxides, metal chalcogenides and others, where increasing selectivity and current density and reducing over-potential, are the major research obstacles within the field [48]. Recently, Saxena et al. reported copper selenide [49] and cobalt telluride [50] that are able to reduce CO_2 to C2 products, such as acetic acid, with greater than 80% Faradaic efficiency (FE) and 75% selectivity at a low applied potential. In a nickel selenide electrocatalyst system, the selectivity of the product can be controlled by the applied potential, and an FE of over 98% can be achieved for acetic acid at lower applied potential [51]. By fine tuning the CO intermediate adsorption energy on the active site using a bimetal copper cobalt selenide electrocatalyst, an FE of 100% towards C2 products such as ethanol and acetic acid was achieved [52]. For comparison, the FE of electrocatalysts used for CO_2 reduction reactions in the literature vary significantly from 3% to over 90% [48], while in a limited number of reports, the FE of biocatalysts varied from 10% for a three enzyme cascade system that produced methanol by cofactor-free direct electron transfer [53] to 23% for a single immobilized enzyme and cofactor regenerating hybrid system that produced formic acid [29]. Both electrocatalytic and bio-electrocatalytic CO_2 reduction research fields are in their infancy, and knowledge can be learned from each to towards a common goal.

1.2. Types of Membranes for CO_2 Capture and Utilization

Membranes are a promising technology platform for CO_2 capture because they are modular, scalable, and compact. This makes them desirable for process intensification and reducing energy costs [54,55]. Membranes encompass many different types of materials and functionality. In this discussion, to distinguish biocatalytic membranes according to their configurations and separation mechanisms, membranes are loosely divided into two categories, based on the physical states of the fluids separated by the membrane: CO_2 gas separation membranes and CO_2 gas–liquid membrane contactors (Figure 1).

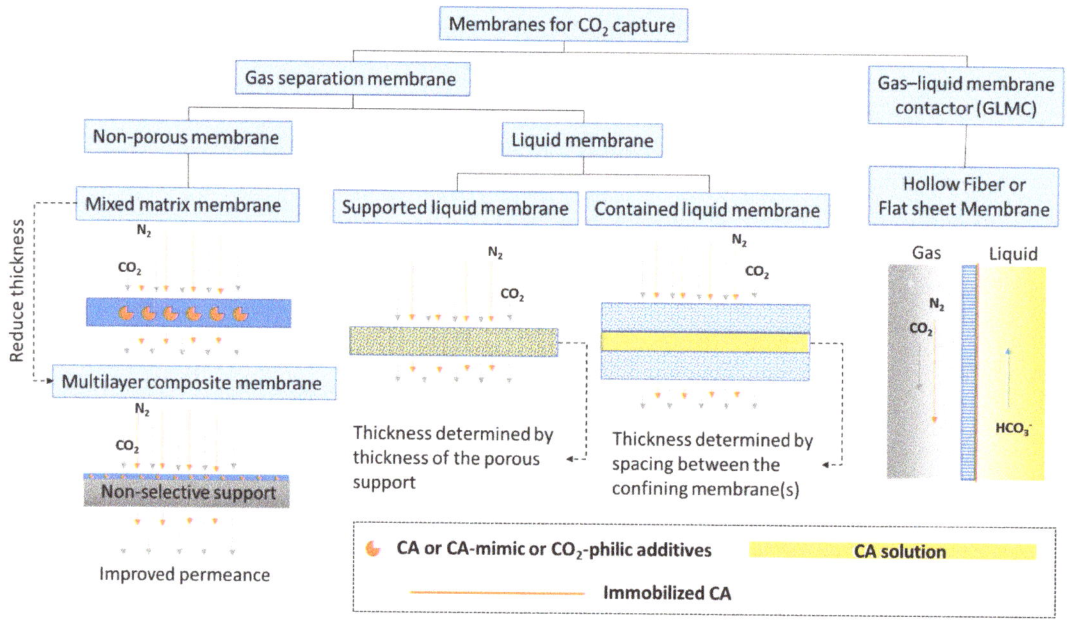

Figure 1. Categories of biocatalytic membranes used for CO_2 capture.

1.2.1. CO_2 Separation Membrane

A membrane that separates two gas phases on either side—CO_2 lean gas mixture on the feed and CO_2 enriched gas phase on the permeate side—is called a CO_2 separation membrane. This category encompasses a large selection of membrane types from non-porous glassy polymer membranes, fixed-site carrier membranes [56], and ultrathin nanocomposite membranes [57], to contained liquid membranes [58]. Research efforts on CO_2 separation membranes have focused on improving performance-limiting membrane properties, such as CO_2 gas permeance and selectivity [59]. New classes of polymer materials, such as polymers of intrinsic microporosity (PIM) [60,61] and ladder polymers [62], have been invented that show superior CO_2 separation properties well above the empirical Robeson upper-bound [63], which classically delineates the trade-off relationship between gas permeability and selectivity. However, physical aging is still an issue that needs to be solved. This problem is common to all glassy polymer membranes, including in the new classes of materials, albeit to a lesser extent owing to the presence of inherent structural porosities. In one case, treatment with super critical CO_2 altered the internal structure of a

PIM, leading to decreased CO_2 permeance [64]. In another case, after being physically aged, ladder polymers showed increased selectivity but decreased permeability [62], indicating a decreased free volume. To alleviate physical aging issues in glassy non-porous polymer membranes, inorganic aging-resistant CO_2-philic components are added to the polymer matrix to form mixed matrix membranes (MMM). Recently, Tan et al. [65] discovered a new method for adding high loadings of zeolite into a polyimide membrane matrix that achieved a CO_2/CH_4 mixed-gas selectivity of ~423 and a CO_2 permeability of ~8300 Barrer at moderate pressure and ambient temperature. To put these numbers in perspective, at a similar CO_2/CH_4 selectivity of 400, the 2008 Robeson upper-bound for the CO_2/CH_4 pair anticipates a CO_2 permeability of only ~1 Barrer [63].

In order to improve the overall sustainability profile of CO_2 separation technologies, biopolymer-based MMM, such as chitosan-based non-porous membranes, have recently emerged as alternatives to conventional non-renewable polymer matrices [66]. Casado-Coterillo et al. [67] fabricated a chitosan MMM filled with metal organic framework (MOF) and non-toxic ionic liquid that achieved a high permeability of 4754–5413 Barrer (or 47–52 GPU) and a CO_2/N_2 selectivity of 12–19. Borgohain et al. [68] synthesized carboxymethyl chitosan as a matrix for compatibilization with scarcely soluble multi-walled carbon nanotubes (MWCNT) to make a thin MMM selective layer (2.7 μm) that exhibited a CO_2 permeance of 43 GPU and a CO_2/N_2 selectivity of 45. The hydrophilicity and free amine groups of the chitosan material could be contributing to the excellent CO_2 transport properties, especially in humidified conditions, compared with the commercial hydrophobic membranes [69]. Owing to their abundance in nature, tailorable functional groups, and excellent membrane forming properties, chitosan [70] and other polysaccharides [71,72], could play an increasing role in the fabrication of novel CO_2 separation membranes.

Another way to improve membrane performance is by making thin film composites (TFC) [73] or integrated multilayer membranes [74], both with ultra-thin CO_2 selective layers for facilitated CO_2 transport. CA and CA mimics have been successfully used to construct both MMM and thin CO_2 selective layers for facilitated CO_2 separation [75,76]. However, these advanced facilitated transport membranes are still at lab-scale and no direct comparison between these and commercial scale CO_2 chemical absorption processes is available in the literature. Nevertheless, a recent techno-economic analysis (TEA) study compared a non-facilitated polymeric membrane process (Membrane Technology and Research, Inc., Newark, CA, USA) [77] to an enzyme-based chemical absorption process (Akermin Inc., St. Louis, MO, USA) and found that the latter is economically more attractive in a simulated CO_2 capture scenario from a 600 MW_e power plant flue gas. This result emphasizes the potential for enzymes to improve energy efficiency of conventional energy intensive processes. Interestingly, the study also predicted that the membrane technology could become more efficient if CO_2 permeance at low pressure (<1.5 bar) could be enhanced. Because CA is already particularly efficient at converting CO_2 to bicarbonate at ambient pressure, developing low pressure facilitated CO_2 transport membranes that utilize the fast enzyme reaction rate is a promising concept.

Liquid membranes that separate two gas phases are also defined as CO_2 separation membranes. CA plays a similar CO_2 hydration facilitator role in liquid membranes, provided there is water present. General types of liquid membranes include supported liquid membranes and contained liquid membranes (Figure 1). Sometimes distinctions are made between supported liquid membranes (SLM) and immobilized liquid membranes (ILM), where in the first case, liquid fills spaces between fibers in the membrane and the second case, liquid fills specific pores in the membrane [78]. However, most of the time, these two nomenclatures are used interchangeably. Disadvantages of common SLMs or ILMs include the formation of gravity-induced downward bulges in the liquid phase (called catenary curves), low tolerance to transmembrane pressure differences, and a high evaporation tendency. All of these problems can be alleviated by contained liquid membrane configurations in which liquid is bound by porous membrane surfaces [78]. Different types of liquid can be used to construct liquid membranes, including hydrogels [79], ionic liquids [80], deep eutectic solvents [81], and aqueous buffers [82]. Both flat sheet and hollow fiber membranes are commonly used. Considering the variety of configurations and liquid types and the large number of associated publications, liquid membranes are further discussed in a separate section from facilitated non-porous polymer membranes.

1.2.2. CO_2 Liquid Contactor Membrane

A membrane that separates a gas phase containing CO_2 from a liquid phase where CO_2 is absorbed, is categorized as a CO_2 liquid contactor membrane (Figure 1). This category emerged as a new hybrid membrane system, called gas–liquid membrane contactors (GLMC), that combines the modularity and high surface area of the membrane with the high selectivity of the chemical absorption process [83,84]. Non-enzymatic GLMC developments have focused on improving membrane stability [85], minimizing pore wetting [86], and selecting the best solvent and activator [9]. Reviews of modeling methods used to analyze the mass transfer in hollow fiber gas–liquid membrane contactors (HFGLMC) for post-combustion carbon capture are available [83]. Improvements to membrane materials were also explored by blending polysulfone (PSf) with PEI, a CO_2-philic polymer. The observed optimal additive ratio for higher capture performance was attributed to chemical affinity, whereas non-optimal conditions inadvertently caused pore wetting and clogging by K_2CO_3 precipitation [87].

Another way to improve GLMC performance is increasing the mass transfer of CO_2 at the gas–liquid interface catalyzed by CA enzymes, which are either immobilized on the membrane [88], dissolved in the solvent [89], or immobilized both on the membrane and on mobile nanoparticles dispersed in the solvent for additional process intensification [90]. A recent TEA study compared a CA-immobilized hollow fiber membrane contactor (HFMC) with benign solvent and vacuum-assisted solvent regeneration with the benchmark case where monoethanolamine (MEA) was used in a conventional packed column process. The projection estimated that at 90% CO_2 capture from a 685 MW_e coal-fired power plant, the enzymatic process achieved a 43% reduction in energy consumption of the capture and compression unit, a 31% reduction in capital cost (CAPEX), and a 28% reduction in operating expenses (OPEX) in comparison with the MEA benchmark [91]. Enzyme-based GLMC is discussed in more detail below.

1.2.3. Other Membrane Structure Functions

The simplest definition of a membrane is a thin layer that acts as a boundary or barrier. This barrier can prevent random mass exchange based on size or physical phase, or can provide protection against harsh environments. Membranes used for CO_2 conversion and utilization applications may require different or added functionality compared with those used for CO_2 capture. For example, as shown in Figure 2, an ultraviolet (UV) protective membrane was used to block UV irradiation and simultaneously retain enzymes (based on their large size) on the biocatalysis side [92], while allowing small cofactor molecules to freely pass between the separate photocatalytic and biocatalytic reaction chambers.

Additionally, membranes provide ample surface area for enzymes to be immobilized, and therefore, can provide high catalytic enhancement. Considering the membrane's separation function, when substrate is delivered as dissolved CO_2-saturated water [35,36], the membrane structure creates a localized environment where the CO_2 conversion reaction can take place continuously in the liquid phase. Membranes can separate either dissolved or immobilized biocatalysts from products [38,39] (exemplified by two schematics under "Separation" in Figure 2). The importance of this seemingly simple function of solid–liquid separation and recovery of enzymes should not be underestimated. An evaluation of using ultrafiltration membranes to separate dissolved enzymes from a CO_2-rich solvent [93], to avoid pumping the enzymes into a high temperature desorber for solvent regeneration, found that even with an enzyme retention rate as high as 99.9%, only 50% of the enzymes are retained after 1 month of operation. Therefore, strategies that prevent enzymes from leaching through or away from membranes can be critical. Biocatalyst retention by immobilization is especially important for operating enzymatic membrane reactors for CO_2 reduction catalyzed by oxidoreductases.

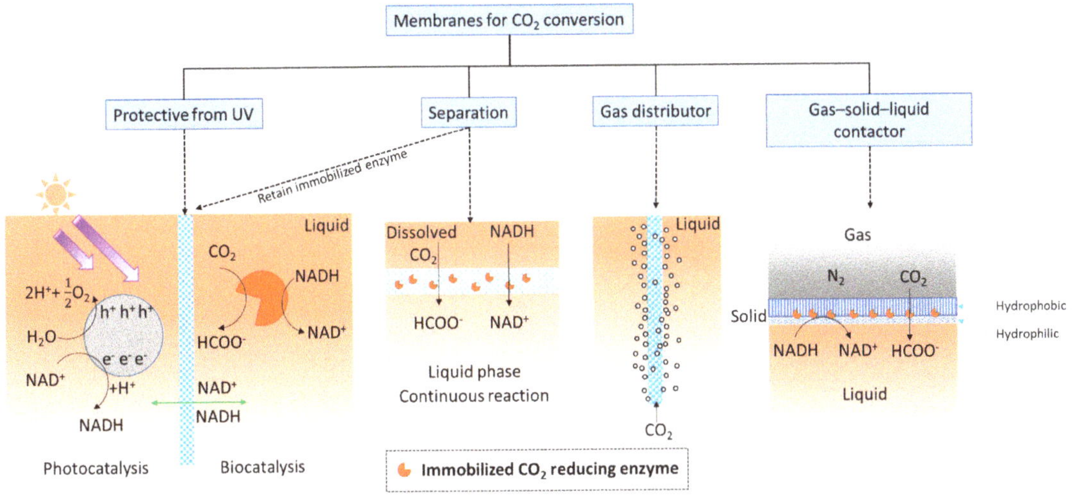

Figure 2. Functionalities of membranes used for biocatalytic CO_2 conversion.

As illustrated under 'gas distributor' in Figure 2, porous membranes, specifically porous hollow fiber membranes, can be used to infuse gaseous CO_2 into the reaction medium [94] to increase the availability of soluble CO_2. This approach is often used in conjunction with adjacent sets of hollow fiber membranes with immobilized enzymes attached [95]. In addition, since gaseous CO_2 is attracted to hydrophobic surfaces, amphiphilic membranes functioning as gas–solid–liquid contactors (Figure 2, right schematic) have recently been developed for converting gaseous CO_2 into water soluble formic acid [96].

Specific examples and the performance of these various membrane functionalities are discussed in detail in Section 5.

2. Facilitated Transport Membranes

2.1. CA vs. CA-Mimic

Published examples of adding CA enzymes directly into the polymer matrix of a separation membrane are rare. The concerns with this approach include enzyme tolerance to membrane fabrication solvents and quality of protein dispersion in the polymer solution. In one example, Zheng et al. [20] compared Pebax-1657 mixed matrix membrane (MMM) doped with either CA-mimic in the form of a Zinc-coordinated metal organic framework (MOF) or with bare CA enzymes. CA-doped MMM fabricated by solvent evaporation showed a 4- to 5-fold improvement compared with the CO_2 permeability of the Pebax-1657 control at a pH range of 7–10. The CA-mimic doped MMM demonstrated an 8- to 9-fold enhancement over a wider pH range of 5–11, which was attributed to the CA-mimic's dual function of resembling the CA enzyme active site and benefiting from a well-defined MOF pore structure for gas transport. In another study, Zhang et al. [75] encapsulated a high loading amount (~24 wt%) of CA enzyme in Zinc-coordinated 2-Methylimidazole metal organic framework (ZIF-8) cavities grown in situ on an oriented halloysite nanotube (HNT) layer supported by a polyacrylonitrile (PAN) porous membrane. The optimal CO_2/N_2 selectivity was 166, which was about 21-fold higher than the negative control membranes containing denatured CA. Because the only difference between the two was the activity of the CA, the enhancement was attributed solely to the catalytic effects of the CA enzyme.

Most of the literature on biomimetic CO_2 facilitated transport membranes only uses CA-mimics for catalytic enhancement. The most common CA-mimics are Zn–cyclen [76],

Cobalt-coordinated Co-2,6-bis(2-benzimidazolyl)pyridine (Co-BBP) [19], Zinc-coordinated histidine-based bolaamphiphile (His-Bola) [97], and Zinc-coordinated bibenzotriazoles (H_2-bibta) [20]. Comparisons of the active sites of CA-mimics with CA enzymes are provided in Figure 3. Although the reaction rates per active site of the CA-mimics are not comparable with those of the CA enzymes, their lower molecular weights make it possible to load a large number of active sites per unit mass of synthetic polymer membrane, which partially offsets the lower activity. Additionally, high thermal stability (>200 °C) and a wide effective pH range make CA-mimics a popular choice for fabricating facilitated transport membranes [76]. This situation may change as ultrastable CA enzymes [21] become more readily available.

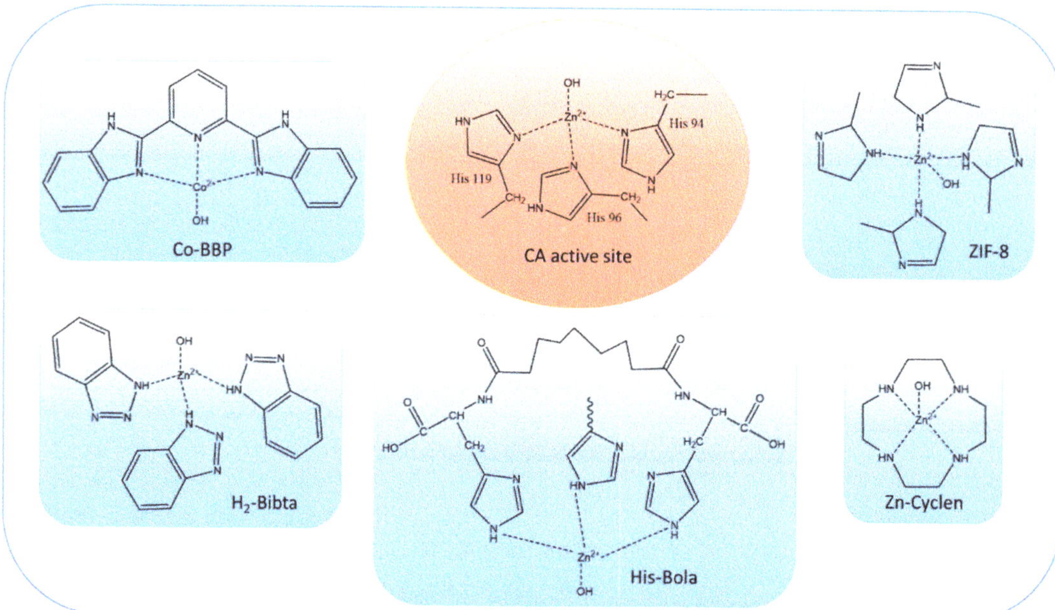

Figure 3. Chemical structures of active sites of CA enzyme (shown in orange oval) and CA-mimics (shown in blue rectangles).

2.2. Membrane Structures

Two common pathways to improve CO_2 separation performance of non-porous membranes are the creation of mixed matrix membranes (MMM) or multilayer composite membranes with a CO_2 selective layer. The former involves the incorporation of solid CO_2-philic additives in a continuous polymer matrix, which combines advantages from both materials. The later reduces membrane thickness to improve CO_2 permeance by supporting a thin CO_2 selective layer on a less selective but more sturdy gas permeable membrane.

Zhang et al. [19] dispersed 1.33 wt% of CoBBP CA-mimic in Pebax-1657 (PEO:PA6 polyamide 60:40 wt%) matrix through a dissolution–evaporation process for MMM and achieved a CO_2 permeability of 675.5 Barrer and a selectivity of 62, which is above the empirical Robeson upper-bound. Nilouyal et al. [97] incorporated up to 9 wt% of Zinc-coordinated His-Bola nanoparticles in Pebax-1657 matrix and obtained a higher CO_2/N_2 selectivity of 158.2 but a lower CO_2 permeability of 188.4 Barrer, which also surpassed the Robeson upper-bound. However, after accounting for their thickness of 70–75 μm, these MMMs had CO_2 permeances in the single digits, at 9 and 3 GPU, respectively. To increase CO_2 permeance, additional CO_2 transport mechanisms are needed. Zheng et al. [20] incorporated a CA-mimic that forms MOF porous structures itself and supplements the catalytic function with additional CO_2 diffusion pathways. Similarly, Wang et al. [98] compounded 28.5 wt% CoBBP CA-mimic with porous organic polymers (POP) that contain inherent microporous structures before mixing them into the Pebax matrix. Both of these examples increased the permeability significantly, with CO_2 permeances of around 30 GPU at a membrane thickness of 50 μm.

Membrane thickness is a bottleneck for achieving high CO_2 permeance. The solution is to construct multi-layer composite membranes with a thin selective layer for providing gas selectivity and thicker gas permeable layers as physical supports for providing mechanical stability. For example, Jahan et al. [99] made cast poly(vinyl alcohol) (PVA) MMM containing 1 wt% crystalline nanocellulose and 5 μmol/g Zn-cyclen on a polysulfone (PSf) ultrafiltration membrane. The thickness of the selective MMM layer was 800 nm with a CO_2 permeance of 126 GPU and a CO_2/CH_4 selectivity of 42, which are significantly higher than the prior examples. Saeed et al. [100] used a similar strategy in combining 1 wt% carbon nanotubes (CNT) as nano-filler with 5 μmol/g Zn-cyclen in PVA to create a selective layer with a thickness of 830 nm supported on PSf membrane. Adding CNT resulted in a higher water swelling capability of the PVA membrane and a high CO_2 permeance of 363 GPU and a CO_2/N_2 selectivity of 120.

In special cases, where CO_2 needs to be separated from smaller gas molecules, such as H_2 and He, the thickness of the membrane cannot be reduced without affecting CO_2 selectivity. For example, the CO_2 permeance of a poly(amidoamine) (PAMAM) dendrimer/poly(ethylene glycol) (PEG) hybrid membrane, which works by a facilitated transport mechanism through amino groups of the PAMAM [101], is lower with an inverse relationship with the membrane thickness ($Q_{CO_2} \propto L^{-0.62}$), whereas hydrogen permeance has an almost inverse relationship with the thickness ($Q_{H_2} \propto L^{-0.95}$). Therefore, the selectivity of CO_2 over H_2 decreased with a reduced membrane thickness. However, when CA enzyme was spray-coated on the sweep side of the hybrid membrane, the membrane's CO_2 permeance became nearly inversely proportional to membrane thickness ($Q_{CO_2} \propto L^{-0.98}$), making reducing membrane thickness a possible strategy for improving CO_2 permeance without sacrificing selectivity. At a membrane thickness of 15 μm, the CO_2 permeance was 490% higher compared with the no-CA membrane. In this example, rather than distributing CA enzymes or CA-mimics homogeneously throughout the polymer matrix, spray-coating CA enzymes only on the membrane surface made it possible to explore the effect of CA enzyme location on the separation properties. For a fixed membrane thickness, CA spray-coated on the feed side alone did not alter CO_2 permeance, nor did CA coated on both the feed and sweep sides perform better than CA coated on the sweep side alone. It was, therefore, concluded that the CO_2 desorption step on the sweep side was the rate limiting step that was assisted by the presence of spray-coated CA enzyme. For membranes that do not contain CO_2 transport facilitating amino groups, CA coated on both sides of the membrane may be necessary for optimal separation performance. Further studies with different membrane chemistries are needed to fully evaluate the benefits and effects of CA placement relative to the membrane, including on the interfaces and dispersed in the matrix.

2.3. Humidity

The mechanism of facilitated CO_2 transport through bicarbonate formation requires participation of water molecules, regardless of whether it is a CA enzyme [75], CA-mimic [76], or amino group facilitated reaction [101]. For membranes with dual transport mechanisms, the dominating mechanism depends on humidity, and higher humidity favors bicarbonate facilitated transport [20]. As an example, the CO_2 permeance of the CA-MOF dual-function MMM increased by 62% from dry to humidified conditions [20]. In most cases, both the CO_2 permeance and selectivity of facilitated transport membranes increase with increasing humidity, and the difference can boost certain membranes to perform CO_2 separation at above the Robeson upper-bound, even when they otherwise do not achieve this under dry conditions [97]. For context, the empirical Robeson upper-bound was initially constructed based on non-facilitated membrane performance and thus serves as a benchmark for indicating the effectiveness of facilitated mechanisms. Moisture is required for all facilitated CO_2 transport membranes described here, and the best performance values reported in Table 1 were all obtained under humidified experimental conditions if not otherwise specified.

Table 1. Facilitated transport membranes with carbonic anhydrase or enzyme mimics for CO_2 separation.

Application	Membrane Configuration	Enzyme, Concentration	Performance	Year, 1st Author, Ref
CO_2 separation from N_2	PVA selective layer containing enzyme mimic supported by PSf ultrafiltration membrane	5 μmol/g (Zn–cyclen /PVA) with 1 wt% CNT	CO_2 permeance: 256–363 GPU CO_2/N_2 selectivity: 107–120	2015–2016 Saeed [76,100]
CO_2 separation from N_2	Biocatalytic composite membranes HNTs/MOF/CA selective layer supported by PAN membrane	24.2 wt% CA in MOF	CO_2 permeance: 24.2 GPU CO_2/N_2 selectivity: 165.5	2017 Zhang [75]
CO_2 separation from N_2	MMM with Cobalt-based CA-mimic CoBBP dispersed in Pebax-1657 (PEO:PA6 polyamide 60:40 wt%)	1.33 wt% CoBBP in Pebax	CO_2 permeability: 675.5 Barrer CO_2 permeance: 9 GPU (75 μm thickness) CO_2/N_2 selectivity: 62	2018 Zhang [19]
CO_2 separation from H_2	PAMAM/PEGDMA/GMA hybrid membrane supported on PES porous support	1 wt% CA loading spray-coated on the hybrid membrane	CO_2 permeance: 14.4 GPU CO_2 permeability: 216 Barrer (15 μm thickness) CO_2/He selectivity: 28.7	2019 Duan [101]
CO_2 separation from N_2	MMM with CoBBP CA-mimic loaded on POP (CoBBP@POP) and together both were loaded in Pebax-1657 matrix	28.5 wt% CoBBP in POP. 5 wt% CoBBP@POP composite in Pebax matrix.	CO_2 permeability: 1620 Barrer CO_2 permeance: 32.4 GPU (50 μm thickness) CO_2/N_2 selectivity: 102	2020 Wang [98]
CO_2 separation from CH_4	PVA selective layer containing CA-mimic supported by PSf ultrafiltration membrane	5 μmol/g (Zn–cyclen /PVA with 1 wt% CNC)	CO_2 permeance: 126 GPU CO_2/CH_4 selectivity: 42	2021 Jahan [99]
CO_2 separation from N_2	MMM with His-NPs CA-mimic loaded in Pebax-1657 matrix	0–9 wt% His-NPs in Pebax-1657 matrix	CO_2 permeability: 188.4 Barrer CO_2 permeance: 2.7 GPU (70 μm thickness) CO_2/N_2 selectivity: 158.2	2022 Nilouyal [97]
CO_2 separation from N_2	MMM with Zinc-coordinated MOF CA-mimic loaded in Pebax-1657 matrix	3% MOF CA-mimic in Pebax-1657	CO_2 permeability: 869 (dry) Barrer 1409 (humid) Barrer CO_2 permeance: 28.2 GPU (50 μm thickness) CO_2/N_2 selectivity: 88.6 (dry) 83 (humid)	2022 Zheng [20]

Acronyms: PVA—poly(vinyl alcohol); PSf—polysulfone; GPU—gas permeation unit: 10^{-6} cm^3(STP)/(cm^2 s cm Hg); CNT—carbon nanotube; PAN—polyacrylonitrile; HNT—halloysite nanotube; MOF—metal organic frameworks; CA—carbonic anhydrase; MMM—mixed matrix membrane; CoBBP—Co-2,6-bis(2-benzimidazolyl)pyridine; PEO—poly(ethylene oxide); PA6—polyamide 6; Barrer—10^{-10} cm^3(STP) cm/(cm^2 s cmHg); PAMAM—poly(amidoamine) dendrimer; PEGDMA—poly(ethylene glycol) dimethacrylate; GMA—glycidyl methacrylate; PES—polyethersulfone; POP—porous organic polymers; CNC—crystalline nanocellulose; His-NPs—Zinc-coordinated self-assembled histidine-based bolaamphiphile nanoparticles.

3. Liquid Membranes

Strictly speaking, liquid membranes (Figure 1) promoted by CA and CA mimics are also considered facilitated transport membranes within the gas separation membrane category, where two gas phases are separated on either side of the membrane. However, the fabrication of liquid membranes is distinctly different from the non-porous facilitated transport membranes detailed in Section 2. Often, commercial porous membranes are used to construct liquid membranes, and the choice of either hydrophobic or hydrophilic materials is based on the type of liquids used and the membrane configurations. The history and recent advances of enzyme-promoted liquid membranes are discussed in this section. Their key parameters and performance metrics are summarized in Table 2.

Table 2. Liquid membranes with carbonic anhydrase for CO_2 separation.

Application	Membrane Configuration	Enzyme, Concentration	Performance	Year, 1st Author, Ref
CO_2 separation from N_2	Microporous PP HFCLM mat with heat exchanger type design (mutually orthogonal fiber orientation)	3 mg/mL CA in 1.0 M $NaHCO_3$	At 10% CO_2 CO_2 permeance: 90 GPU CO_2/N_2 selectivity: 234	2006 Bao [78]
CO_2 separation from air	Microporous PP HFCLM bundle with feed and sweep fibers intimately commingled	10 mg/L CA in poly(acrylic acid-co-acrylamide) hydrogel	Able to reduce CO_2 from 0.52% to 0.09%	2008 Cheng [79]
CO_2 separation from air	Microporous PVDF HFCLM bundle with feed and sweep fibers intimately commingled	121.8 W-A U/L CA displayed on the surface of *E. coli* suspended in water	40% increase in CO_2 removal rate, 2 times more stable than free CA	2011 Fan [102]
CO_2 separation from N_2	SLM with enzyme solution impregnating hydrophilic PVDF membrane; hybrid nylon-silica CLM sandwiched between two hydrophobic PVDF membranes	0.2 mg/mL CA in 1 M $NaHCO_3$ pH~8	CO_2 permeance: 108 GPU, silica xerogel provides additional catalytic benefit	2011 Favre [103]
CO_2 separation from N_2	SILM with porous hydrophobic PVDF membrane	0.01 wt% CA in hydrophobic $[C_4MIM][Tf_2N]$ ionic liquid or PEG 300	Max CO_2/N_2 selectivity: 48; enzyme enhancement is more profound at higher water content	2012 Neves [104]
CO_2 separation from N_2, H_2, CH_4	SILM with hydrophobic PVDF microfiltration membrane	5 mg/mL in hydrophobic $[C_4MIM][Tf_2N]$ ionic liquid	Selectivity: CO_2/N_2:30.3 CO_2/CH_4:19.9 CO_2/H_2:11.2	2016 Bednar [80]
CO_2 separation from N_2	SLM with porous hydrophilic cellulose acetate membrane reinforced by pectin	2 mg CA/mL in Tris buffer (20Mm, pH 8.3)	CO_2 permeability: 93 Barrer CO_2 permeance: 0.75 GPU (120 µm thickness) CO_2/N_2 selectivity: 54	2018 Nemestóthy [82]
CO_2 separation from N_2	ILM within 8 nm hydrophilic silica mesopores and thickness of 18 nm	2 CA per nanopore; effective conc. of 100 mg CA mL^{-1}	CO_2 permeance: 2600 GPU Selectivity: CO_2/N_2:788; CO_2/H_2:1500	2018 Fu [105]
CO_2 separation from N_2	SLM with DES filling hydrophilic PTFE microfiltration membrane	0.5 mg CA/g DES (choline chloride and levulinic acid)	CO_2 permeability: 78 Barrer CO_2/N_2 selectivity: 32 Adding CA failed to enhance selectivity	2021 de Castro [81]
CO_2 separation from N_2 and CH_4	SLM with DES filling hydrophilic PTFE microfiltration membrane	0.1 mg CA/mL DES (choline chloride and urea)	CO_2 permeability: 140 Barrer (w/CA) Selectivity: CO_2/N_2: below RUB CO_2/CH_4: above (w/o CA) and on (w/CA) RUB	2021 Craveiro [106]
CO_2 separation from CH_4	SLM with water-in-oil emulsion filling porous hydrophobic PVDF membrane	1 wt% disperse phase (0.5 g CA/L K_2CO_3 pH 11, 5% PEG 300) in corn oil with 2 wt% Tween 80	Permeability of CO_2 increased by ~15% and CO_2/CH_4 selectivity increased by 2.9-fold with CA	2022 Mondal [107]
CO_2 separation from N_2 and CH_4	SLM with hydrophilic PTFE microfiltration membrane	0.5 mg CA/g solvent (12.5/14.5/75.0 wt% ChOH/water/glycerol)	CO_2 permeability: 81 Barrer Selectivity: CO_2/N_2: 90.5	2022 Castro [108]

Acronyms: PP—polypropylene; HFCLM—hollow fiber contained liquid membrane; CA—carbonic anhydrase; GPU—gas permeation unit: 10^{-6} cm^3(STP)/(cm^2 s cmHg); PVDF—polyvinylidene fluoride; W-A—Wilbur and Anderson; SLM—supported liquid membrane; CLM—contained liquid membrane; SILM—supported ionic liquid membrane; $[C_4MIM][Tf_2N]$—1-butyl-3-methylimidazolium bis(trifluoromethanesulfonyl)imide; PEG—polyethylene glycol; ILM—immobilized liquid membrane; DES—deep eutectic solvents; RUB—Robeson upper-bound; ChOH—choline hydroxide.

3.1. Early Developments in CA-Promoted Supported Liquid Membrane (SLM)

The concept of using CA for enhancing CO_2 transport through a thin liquid membrane dates back to the late 1960s. Enns [109] demonstrated that transport enhancement across a membrane resulted from a dominating bicarbonate diffusion mechanism because at pH values higher than 6.1, bicarbonate concentration was higher than dissolved CO_2, and increased bicarbonate concentration was facilitated by the rapid interconversion between dissolved CO_2 molecules and bicarbonate ions catalyzed by CA. The supported liquid membrane (SLM) used in these tests was created by saturating a Millipore filter with nominal pore size of 0.45 μm, void space of 80%, and a thickness of 150 μm with 25 mM aqueous sodium bicarbonate solutions containing 1–10 mg/mL dissolved CA. In SLM, there is a continuous liquid phase from one side of the membrane to the other. Capillary forces hold the liquid within the open pores of the solid support. CO_2 transport enhancement of more than two orders of magnitude was possible at pH 9.0 with the highest enzyme concentration. Shortly after, Ward et al. [12] described an "immobilized liquid membrane", also a flat sheet supported liquid membrane (SLM), in which a porous cellulose acetate film was saturated with 2 N aqueous potassium bicarbonate containing 2 mg/mL of CA. An initial 6-fold enhancement in CO_2 transport was observed, but this decreased to zero after several days, indicating low stability of the liquid membrane and the CA being used. Therefore, improving the stability of the liquid membrane by reducing solvent evaporation is a primary research objective and requirement for CO_2 separation applications.

3.2. SLM with Non-Volatile Liquids

One of the strategies for retaining enzyme activity and reducing evaporative solvent loss in SLM is the use of non-volatile liquids and thermostable CA strains [110]. Neves et al. [104] chose a hydrophobic ionic liquid supported by a hydrophobic porous PVDF membrane. The addition of 0.01 wt% CA in the ionic liquid showed an enhancement effect only at higher water content, signifying a compromise between optimizing the ionic liquid internal structure and benefiting from enzymatic CO_2 hydration enhancement. The same was true when a higher CA concentration of 5 mg/mL was added and maximum CO_2 selectivity versus other gases remained low [80]. Recently, deep eutectic solvents (DES), such as the choline chloride and levulinic acid pair, have been explored to form stable SLM. However, adding CA into DES failed to improve CO_2 selectivity over N_2 [81] and in certain cases, decreased its selectivity over CH_4 [106]. Castro et al. [108] prepared a choline hydroxide-based alkaline solvent containing 14.5% water and 75% glycerol supported on a hydrophilic PTFE microfiltration membrane. With the help of 0.5 mg CA/g solvent and the presence of a large percentage of water, CO_2/N_2 selectivity increased to 90.5. Another promising case involves the formation of a stable emulsion for filling the porous membrane. Mondal et al. [107] formulated a water-in-oil emulsion with 1 wt% aqueous disperse phase in corn oil stabilized by 2 wt% Tween 80 and filled the pores of the porous hydrophobic PVDF membrane. The aqueous phase comprised 0.5 g CA per liter potassium carbonate solvent and 5 wt% PEG 300 for enzyme stabilization. With the help of the CA enzyme, CO_2 permeability increased by ~15% and CO_2/CH_4 selectivity increased by 2.9-fold.

3.3. CA-Promoted Contained Liquid Membrane (CLM)

As interest in capturing CO_2 from air and from power plant flue gas grew, new membrane configurations emerged to overcome supported liquid membrane instability. Contrary to SLMs, where hydrophilicity/hydrophobicity of the liquid and support should match in order for the liquid to fill the pores or space in between the fibers of the support [82], contained liquid membranes (CLM) use a mismatch of the hydrophilicity/hydrophobicity of the liquid and the membrane surfaces to create physical barriers that contain the liquid in a confined space (e.g., between flat membrane sheets, or between adjacent hollow fibers (HF)), forming a liquid membrane. Because of the additional physical barriers between the liquid and gas phases, CLMs tend to be less prone to the instability caused by solvent evaporation.

Bao et al. [78] constructed a CLM by surrounding a microporous PP woven HF mat, with mutually orthogonal fiber orientations between hollow fibers containing feed and sweep gases (arranged like a heat exchanger), using different CO_2 absorption liquids. The CLM with the space between the feed and sweep fibers filled with 1.0 M $NaHCO_3$ containing 3 mg/mL dissolved CA performed better than that filled with 20 wt% diethanolamine (DEA). Alternatively, Trachtenberg et al. (Carbozyme Inc., Monmouth Junction, NJ, USA) [58] immobilized CA on the liquid side wall of both feed and sweep hollow fiber microporous membranes in a contained liquid membrane (CLM) configuration to intensify the CO_2 mass transfer between the gas and liquid and back from liquid to gas at the gas–liquid interfaces. Immobilization alleviated solvent evaporation and enzyme instability, and 85% CO_2 removal was achieved in a preliminary test using 15.4% CO_2, a concentration typical for flue gas from coal-fired power plants. Zhang et al. [111] immobilized CA in a poly(acrylic acid-co-acrylamide)/hydrotalcite (PAA-AAm/HT) nanocomposite hydrogel, in the interstitial space between feed and sweep hollow fiber membranes, and reported the immobilized CA retained over 76% activity. However, the bicarbonate diffusion rate through the hydrogel phase was found to be the rate-limiting step of the overall process. The same was concluded about enzyme-based facilitated transport contained liquid membranes (EBCLM), where increases in liquid thickness led to increases in CO_2 selectivity but decreases in CO_2 permeance [112].

3.4. Liquid Membrane Thickness

Whether it is a SLM or CLM, the thickness of the liquid membrane, which is determined by the thickness of the support membrane or the distance between two surfaces of the support membrane, greatly influences the CO_2 permeance of the liquid membrane. Typical liquid membrane thicknesses reported in the literature range from 10s to 100s μm, which could result in low CO_2 permeance even if their permeabilities are high. To decrease the liquid membrane thickness and greatly improve CO_2 permeance, Fu et al. [105] fabricated an inorganic membrane with close-packed arrays of silica mesopores with a diameter of 8 nm and total thickness (depth) of 1 μm, with the top 18 nm being hydrophilic and the rest being hydrophobic. The hydrophilic portions of the pores were filled with CA solution through capillary forces with an average of 2 CA enzymes per 18 nm thick liquid film in each mesopore. This drastic reduction in liquid layer thickness and high nominal CA concentration inside the channel resulted in a remarkably high CO_2 permeance of 2600 GPU, and high selectivities of 788 and 1500 for CO_2/N_2 and CO_2/H_2, respectively, were achieved at room temperature and atmospheric pressure. This example experimentally verified the hypothesis that a reduced liquid layer thickness coupled with high enzyme loading can produce a liquid membrane with outstanding permeance and selectivity. Such encouraging results argue for more studies to demonstrate this phenomenon with other types of support membranes, which will require advancements in respective material processing techniques for fabricating well-controlled nano-sized structures.

4. Gas–Liquid Membrane Contactor

4.1. Advantages Compared with Conventional Gas Separation Membrane and Chemical Absorption

An important practical challenge for conventional CO_2 separation membranes lies in the limited partial pressure driving force at atmospheric application conditions. The CO_2 concentration in the flue gas of a coal-fired power plant is only ~14% and is even lower in natural gas power plant emissions at only ~5% [113]. Both are emitted at near atmospheric pressure, leading to a CO_2 partial pressure of only 0.05–0.14 bar. Gas permeation occurs when the partial pressure of CO_2 on the feed side is greater than the partial pressure of CO_2 on the permeate side:

$$n_f \times p_f \geq n_p \times p_p \tag{3}$$

where, n_f is the molar concentration of CO_2 in the feed, p_f is the feed pressure, n_p is the molar concentration of CO_2 in the permeate, and p_p is the permeate pressure. Rearrangement of Equation (3) shows that the maximum separation that can be achieved by a membrane is limited by the ratio of the feed and permeate pressures:

$$\frac{p_f}{p_p} \geq \frac{n_p}{n_f} \quad (4)$$

A feed to permeate pressure ratio greater than one is needed to drive the CO_2 enriching process, no matter how selective the membrane is, and the compression and vacuum cost could easily make the whole process unaffordable [8].

Chemical reactive absorption processes using primary amine solvents have been the benchmark technology for capturing CO_2 at atmospheric pressure owing to the high CO_2 removal rate. However, broad adoption of the technology has been impeded by corresponding high regeneration energy consumption (high OPEX) and large equipment size (high CAPEX) [114]. One solution for reducing the high energy penalty is by using lower energy solvents such as sterically hindered amines, tertiary amines, and inorganic carbonate-based solvents [115]. However, the CO_2 absorption rates of these solvents are often low despite having higher absorption capacity on a mole basis than the primary amine benchmark, monoethanolamine (MEA) [116]. In combination with CA enzymes, however, low energy solvents have the potential to meet MEA absorption efficiency [117].

In addition, the CO_2 absorption rate can be improved with mass transfer intensification strategies, such as the use of gas–liquid membrane contactors (GLMC), which provide a well-defined interfacial area that is orders of magnitude higher than conventional packed columns [118] along with modularity and a small footprint. Additional advantages of using GLMC stem from its non-dispersive nature which helps avoid common problems in conventional packed columns, such as flooding, foaming, channeling, and entrainment [119]. GLMC modules are typically assembled as a bundle of hydrophobic microporous hollow fibers packed parallel in a shell. Gas and liquid usually flow counter-currently on opposite sides of the membrane (Figure 1). Leimbrink et al. [118] compared the CO_2 absorption performance of conventional packed column, rotating packed beds, and a GLMC in 30% MDEA, with and without 0.2 wt% CA. Without CA, the three contacting devices had similar CO_2 absorption per effective interfacial area. With CA, although the membrane contactor showed lower enhancement from the CA enzyme (due to the large specific surface area of the membrane contactor making it seem less effective from enzyme enhancement), it achieved the highest CO_2 absorption per contactor volume, requiring only a quarter of the size to achieve the same absorption required for a packed column process.

4.2. Developments in CA-Promoted GLMC

Interest in improving the CO_2 removal rate of artificial lung devices prompted the early development of enzyme immobilized GLMC for CO_2 transport. In contrast to conventional GLMC used for CO_2 absorption, the essential function of an artificial lung is to act similar to a desorber contactor, where CO_2, delivered as a liquid flow in the form of dissolved bicarbonate ions, is quickly converted to CO_2 gas and carried away by the oxygen sweep gas. The oxygen sweep gas acts as the low CO_2 partial pressure driving force for desorption. This desorption reaction is catalyzed by CA immobilized at the surface of the hollow fiber membranes. Federspiel and coworkers [13,120–123] utilized plasma-based surface modification techniques to introduce hydroxyl or amino functional groups on hollow fiber membranes (HFM) followed by surface chemical activation using cyanogen bromide or glutaraldehyde, both of which are reactive toward lysine amino groups in CA. Enhancements of 115% and 37% in CO_2 removal rates were observed from buffer and blood, respectively, after optimizing surface treatment and enzyme immobilization conditions, including the use of chitosan as a tether polymer. Further increases in enzyme activity, by use of a highly active recombinant human CA, did not yield additional improvement to the CO_2 removal rate due to the lack of sufficient driving force at physiological conditions for CA to push the reaction toward bicarbonate dehydration at the fluid boundary layer. This emphasizes that CA offers a benefit when the reaction is kinetically limited in the presence of a driving force but does not overcome equilibrium dominated conditions. However, when the local equilibrium was altered by introducing acidity at the boundary layer by using an acidic sweep gas, a 109% increase in the CO_2 removal rate was achieved while still maintaining the bulk blood at physiological pH. This further emphasizes that altering the local equilibrium is a fundamental strategy by which enhanced CO_2 mass transfer can be achieved in the presence of CA. While GLMC has only been explored somewhat for CO_2 stripping [124,125], advances in artificial lungs with immobilized CA on HFM could potentially provide alternatives to the high temperature CO_2 stripping process which consumes the most energy in conventional chemical absorption processes.

The research field of CA-immobilized GLMC for CO_2 absorption is expanding [126] as more attention is placed on the negative effects of greenhouse gas emissions on our environment and efforts increase to find more efficient and sustainable solutions to mitigate this issue. Representative research studies with their materials selection and key performance outcomes are summarized in Table 3. The dominant research themes include membrane surface modification, enzyme immobilization, absorption solvent development, and reducing mass transfer barriers.

Table 3. Gas liquid membrane contactor with carbonic anhydrase for CO_2 absorption and desorption.

Application	Membrane Configuration	Enzyme, Concentration	Performance	Year, 1st Author, Ref
Artificial lungs CO_2 desorption	PMP HFGLMC	Immobilized CA up to 88% theoretical monolayer coverage, 0.3 U esterase activity	Rates of CO_2 exchange from buffer increased by 75% with immobilized CA	2007 Kaar [13]
Artificial lungs CO_2 desorption	PMP and PP HFGLMC	Immobilized CA 0.99–8.8 U esterase activity	CO_2 removal rate increased by 115% and 37% from buffer and from blood, respectively	2012–2015 Arazawa [120–123]
CO_2 absorption	PP flat sheet membrane with LbL polyelectrolytes PEI/PSS/PAH/MSNP	440 μg CA cm^{-2} per layer tested up to 3 layers	CO_2 hydration rate of 19 ± 4 μmol cm^{-2} min^{-1} per layer tested up to 3 layers using CO_2-saturated buffer	2015 Yong [88]
CO_2 absorption	Hydrophobic PVDF flat sheet membrane with TiO_2 coating	700 μg CA cm^{-2}	CO_2 hydration rate of 140 μmol cm^{-2} min^{-1} nominal membrane area using CO_2-saturated buffer	2015 Hou [127]
CO_2 absorption	PVDF flat sheet Janus membrane with fluorosilane-treated superhydrophobic and CNT-coated hydrophilic sides	165 ± 22 μg CA cm^{-2}	CO_2 hydration rate of 1.32 μmol cm^{-2} min^{-1} from 100% CO_2 gas to pure water	2015 Hou [128]
CO_2 absorption	PP- or TiO_2-coated superhydrophobic PP HFGLMC	200 μg immobilized CA (on TiO_2 NP)/mL suspended in absorption buffer	CO_2 hydration rate of 0.96 μmol cm^{-2} min^{-1} from 20% CO_2 gas mixture to buffer	2016 Hou [129]
CO_2 absorption	GLMC with a tubular porous glass membrane with hydrophobic coating on the outer skin	10 mM enzyme mimic in 0.5 M K_2CO_3	10.8 μmol cm^{-2} min^{-1} from 10% CO_2 gas mixture to 1 M NaOH; 10-fold increase in rate constant using 10 mM enzyme mimic in 0.5 M K_2CO_3	2016 Saeed [130]
CO_2 absorption	HFGLMC with porous PP or non-porous Psf with PDMS coating; LbL polyelectrolytes PEI/PSS/PAH for enzyme adsorption	Three tri-layers (PSS/PAH/CA) and about 0.15 mg CA in HFGLMC.	3-fold improvement in CO_2 absorption rate in 30 wt% K_2CO_3 with immobilized CA which retained > 80% activity after exposure to common contaminant from flue gas but did not tolerate high pH combined with high temperature	2016–2017 Yong [131,132]

Table 3. *Cont.*

Application	Membrane Configuration	Enzyme, Concentration	Performance	Year, 1st Author, Ref
CO_2 absorption	HFGLMC based on hydrophobic porous PP (bulk and pores) and a dense hydrophilic PVDF layer and PIL coating on flue gas side	1 wt% CA concentrate dissolved in 30% MDEA or MEA.	2.2- and 1.7-fold enzyme enhancement in 30% MDEA with (CO_2 flux 0.41 μmol cm^{-2} min^{-1} from 15% CO_2) or without PIL coating, respectively; negative effect for adding CA in 30% MEA	2017 Kim [89]
CO_2 absorption from CO_2/Xe mixture	Dense flat sheet PDMS GLMC separating gas and liquid phase microfluidic channels with alveolar design	0.1 mg CA/g of CP ionic liquid with water activity of 0.753	Enzyme has no effect on Xe transport but has 1.9-fold enhancement for CO_2 absorption	2018 Malankowska [133]
CO_2 absorption	PVDF HFGLMC with co-deposited PDA/PEI for enzyme immobilization	498 U esterase activity per m^2 membrane	15 μmol cm^{-2} min^{-1}, 150% higher than non-biocatalytic membrane	2019 Xu [134]
CO_2 absorption	Electrospun PSMA nanofiber membrane as enzyme carrier and gas–liquid contacting surface positioned by flotation device	10 mg CA/mg nanofiber membrane	CO_2 hydration rate 8.9 μmol cm^{-2} min^{-1} from 100% CO_2 gas	2020 Kim [135]
CO_2 absorption	MOF grown on Al_2O_3 membrane filter for enzyme adsorption	0.1 mg CA/membrane or 75 μg CA/cm^2 nominal area	CO_2 hydration rate 108 μmol cm^{-2} min^{-1} from 5% CO_2 gas into water	2021 Liu [136]
CO_2 absorption	Flat sheet PP GLMC with co-deposited PEI/PDA for enzyme immobilization	94.3 μg CA/cm^2	CO_2 hydration rate 1.74 μmol cm^{-2} min^{-1} from 15% CO_2 into 100 mM Tris buffer	2021 Rasouli [137]
CO_2 absorption	Biocatalytic Flat sheet PP GLMC and MNP both were co-deposited with PEI/PDA and used for enzyme immobilization	6.49–65.44 mg CA/$L_{reactor}$	CO_2 hydration rate 1.7 μmol cm^{-2} min^{-1} from 15% CO_2 into 100 mM Tris buffer	2022 Rasouli [90]

Acronyms: PMP—poly(methyl pentene); HFGLMC—hollow fiber gas liquid membrane contactor; CA—carbonic anhydrase; PP—polypropylene; LbL—layer-by-layer assembly; PEI—polyethylenimine; PSS—poly(styrene sulfonate); PAH—poly(allylamine hydrochloride); MSNP—mesoporous silica nanoparticle; PVDF—polyvinylidene fluoride; CNT—carbon nanotube; NP—nanoparticle; GLMC—gas liquid membrane contactor; PSf—polysulfone; PIL—poly(ionic liquids); MDEA—methyldiethanolamine; MEA—monoethanolamine; CP—cholinium propionate; PDA—polydopamine; PSMA—poly(styrene-co-maleic anhydride); Al_2O_3—aluminum oxide; MNP—magnetic nanoparticle.

4.3. Materials and Surface Modifications

The role of a membrane in GLMC is to separate the liquid phase from the gas phase and provide well-defined non-dispersive contacting areas between the two phases. Therefore, the most commonly used type of membranes are hydrophobic porous membranes, including, poly(methyl pentene) (PMP) [13], polypropylene (PP) [88], and polyvinylidene fluoride (PVDF) [134]. Serving the same function, tubular porous glass membranes can also be used after applying hydrophobic coatings on the outer membrane skin [130]. Surface modifications, such as coating with sol-gel TiO_2, have been implemented on PVDF membranes to improve enzyme compatibility [127]. However, decreased water contact angle and liquid entry pressure were observed, and severe pore blockages occurred after multiple sol-gel coating cycles. Superhydrophobic coating, 1H,1H,2H,2H-perfluorodecyltriethoxysilane (PDTS), was applied on a TiO_2 functionalized PP membrane, leading to improved operational stability while inadvertently also increasing membrane resistance [129].

In addition to deliberate surface modifications carried out by chemical reagents and polymer coatings, the surface properties of membranes, including the parts of pores exposed to enzyme immobilization solutions, are drastically altered by immobilized enzymes, often leading to membrane pore wetting. In one study, at a high enzyme loading, the surface pore openings were narrowed and the water contact angle of a pristine PP membrane dropped from 131° to 78° after enzyme immobilization [90]. However, the extra mass transfer resistance caused by partially wetted pores was overcome by enzymatic CO_2 hydration enhancement caused by CA enzymes attached inside the wetted pores [137]. A biocatalytic GLMC operated in fully pore-wetted conditions has been described where CA was adsorbed on hygroscopic and catalytic MOF grown on the surface of a hydrophilic Al_2O_3 membrane, where a CO_2 hydration rate of 108 μmol cm^{-2} min^{-1} from 5% CO_2 gas into water was achieved [136], but the buffering capacity of the system was not explained. The additional mass transfer resistance of the wetted pores was likely overcome by the combined catalytic hydration enhancements from CA and MOF. A recent modelling study concurred with the observation that mass transfer resistance in wetted pores can be reduced by catalyzed CO_2 hydration [138].

To overcome pore wetting caused by immobilized enzymes, a "Janus" configuration was used to modify a PVDF membrane, with hydrophilic carbon nanotubes (CNT) on one side and superhydrophobic fluorosilane on the other side of the membrane [128]. CA was immobilized through physical adsorption onto the hydrophilic side to prevent interference with the pore structure and properties of the hydrophobic membrane. In a pressure drop test, where the hydrophilic side faced water and the superhydrophobic side faced the overhead gas chamber filled with 100% CO_2 gas to 1 atm, the CA immobilized membrane achieved up to a 2-fold increase in CO_2 hydration rate compared with the no-enzyme control membranes. The advantage of immobilizing CA on a Janus membrane located at the gas–liquid interface was demonstrated by the fact that over 30 mg of free CA had to be dissolved in the liquid bulk to achieve a similar CO_2 hydration rate as 2.97 mg CA immobilized on the membrane. Because the mass transfer at the gas and liquid interface is the rate-limiting step, the immobilized enzyme concentrated at the interface was able to catalyze the CO_2 hydration more efficiently.

In addition to hydrophobic porous membranes, non-porous CO_2 permeable membranes were also explored as GLMC for CO_2 absorption. CO_2 absorption into 30% K_2CO_3 solvent using a non-porous polysulfone (PSf) HFM coated with a layer of PDMS [132] was 70–90% that of a porous PP HFM, both incorporating immobilized CA through layer-by-layer electrostatic adsorption [131]. A 60-μm-thick free-standing non-porous PDMS membrane used in microfluidic devices to separate anesthesia gases from an ionic liquid (IL)-based CO_2 absorption solvent exhibited a 1.9-fold increase in CO_2 affinity when 0.1 mg CA/ g IL was added, while Xenon permeability was not affected [133]. A seemingly counter intuitive example showing promise was based on a porous hydrophobic PP membrane with a dense non-porous hydrophilic PVDF skin that was surface-coated with a CO_2 selective, hydrophilic, and anti-fouling poly(ionic liquid) (PIL) layer facing the flue gas side [89]. The added resistance from the hydrophilic PVDF dense non-porous skin layer was insignificant compared with the improved affinity of CO_2 brought about by the PIL layer, which resulted in a synergy with dissolved enzyme in the solvent and an overall improvement in CO_2 absorption.

These encouraging examples show that studies on modification of the membrane surface with control over hydrophilicity/hydrophobicity and the location of the immobilized enzymes and CO_2-philic additives will continue to evolve in positive directions. At the current stage, larger scale demonstration studies are urgently needed to prove the feasibility and longevity profiles of enzymatic GLMC processes for commercialization.

4.4. Enzyme Immobilization

The basic principles of enzyme immobilization outlined in the introduction are applicable to membrane immobilization. For GLMC applications, enzymes are exclusively immobilized on the surface of the pre-made carriers, either stationary on the membranes [131,134] or mobile on nanoparticles that are dispersed in the absorption solvent [90,129]. Because GLMCs are positioned at the gas–liquid interface where the limiting step in mass transfer occurs it makes sense to expose the immobilized enzyme as close to this interface as possible. This requires enzymes to be adsorbed to or immobilized on the liquid-facing membrane side or requires enzymes to stay mobile in the liquid and thus have chances to approach the gas–liquid interface as the liquid flows. These principles were combined in a recent report [135] where CA enzymes were immobilized on an electrospun poly(styrene-co-maleic anhydride) PSMA nanofiber membrane that was floated at the air–liquid interface assisted by 3D printed flotation devices, and the actual air and liquid interface was refreshed frequently as the liquid was agitated. Such a configuration could be useful outside of the column-based absorption systems, such as for enhancing CO_2 uptake in natural systems such as in ponds and lakes.

One important parameter that studies across the field strive to improve is the enzyme loading. Enzyme loading is commonly reported as the mass of enzyme (μg or mg) or esterase activity (U) of CA, either as a total number for the module assembly [13] or divided by the nominal area of the membrane [136] or volume of the reactor [90]. Various strategies can be used to increase enzyme loading. For example, enzyme loading increased linearly with the number of alternating enzyme layers in a layer-by-layer (LbL) electrostatic adsorption technique [131,132]. Enzyme loading can also be optimized by moderating chemical reagent ratios to control surface functional group density [134] or by optimizing instrument power and duration settings, such as plasma radio frequency glow discharge [120–123]. A monolayer of enzyme coverage, estimated through geometrical calculations, can be compared with the obtained enzyme loading [13]. However, beyond a certain point, the CO_2 hydration rate of the immobilized enzyme no longer follows a linear correlation with the total amount of enzyme detectable by the esterase activity assay [128]. This is because only the surface-exposed enzymes are able to catalyze the extremely fast CO_2 hydration reaction, while enzymes buried deeper under the surface are mass transfer limited and, therefore, not able to contribute to the catalytic effect. Nevertheless, higher enzyme loading could indeed improve product longevity because fresh layers of enzyme could be exposed over time to continue the catalytic enhancement [128].

It is widely acknowledged that the immobilized enzyme orientation on surfaces can affect its activity, and the impact of such effects with immobilized CA enzymes have already been studied on simple geometries such as ultra-flat template-stripped gold (TSG) [139]. However, due to the more complex geometry of membranes and the non-specificity of chemical bonds or physical interactions involved in many immobilization approaches, experiments on the controlled orientation of immobilized CA on membranes and the effect on CO_2 absorption performance have not yet been reported. This is an important research direction as the fields of enzyme immobilization and protein engineering converge toward orienting immobilized enzymes with enhanced activity and stability.

4.5. Solvents and Form of Substrate

Initial evaluations of biocatalytic GLMC measured the enzymatic CO_2 hydration rate enhancement with the membrane immersed in CO_2-saturated buffers [88,127]. However, this type of configuration circumvented the critical and rate-limiting mass transfer step of CO_2 from the gas phase to the liquid phase, which is a critical performance function of GLMC that should be evaluated. More realistic configurations, with membranes positioned at gas–liquid interfaces, were later implemented with 100% CO_2 gas [128] and soon after with more realistic lower concentration CO_2 gas mixtures [129].

Common liquids used as CO_2 absorption solvents for lab scale testing are water [136] and low concentration aqueous buffers [137]. The advantages of using these solvents, include abundancy, non-corrosiveness, and less tendency to wet the hydrophobic membrane. However, a drawback is their low CO_2 capture capacity, which inevitably requires the use of higher amounts of liquid to absorb more CO_2. Concentrated aqueous solutions of potassium carbonate are relatively benign and are known to both benefit from CA kinetic enhancement and have a useful carbon loading capacity [14]. An aqueous 30 wt% K_2CO_3 solvent was used in an HFGLMC apparatus where three layers of CA were electrostatically immobilized, resulting a three-fold improvement in CO_2 absorption rate compared with the non-enzyme control membrane [131,132]. The immobilized CA retained over 80%

activity after a 200 h exposure to common contaminants from flue gas and to the high pH environment brought about by the high solute concentration.

The low regeneration energy solvent 30% methyldiethanolamine (MDEA) was compared with the benchmark solvent 30% monoethanolamine (MEA), both with 1 wt% dissolved CA concentrate, in an HFGLMC configuration [89]. Dissolved CA contributed up to a 2.2-fold CO_2 absorption rate enhancement in 30% MDEA, while showing a negative effect for 30% MEA. The explanation is that MEA reacts directly with CO_2 to rapidly form carbamate molecules, which is not enhanced further by CA, which instead acts through a bicarbonate formation mechanism. A less common category of absorption solvent for GLMC, ionic liquid, was also studied with dissolved CA [133]. The water content was essential for detecting an enzyme enhancement effect because the hydration reaction requires 1 mole of water per mole of CO_2 converted to bicarbonate.

In summary, CA-enhanced GLMCs, with both immobilized and dissolved CA, offer rate enhancement benefits from the biocatalyst and, thus, are able to use aqueous benign solvents, which greatly reduces membrane wetting and improves membrane longevity. Because pumping and heating large volumes of water take a lot of energy, future studies that couple the CO_2 capture in natural liquid sources with direct utilization have the potential to avoid the desorption cost and make this process more desirable.

5. Enzyme Membrane Reactor

Beyond CO_2 capture, membranes play a vital role in the design of efficient enzyme membrane reactors (EMR) for carrying out CO_2 utilization chemical conversion reactions. These processes benefit from the combined separation function of the membrane together with continuous-flow chemical reactors. Discussions in this section focus on intense research on the design and application of EMR for CO_2 reduction reactions that incorporate oxidoreductase enzymes and representative studies are summarized in Table 4.

Table 4. CO_2 conversion to chemicals with oxidoreductase.

Application	Membrane Configuration	Enzyme, Concentration	Cofactor Regeneration Electron Transfer System	Year, 1st Author, Ref
CO_2 conversion to formic acid	Ceramic tubular membrane as UV-light blocker	FDH; DAH	UV > TiO_2 (EtOH as hole quencher) > MV > DAH > NADH > FDH	2005 Kurayama [92]
CO_2 conversion to methanol	Flat sheet polymeric membranes with immobilized enzymes by direct membrane fouling	FDH; FaldDH; ADH	NADH > FDH; FaldDH; ADH	2015 Luo [37]
CO_2 conversion to formic acid	Hydrophobic HFM as gas distributor and PAA-grafted PE HFM as enzyme carrier	FDH	NADH > FDH	2016 Wang [95]
Formaldehyde conversion to methanol	Hydrophilic flat sheet macroporous (200 nm) SiC membrane pretreated with NaOH and surface functionalized with PEI or APTES as enzyme carrier	ADH	NADH > ADH	2018 Zeuner [140]
CO_2 conversion to formic acid	PAA-grafted PE HFM modified by PEI through electrostatic interaction as CO_2-philic surface	FDH	NADH > FDH	2018 Wang [141]
CO_2 conversion to formic acid	Electrospun PS nanofiber membrane surface modified by acid treatment, APTES, and GA activation as enzyme carrier	FDH	Cu foam electrode > NADH > FDH	2018 Barin [142]
CO_2 conversion to methanol	PVDF porous membrane functionalized by dead-end filtration of MOFs containing enzymes and cofactor	FDH; FaldDH; ADH; GDH	L-glutamate > GDH > NADH > FDH; FaldDH; ADH	2019 Zhu [143]

Table 4. *Cont.*

Application	Membrane Configuration	Enzyme, Concentration	Cofactor Regeneration Electron Transfer System	Year, 1st Author, Ref
CO_2 conversion to formic acid	Porous HFM used as both gas distributor and enzyme carrier	FDH	UV > TiO_2 (EDTA as hole quencher) > $[Cp{*}Rh(bpy)(H2O)]^{2+}$ > NADH > FDH	2020 Gu [144]
CO_2 conversion to formic acid	PAA-grafted PE HFM modified by PEI compared with PEI/PDA co-deposited SiO_2 microsphere as enzyme carriers	FDH	NADH > FDH	2021 Guo [145]
CO_2 absorption and conversion to formic acid	PP or ceramic GLMC modified by PEI/PDA and in situ grown MOFs encapsulating enzymes	CA; FDH	NADH > FDH	2021 Chai [146]
CO_2 conversion to formic acid	Ultrafiltration membrane with hydrophobic PP support layer and hydrophilic regenerated cellulose skin layer for enzyme immobilization	FDH	UV > MIL-125-Py-Rh (TEOA as hole quencher) > NADH > FDH	2022 Lin [147]
CO_2 conversion to formic acid	Ultrafiltration membrane with hydrophobic PP support layer and hydrophilic regenerated cellulose skin layer for enzyme immobilization	FDH	UV > Rh_{m3}-N-PCN (TEOA as hole quencher) > NADH > FDH	2022 Zhang [96]
CO_2 conversion to formaldehyde	PE hollow fiber membrane was used as the enzyme-bearing reactor and gas distributor	FDH; FaldDH	UV > TiO_2 (EDTA or H_2O as hole quencher) > $[Cp{*}Rh(bpy)(H_2O)]^{2+}$ > NADH > FDH; FaldDH	2022 Guo [94]

Acronyms: UV—ultraviolet light; FDH—formate dehydrogenase; MV—methyl viologen; DAH—diaphorase; EtOH—ethanol; FaldDH—formaldehyde dehydrogenase; ADH—alcohol dehydrogenase; HFM—hollow fiber membrane; PAA—poly(acrylic acid); PE—polyethylene; SiC—silicon carbide; NaOH—sodium hydroxide; PEI—polyethylenimine; APTES—(3-aminopropyl)triethoxysilane; GA—glutaraldehyde; MOF—metal organic frameworks; GDH—glutamic dehydrogenase; EDTA—ethylenediaminetetraacetic acid; $[Cp{*}Rh(bpy)(H_2O)]^{2+}$—pentamethylcyclopentadienyl rhodium bipyridine complex; GLMC—gas–liquid membrane contactor; PDA—polydopamine; TEOA—triethanolamine; MIL-125-Py-Rh—a Ti-based MOF with anchored Rh complex; Rh_{m3}-N-PCN—Rhodium complex covalently grafted on amine functionalized polymeric carbon nitride.

5.1. Location of the Immobilized Enzymes

Biocatalyst stability, recovery and reuse are improved by immobilizing enzymes on membrane supports, utilizing both their high surface area and separation functions. Common immobilization strategies involve modifying porous membrane surfaces using UV-grafting [141], acid [142] or base [140] treatment, silanization [140], polydopamine (PDA) and PEI deposition [145]. The carboxylic acid or amine surfaces are further activated with carbodiimide coupling reagent [95] or glutaraldehyde [142], respectively, rendering the surface covalently reactive toward amino groups on the enzyme. Enzymes immobilized on membrane surfaces have alleviated mass transfer limitations and thus higher activity.

Leveraging the separation function of a membrane, direct membrane fouling [37], also known as dead-end filtration [143], has been a popular non-chemical alternative for enzyme immobilization. Enzyme cascades can be easily constructed by direct membrane fouling methods. Luo et al. [37] immobilized formate dehydrogenase (FDH), formaldehyde dehydrogenase (FaldDH), and alcohol dehydrogenase (ADH), both as a co-localized mixtures and as separate layers, in sequences using direct membrane fouling techniques. While the direct cascade concept of reducing product inhibition and enhancing substrate shuttling is appealing, the presence of a bottleneck enzyme in the cascade, FaldDH in this case, requires a threshold formic acid concentration to move the reaction in the reduction direction, making co-localization less effective. Rather, a cascade consisting of separate enzyme layers is more conducive to higher final product yield because of the flexibility this gives in optimizing reaction steps separately. Zhu et al. [143] encapsulated individual enzymes in metal organic frameworks (MOF) and loaded them either randomly or in sequence into the pores of a support membrane by dead-end filtration. Consistent with the foregoing study, they found that the ordered multi-enzyme cascade system achieved the highest methanol yield.

Another aspect of localizing enzymes by immobilization is to protect enzymes from harsh environments (Figure 2). Kurayama et al. [92] used a ceramic membrane to separate the enzymatic reaction step from the photocatalytic chamber used for cofactor regeneration. They found that when ultra-violet (UV) light shone on the whole system and directly on the enzymes, no formic acid production occurred, indicating the critical importance of the UV blocking function. Alternatively, Guo et al. [94] separated the NADH-depleted liquid and pumped it to a separate quartz vessel for photo-regeneration, away from the enzymatic reaction chamber. Tian et al. [148] encapsulated enzyme in zeolitic imidazolate-based MOF to protect FDH from possible deactivation by reactive oxygen species (ROS) generated in the photocatalytic reaction. Chai et al. [146] used thermal stable MOFs as protective coatings to encapsulate FDH and protect it from high temperature flue gas conditions. While the free enzyme denatured immediately, the MOF encapsulated enzyme maintained activity even after 4 h at 100 °C. However, they pointed out that the NADH cofactor has larger dimensions than the pore sizes of the MOF, thus limiting the effectiveness of the immobilized enzymes to those exposed at the surfaces. This issue was investigated in another study [143], where FDH, FaldDH, or ADH were paired with and co-immobilized with glutamate dehydrogenase (GDH) for enzymatic NADH regeneration. The methanol production rate increased in the cases where NADH was also co-immobilized in the MOF structures. It seems that although the well-defined native MOF pores do not fit the large NADH, the pre-encapsulated NADH generated larger cavities that were able to accommodate it for a faster reaction.

5.2. Roles of Membrane in Substrate Uptake

Due to its low aqueous solubility, delivering CO_2 to the active site of enzymes becomes a rate-limiting step. For many of the early studies, CO_2-saturated water or buffer [37,92] was used to circumvent the gas to liquid mass transfer limitation. For continuous delivery as a gas, CO_2 was bubbled into the reaction chamber [142]. CO_2 gas can also be infused into the liquid by using hollow fiber membranes (HFM) as a CO_2 gas distributor [95], utilizing both the high surface areas and porous structures of the membranes to generate small gas bubbles for faster dissolution. For batch reactions, CO_2 can be blown into the liquid and pressurized in the head space to a certain pressure before sealing the reaction chamber [145].

Another strategy to increase CO_2 uptake in the reaction solution, borrowing knowledge from the CO_2 reactive absorption process, is the use of amine-based CO_2-philic materials. Wang et al. [141] found that a PEI modified HFM surface was able to bring in more CO_2 from the gas phase in the form of bicarbonate ions and improve formic acid production of dissolved FDH. This implies that CA could also assist FDH to reach a higher productivity. Chai et al. [146] constructed a GLMC with immobilized CA and FDH, combining CO_2 capture with CO_2 conversion in a single step. By positioning the GLMC at the gas–liquid interface with the hydrophilic side facing the liquid, the biocatalytic membrane produced 5.6 µmol formic acid after 4 h of reaction using 20% CO_2 gas.

In addition, the hydrophilicity and hydrophobicity of the membrane have an influence on CO_2 gas accessibility to the enzyme. Lin et al. [147] measured the CO_2 gas bubble adhesion force on both hydrophobic and hydrophilic membranes and concluded that hydrophobic membranes attracts CO_2 gas bubbles better. FDH immobilized on a hydrophobic layer achieved higher formic acid yield than that immobilized on a hydrophilic layer.

5.3. Cofactor Regeneration

While many studies used the natural cofactor NADH as the electron donor [95,140,141,145,146] as a proof-of-concept, the continuous supply of NADH to run the reaction can be costly and is uneconomical for real-world CO_2 reduction applications. To overcome this limitation, NADH regeneration has been the focal point of many recent research publications. There are three main approaches for NADH regeneration, namely, biocatalytic [143], photocatalytic [147], and electrochemical [142].

Biocatalytic regeneration involves the use of another enzyme and its substrate as a natural reducing electron source to reduce NAD^+ to NADH. Zhu et al. [143] co-immobilized glutamic dehydrogenase (GDH) with FDH, FaldDH, and ADH. A continuous supply of relatively inexpensive L-glutamate as the reductant powered the enzymatic cascade reduction reactions from CO_2 to methanol.

Photocatalytic regeneration uses semiconductors or organic dye-based photosensitizers to harvest light energy for creating high energy electrons that can be transferred to NAD^+ for its reduction. However, direct electron transfer from the photosensitizer to NAD^+ is inefficient, and an electron mediator is usually needed. Kurayama et al. [92] paired TiO_2 particle photo-catalysts with oxidized methyl viologen (MV^{2+}) as the electron mediator with the enzyme diaphorase (DAH) that accepts the reduced form of MV^+ as the reducing equivalent in reducing NAD^+ to NADH. The results showed that the concentrations of both MV and NADH affected the overall CO_2 conversion rate, and that the DAH catalytic reaction was the rate-limiting step. The requirement of an additional enzyme to transfer the reducing equivalents from an electron mediator to NAD^+ is disadvantageous as it adds additional bottlenecks, such as the DAH catalytic process.

Electrochemical approaches aim to overcome cofactor-dependent reaction limitations. A Rhodium-based compound named pentamethylcyclopentadienyl rhodium bipyridine complex, $[Cp{*}Rh(bpy)\ (H_2O)]^{2+}$, was reported to directly interact either with nicotinamide cofactors or with a reagentless source of reduction equivalents, such as a cathode [149]. Gu et al. [144] employed soluble homogenous $[Cp{*}Rh(bpy)\ (H_2O)]^{2+}$ along with a heterogeneous TiO_2 photo-catalyst and achieved a maximum turnover number of 125 after 4.5 h for NADH regeneration. Guo et al. [94] utilized a similar regeneration system for the synthesis of formaldehyde from CO_2 with both FDH and FaldDH immobilized on hollow fiber membrane. They compared the pH and sacrificial electron donor and found that ethylenediaminetetraacetic acid (EDTA) produced much higher formaldehyde yield than water over a pH of 5–7.5.

The problem with using homogeneous (dissolved) electron mediators is that they travel with the reaction liquid throughout the system and thus require a large amount of mediator to be used to achieve an effective concentration. The solution to this is to confine or immobilize the electron mediator in a localized domain. Tian et al. [148] constructed an elaborate compartmentalized photocatalyst-enzyme system inspired by the thylakoid membrane (Figure 4). The electron–hole pair separation ability of an inorganic photosensitizer graphitic carbon nitride C_3N_4 was improved by incorporating aromatic thiophene into network structures, and the selective electron transfer to NAD^+ was mediated by a Cobalt complex coordinated with bipyridine covalently bonded to the surface modifier polyethylenimine (PEI) for efficient NADH regeneration. Lin et al. [147] synthesized a Ti-based MOF with an electron-transferring Rh complex anchored to the light-harvesting iminopyridine unit of the MOF and found an NADH generation yield of 66.4% in 60 min. Zhang et al. [96] covalently grafted a Rh complex onto amino-functionalized polymeric carbon nitride photo-catalyst for efficient regeneration of NADH with a conversion of 66% in 20 min.

A common drawback for all photocatalytic systems is the use of a sacrificial electron donor, after photo excitation and electron transfer, to deliver electrons to fill positively charged holes and regenerate the photo-catalysts. This electron donor can be abundant compounds, such as water, but the most efficient ones, such as EDTA and triethanolamine (TEOA), would cause additional cost and chemical waste issues. Recently, some metal-dependent FDHs have been reported to accept artificial electron donors by mediated electron transfer (MET) [150], circumventing the use of the NADH cofactor. Obtaining electrons directly from electrodes through direct electron transfer (DET) is an appealing concept but it usually comes at the cost of lower interfacial electron transfer rates due to the need for direct contact between electrode and enzyme, thus limiting effective enzyme loading to only a monolayer [151]. However, as of now, neither of these NADH-free mechanisms has been used in conjunction with membrane technologies, and future work on utilizing high surface area conductive membranes could potentially overcome the enzyme loading limitation and result in highly efficient integrated electrochemical-enzymatic CO_2 reduction technologies.

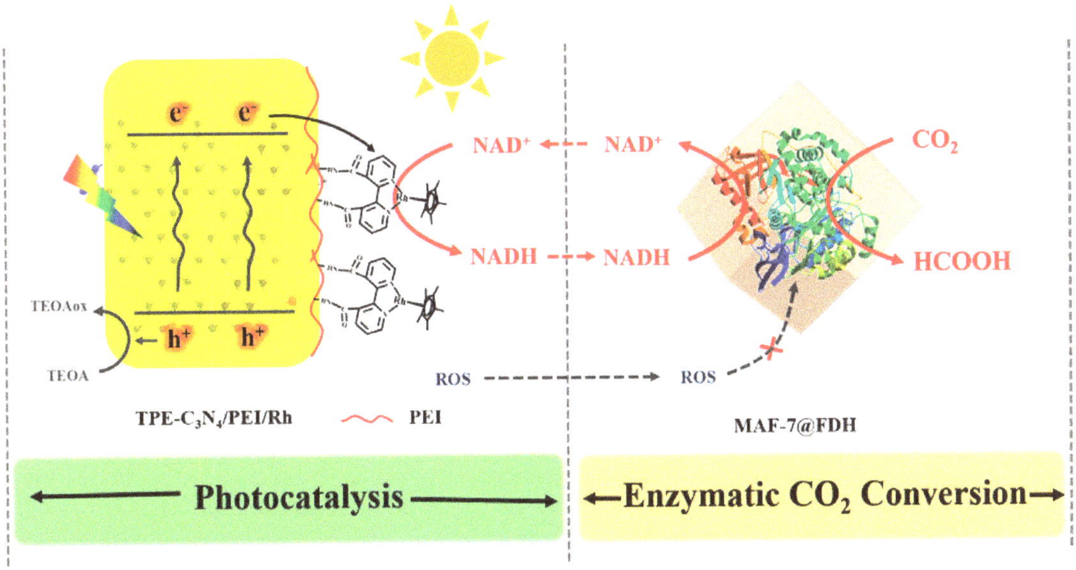

Figure 4. Modified graphitic carbon nitride decorated with a Cobalt complex for photocatalytic NADH regeneration (Reprinted with permission from Ref. [148]. Copyright 2020, American Chemical Society).

5.4. Long-Term Stability of EMR

The long-term stability of EMR includes storage and operational stabilities, and both need to be addressed before large scale-up and commercialization is possible. A consistent finding across studies is that the storage stabilities of immobilized enzymes are greatly improved over their free dissolved enzyme counterparts. For example, FDH covalently attached on HFM retained 83% and 67% activity after 30 and 60 days, respectively, in buffer at 4 °C, compared with 48% activity retained for the dissolved FDH after only 14 days under those conditions [95]. FDH immobilized on electrospun polystyrene nanofiber retained 41% activity after 20 days storage in buffer at 4 °C [142], and close to 100% activity retention was reported for CA and FDH encapsulated in MOF membranes after 20 days storage in ambient air [146]. Most studies report operational stabilities as the "number of recycles", which is usually limited to 10–20 cycles with total operation times of several to tens of hours [37,142]. For example, ADH immobilized on ceramic silicon carbide membranes retained less than 20% of the initial activity after 17 cycles of reuse [140]. However, high activity retention over many days of operation will be needed to fully validate the operational stability of EMRs. Lin et al. compared FDH immobilized either on the hydrophobic support layer or the hydrophilic skin layer of a photo-biocatalytic membrane system over five 24 h cycles of operation (120 h total) and found that although the former generated more formic acid, both configurations were able to retain all of their initial productivity at the end of test [147]. Guo et al. comprehensively tested operational stability by running their reaction continuously for 48 h, finding that a photo-enzyme coupled system was significantly more stable than photocatalysis (UV/TiO_2 reduction of CO_2) or biocatalysis (FDH and FaldDH without NADH regeneration) alone [94]. This type of continuous operation testing is highly important in future studies. Analogous to the CO_2 reduction reaction, the enzymatic fuel cell field has long been battling with short lifetimes of bio-electrodes and instabilities of electron mediators [152]. By carefully fine tuning the enzyme immobilization matrix, lifetimes of more than 45 days or even 200 days of continuous operation are possible [153–155]. Certainly, the experiences [156,157] gained in biofuel cell research should be adopted for the fabrication of stable biocathodes for CO_2 reduction reactions, and we expect more long-term stability tests will be carried out with this goal.

6. Conclusions and Future Perspectives

Biocatalytic membranes are a promising technology category for CO_2 capture and utilization that combine high reaction rates and enzyme selectivity with high surface area and separation functions of the membrane. CA and its mimics have been used in both immobilized and in dissolved forms in conjunction with membranes, where the presence of water molecules is essential for the facilitated CO_2 transport mechanism to function, regardless of the membrane type. For mixed matrix membranes and liquid membranes, membrane thickness is the bottleneck for achieving high CO_2 permeance. Recent trends in constructing ultra-thin CA immobilized selective layers or liquid-immobilizing porous structures, both supported by mechanically stable non-selective layers, for high permeance CO_2 separation, are expected to continue as more sophisticated fabrication methods are devised for reducing the membrane thickness.

CA-enhanced gas–liquid membrane contactors (GLMC), with both immobilized and dissolved CA, offer rate enhancement benefits for the biocatalyst and, thus, are able to use benign aqueous solvents, which greatly reduces membrane wetting and improves membrane longevity. Although issues such as enzyme-induced membrane wetting, surface hydrophilicity and pore blockages can occur, the additional mass transfer resistance of the wetted pores can still be overcome by the catalyzed CO_2 hydration. Because the mass transfer at the gas–liquid interface is the rate-limiting step, immobilized enzymes concentrated at this interface are able to catalyze the CO_2 hydration reaction more efficiently. Modifications of membrane surfaces with increased control over hydrophilicity/hydrophobicity and the location of immobilized enzymes and CO_2-philic additives will continue to evolve and improve reaction performance. In addition, only limited exploration of GLMC for CO_2 stripping has occurred. Advances in artificial lungs with immobilized CA on hollow fiber membranes (HFM) could inspire alternatives to high temperature CO_2 stripping processes which are responsible for high energy consumption.

While it is widely acknowledged that the orientation of enzymes on surfaces affects the activity of immobilized enzymes, complex membrane geometries and non-specificity of chemical bonds or physical interactions involved in many immobilization approaches have hindered experiments on controlled orientation of immobilized CA on membranes. Thus, the impact of oriented CA on CO_2 absorption performance is yet to be reported. This is an important research direction as the fields of enzyme immobilization and protein engineering converge toward orienting immobilized enzymes with enhanced activity and stability. Likewise, enzyme cascades for CO_2 conversion, currently constructed by direct membrane fouling, would benefit from improved control over enzyme placement and orientation. The presence of a bottleneck enzyme in the cascade requires a minimum threshold substrate concentration to move the reaction in the reduction direction, making co-localization of enzyme pairs potentially less effective, though controlling the molar ratio of enzymes in proximity to each other may help alleviate this issue. Alternatively, cascade reactions consisting of separate enzyme layers could achieve higher final product yields, due to the corresponding flexibility in optimizing each separate reaction step.

Poor aqueous solubility of CO_2 is a rate limitation for delivering CO_2 to enzyme active sites for CO_2 conversion. Utilizing CA-promoted CO_2 hydration reactions together with HFM CO_2 gas infusion is one way to supply higher concentrations of carbon in the form of bicarbonate to solute selective membranes or to CO_2 reducing enzymes. Furthermore, coupling CO_2 capture in natural liquid sources with direct utilization has the potential to completely avoid desorption costs and make this process more desirable. New gas–liquid–solid tri-phase contactor configurations have the potential to further integrate and improve the capture and conversion processes.

A critical challenge in the application of CO_2 reducing enzymes for CO_2 conversion is the delivery of reducing equivalents, i.e., electrons and protons. While methods to regenerate the natural cofactor NADH have been the focus of many studies in this field, and proofs-of-concept for biocatalytic, photocatalytic, and electrochemical NADH regeneration methods have been reported, more research is needed to develop and scale up these important processes. In other areas of biocatalytic reactor research, certain metal-dependent FDHs are reported to accept artificial electron donors by mediated electron transfer (MET) or obtain electrons directly from the electrode through direct electron transfer (DET). Since neither of these NADH-free mechanisms has yet been used in conjunction with membrane technologies, future work on utilizing high surface area conductive membranes should be conducted. Such efforts could potentially overcome enzyme loading limitations and result in highly efficient integrated electrochemical-enzymatic CO_2 reduction technologies that are urgently needed to address the climate change crisis.

Author Contributions: Writing—original draft preparation, J.S.; writing—review and editing, S.S. All authors have read and agreed to the published version of the manuscript.

Funding: This work was made possible by funding from North Carolina State University, the Novo Nordisk Foundation, and the Alliance for Sustainable Energy, LLC, Managing and Operating Contractor for the National Renewable Energy Laboratory for the U.S. Department of Energy.

Institutional Review Board Statement: Not applicable.

Informed Consent Statement: Not applicable.

Data Availability Statement: Not applicable.

Conflicts of Interest: The authors declare no conflict of interest.

References

1. Friedlingstein, P.; O'Sullivan, M.; Jones, M.W.; Andrew, R.M.; Hauck, J.; Olsen, A.; Peters, G.P.; Peters, W.; Pongratz, J.; Sitch, S.; et al. Global Carbon Budget 2020. *Earth Syst. Sci. Data* **2020**, *12*, 3269–3340. [CrossRef]
2. Shakun, J.D.; Clark, P.U.; He, F.; Marcott, S.A.; Mix, A.C.; Liu, Z.; Otto-Bliesner, B.; Schmittner, A.; Bard, E. Global Warming Preceded by Increasing Carbon Dioxide Concentrations during the Last Deglaciation. *Nature* **2012**, *484*, 49–54. [CrossRef] [PubMed]
3. Lindsey, R. Climate Change: Atmospheric Carbon Dioxide. Available online: https://www.climate.gov/news-features/understanding-climate/climate-change-atmospheric-carbon-dioxide (accessed on 18 July 2022).
4. Lenton, T.; Rockström, J.; Gaffney, O.; Rahmstorf, S.; Richardson, K.; Steffen, W.; Shellnhuber, H.J. Climate Tipping Points—Too Risky to Bet Against. *Nature* **2019**, *575*, 592–595. [CrossRef]
5. Lebling, K.; Leslie-Bole, H.; Psarras, P.; Bridgwater, E.; Byrum, Z.; Pilorgé, H. Direct Air Capture: Assessing Impacts to Enable Responsible Scaling. *World Resour. Inst.* **2022**, 1–28. [CrossRef]
6. Peplow, M. The Race to Upcycle CO_2 into Fuels, Concrete and More. *Nature* **2022**, *603*, 780–783. [CrossRef]
7. Mondal, M.K.; Balsora, H.K.; Varshney, P. Progress and Trends in CO_2 Capture/Separation Technologies: A Review. *Energy* **2012**, *46*, 431–441. [CrossRef]
8. Merkel, T.C.; Lin, H.; Wei, X.; Baker, R. Power Plant Post-Combustion Carbon Dioxide Capture: An Opportunity for Membranes. *J. Memb. Sci.* **2010**, *359*, 126–139. [CrossRef]
9. Witek-Krowiak, A.; Dawiec, A.; Modelski, S.; Podstawczyk, D. Carbon Dioxide Removal in a Membrane Contactor—Selection of Absorptive Liquid/Membrane System. *Int. J. Chem. Eng. Appl.* **2012**, *3*, 391–395. [CrossRef]
10. Figueroa, J.D.; Fout, T.; Plasynski, S.; McIlvried, H.; Srivastava, R.D. Advances in CO_2 Capture Technology—The U.S. Department of Energy's Carbon Sequestration Program. *Int. J. Greenh. Gas Control* **2008**, *2*, 9–20. [CrossRef]
11. Khalifah, R.G. The Carbon Dioxide Hydration Activity of Carbonic Anhydrase. I. Stop-Flow Kinetic Studies on the Native Human Isoenzymes B and C. *J. Biol. Chem.* **1971**, *246*, 2561–2573. [CrossRef]
12. Ward, W.J.; Robb, W.L. Carbon Dioxide-Oxygen Separation: Facilitated Transport of Carbon Dioxide across a Liquid Film. *Science* **1967**, *156*, 1481–1484. [CrossRef] [PubMed]
13. Kaar, J.L.; Oh, H.I.; Russell, A.J.; Federspiel, W.J. Towards Improved Artificial Lungs through Biocatalysis. *Biomaterials* **2007**, *28*, 3131–3139. [CrossRef] [PubMed]

14. Salmon, S.; House, A. Enzyme-Catalyzed Solvents for CO_2 Separation. In *Novel Materials for Carbon Dioxide Mitigation Technology*; Novel Materials for Carbon Dioxide Mitigation Technology; Elsevier: Amsterdam, The Netherlands, 2015; pp. 23–86.
15. Qi, G.; Liu, K.; House, A.; Salmon, S.; Ambedkar, B.; Frimpong, R.A.; Remias, J.E.; Liu, K. Laboratory to Bench-Scale Evaluation of an Integrated CO_2 Capture System Using a Thermostable Carbonic Anhydrase Promoted K2CO3 Solvent with Low Temperature Vacuum Stripping. *Appl. Energy* **2018**, *209*, 180–189. [CrossRef]
16. Shen, J.; Yuan, Y.; Salmon, S. Carbonic Anhydrase Immobilized on Textile Structured Packing Using Chitosan Entrapment for CO_2 Capture. *ACS Sustain. Chem. Eng.* **2022**, *10*, 7772–7785. [CrossRef]
17. Shen, J.; Yuan, Y.; Salmon, S. Durable and Versatile Immobilized Carbonic Anhydrase on Textile Structured Packing for CO_2 Capture. *Catalysts* **2022**, *12*, 1108. [CrossRef]
18. Niazi, M.B.K.; Jahan, Z.; Ahmed, A.; Rafiq, S.; Jamil, F.; Gregersen, Ø.W. Effect of Zn-Cyclen Mimic Enzyme on Mechanical, Thermal and Swelling Properties of Cellulose Nanocrystals/PVA Nanocomposite Membranes. *J. Polym. Environ.* **2020**, *28*, 1921–1933. [CrossRef]
19. Zhang, Y.; Wang, H.; Zhou, S.; Wang, J.; He, X.; Liu, J.; Zhang, Y. Biomimetic Material Functionalized Mixed Matrix Membranes for Enhanced Carbon Dioxide Capture. *J. Mater. Chem. A* **2018**, *6*, 15585–15592. [CrossRef]
20. Zheng, W.; Tian, Z.; Wang, Z.; Peng, D.; Zhang, Y.; Wang, J.; Zhang, Y. Dual-Function Biomimetic Carrier Based Facilitated Transport Mixed Matrix Membranes with High Stability for Efficient CO_2/N_2 Separation. *Sep. Purif. Technol.* **2022**, *285*, 120371. [CrossRef]
21. Alvizo, O.; Nguyen, L.J.; Savile, C.K.; Bresson, J.A.; Lakhapatri, S.L.; Solis, E.O.P.; Fox, R.J.; Broering, J.M.; Benoit, M.R.; Zimmerman, S.A.; et al. Directed Evolution of an Ultrastable Carbonic Anhydrase for Highly Efficient Carbon Capture from Flue Gas. *Proc. Natl. Acad. Sci. USA* **2014**, *111*, 16436–16441. [CrossRef]
22. Reardon, J.; Bucholz, T.; Hulvey, M.; Tuttle, J.; Shaffer, A.; Pulvirenti, D.; Weber, L.; Killian, K.; Zaks, A. Low Energy CO_2 Capture Enabled by Biocatalyst Delivery System. *Energy Procedia* **2014**, *63*, 301–321. [CrossRef]
23. Smit, B.; Park, A.H.A.; Gadikota, G. The Grand Challenges in Carbon Capture, Utilization, and Storage. *Front. Energy Res.* **2014**, *2*, 2013–2015. [CrossRef]
24. Bulushev, D.A.; Ross, J.R.H. Towards Sustainable Production of Formic Acid. *ChemSusChem* **2018**, *11*, 821–836. [CrossRef] [PubMed]
25. Grubel, K.; Jeong, H.; Yoon, C.W.; Autrey, T. Challenges and Opportunities for Using Formate to Store, Transport, and Use Hydrogen. *J. Energy Chem.* **2020**, *41*, 216–224. [CrossRef]
26. Bar-Even, A. Formate Assimilation: The Metabolic Architecture of Natural and Synthetic Pathways. *Biochemistry* **2016**, *55*, 3851–3863. [CrossRef] [PubMed]
27. Mao, W.; Yuan, Q.; Qi, H.; Wang, Z.; Ma, H.; Chen, T. Recent Progress in Metabolic Engineering of Microbial Formate Assimilation. *Appl. Microbiol. Biotechnol.* **2020**, *104*, 6905–6917. [CrossRef]
28. Marpani, F.; Pinelo, M.; Meyer, A.S. Enzymatic Conversion of CO_2 to CH_3OH via Reverse Dehydrogenase Cascade Biocatalysis: Quantitative Comparison of Efficiencies of Immobilized Enzyme Systems. *Biochem. Eng. J.* **2017**, *127*, 217–228. [CrossRef]
29. Song, H.; Ma, C.; Liu, P.; You, C.; Lin, J.; Zhu, Z. A Hybrid CO_2 Electroreduction System Mediated by Enzyme-Cofactor Conjugates Coupled with Cu Nanoparticle-Catalyzed Cofactor Regeneration. *J. CO_2 Util.* **2019**, *34*, 568–575. [CrossRef]
30. Chen, Y.; Li, P.; Noh, H.; Kung, C.; Buru, C.T.; Wang, X.; Zhang, X.; Farha, O.K. Stabilization of Formate Dehydrogenase in a Metal–Organic Framework for Bioelectrocatalytic Reduction of CO_2. *Angew. Chem.* **2019**, *131*, 7764–7768. [CrossRef]
31. Liao, Q.; Liu, W.; Meng, Z. Strategies for Overcoming the Limitations of Enzymatic Carbon Dioxide Reduction. *Biotechnol. Adv.* **2022**, *60*, 108024. [CrossRef]
32. Luo, J.; Song, S.; Zhang, H.; Zhang, H.; Zhang, J.; Wan, Y. Biocatalytic Membrane: Go Far beyond Enzyme Immobilization. *Eng. Life Sci.* **2020**, *20*, 441–450. [CrossRef]
33. Lim, F.N.A.R.; Marpani, F.; Dilol, V.E.A.; Pauzi, S.M.; Othman, N.H.; Alias, N.H.; Him, N.R.N.; Luo, J.; Rahman, N.A. A Review on the Design and Performance of Enzyme-Aided Catalysis of Carbon Dioxide in Membrane, Electrochemical Cell and Photocatalytic Reactors. *Membranes* **2022**, *12*, 28. [CrossRef]
34. Shen, J.; Zhang, S.; Fang, X.; Salmon, S. Advances in 3D Gel Printing for Enzyme Immobilization. *Gels* **2022**, *8*, 460. [CrossRef]
35. Sun, J.; Wei, L.; Wang, Y.; Zhao, Z.; Liu, W. Immobilization of Carbonic Anhydrase on Polyvinylidene Fluoride Membranes. *Biotechnol. Appl. Biochem.* **2018**, *65*, 362–371. [CrossRef] [PubMed]
36. Sun, J.; Wang, C.; Wang, Y.; Ji, S.; Liu, W. Immobilization of Carbonic Anhydrase on Polyethylenimine/Dopamine Codeposited Membranes. *J. Appl. Polym. Sci.* **2019**, *136*, 1–9. [CrossRef]
37. Luo, J.; Meyer, A.S.; Mateiu, R.V.; Pinelo, M. Cascade Catalysis in Membranes with Enzyme Immobilization for Multi-Enzymatic Conversion of CO_2 to Methanol. *N. Biotechnol.* **2015**, *32*, 319–327. [CrossRef] [PubMed]
38. Ren, S.; Li, C.; Tan, Z.; Hou, Y.; Jia, S.; Cui, J. Carbonic Anhydrase@ZIF-8 Hydrogel Composite Membrane with Improved Recycling and Stability for Efficient CO_2 Capture. *J. Agric. Food Chem.* **2019**, *67*, 3372–3379. [CrossRef]
39. Wen, H.; Zhang, L.; Du, Y.; Wang, Z.; Jiang, Y.; Bian, H.; Cui, J.; Jia, S. Bimetal Based Inorganic-Carbonic Anhydrase Hybrid Hydrogel Membrane for CO_2 Capture. *J. CO_2 Util.* **2020**, *39*, 101171. [CrossRef]
40. Moehlenbrock, M.J.; Minteer, S.D. Introduction to the Field of Enzyme Immobilization and Stabilization. In *Enzyme Stabilization and Immobilization: Methods and Protocols*; Minteer, S.D., Ed.; Humana Press: New York, NY, USA, 2017; Volume 1504, pp. 1–7. ISBN 978-1-60761-894-2.

41. Guisan, J.M. New Opportunities for Immobilization of Enzymes. In *Immobilization of Enzymes and Cells: Third Edition*; Guisan, J.M., Ed.; Methods in Molecular Biology; Humana Press: Totowa, NJ, USA, 2013; Volume 1051, pp. 1–13. ISBN 978-1-62703-549-1.
42. Rasouli, H.; Nguyen, K.; Iliuta, M.C. Recent Advancements in Carbonic Anhydrase Immobilization and Its Implementation in CO_2 Capture Technologies: A Review. *Sep. Purif. Technol.* **2022**, *296*, 121299. [CrossRef]
43. Molina-Fernández, C.; Luis, P. Immobilization of Carbonic Anhydrase for CO_2 Capture and Its Industrial Implementation: A Review. *J. CO_2 Util.* **2021**, *47*, 101475. [CrossRef]
44. Russo, M.E.; Capasso, C.; Marzocchella, A.; Salatino, P. Immobilization of Carbonic Anhydrase for CO_2 Capture and Utilization. *Appl. Microbiol. Biotechnol.* **2022**, *106*, 3419–3430. [CrossRef]
45. Hedstrom, L. Enzyme Specificity and Selectivity. In *Encyclopedia of Life Sciences*; Wiley: Chichester, UK, 2010; pp. 1–8. ISBN 9780470015902. [CrossRef]
46. Bender, D.A. Tricarboxylic Acid Cycle. In *Encyclopedia of Food Sciences and Nutrition*; Elsevier: Amsterdam, The Netherlands, 2003; pp. 5851–5856. ISBN 9780123786319.
47. Yates, N.D.J.; Fascione, M.A.; Parkin, A. Methodologies for "Wiring" Redox Proteins/Enzymes to Electrode Surfaces. *Chem.—A Eur. J.* **2018**, *24*, 12164–12182. [CrossRef] [PubMed]
48. Liu, Y.; Li, F.; Zhang, X.; Ji, X. Recent Progress on Electrochemical Reduction of CO_2 to Methanol. *Curr. Opin. Green Sustain. Chem.* **2020**, *23*, 10–17. [CrossRef]
49. Saxena, A.; Liyanage, W.; Masud, J.; Kapila, S.; Nath, M. Selective Electroreduction of CO_2 to Carbon-Rich Products with a Simple Binary Copper Selenide Electrocatalyst. *J. Mater. Chem. A* **2021**, *9*, 7150–7161. [CrossRef]
50. Saxena, A.; Singh, H.; Nath, M. Cobalt Telluride Electrocatalyst for Selective Electroreduction of CO_2 to Value-Added Chemicals. *Mater. Renew. Sustain. Energy* **2022**, *11*, 115–129. [CrossRef]
51. Saxena, A.; Liyanage, W.P.R.; Kapila, S.; Nath, M. Nickel Selenide as an Efficient Electrocatalyst for Selective Reduction of Carbon Dioxide to Carbon-Rich Products. *Catal. Sci. Technol.* **2022**, *12*, 4727–4739. [CrossRef]
52. Saxena, A.; Kapila, S.; Medvedeva, J.E.; Nath, M. Copper Cobalt Selenide as a Bifunctional Electrocatalyst for the Selective Reduction of CO_2 to Carbon-Rich Products and Alcohol Oxidation. *ACS Appl. Mater. Interfaces* **2023**, *15*, 14433–14446. [CrossRef]
53. Schlager, S.; Dumitru, L.M.; Haberbauer, M.; Fuchsbauer, A.; Neugebauer, H.; Hiemetsberger, D.; Wagner, A.; Portenkirchner, E.; Sariciftci, N.S. Electrochemical Reduction of Carbon Dioxide to Methanol by Direct Injection of Electrons into Immobilized Enzymes on a Modified Electrode. *ChemSusChem* **2016**, *9*, 631–635. [CrossRef]
54. He, X. A Review of Material Development in the Field of Carbon Capture and the Application of Membrane-Based Processes in Power Plants and Energy-Intensive Industries. *Energy. Sustain. Soc.* **2018**, *8*, 34. [CrossRef]
55. Bernardo, P.; Drioli, E.; Golemme, G. Membrane Gas Separation: A Review/State of the Art. *Ind. Eng. Chem. Res.* **2009**, *48*, 4638–4663. [CrossRef]
56. He, X.; Lindbråthen, A.; Kim, T.J.; Hägg, M.B. Pilot Testing on Fixed-Site-Carrier Membranes for CO_2 Capture from Flue Gas. *Int. J. Greenh. Gas Control* **2017**, *64*, 323–332. [CrossRef]
57. Merkel, T.; Kniep, J.; Wei, X.; Carlisle, T.; White, S.; Pande, S.; Fulton, D.; Watson, R.; Hoffman, T.; Freeman, B.; et al. *Pilot Testing of a Membrane System for Postcombustion CO_2 Capture*; National Energy Technology Laboratory: Pittsburgh, PA, USA; Morgantown, WV, USA, 2015.
58. Trachtenberg, M.C.; Cowan, R.M.; Smith, D.A.; Horazak, D.A.; Jensen, M.D.; Laumb, J.D.; Vucelic, A.P.; Chen, H.; Wang, L.; Wu, X. Membrane-Based, Enzyme-Facilitated, Efficient Carbon Dioxide Capture. *Energy Procedia* **2009**, *1*, 353–360. [CrossRef]
59. Zhou, F.; Tien, H.N.; Xu, W.L.; Chen, J.T.; Liu, Q.; Hicks, E.; Fathizadeh, M.; Li, S.; Yu, M. Ultrathin Graphene Oxide-Based Hollow Fiber Membranes with Brush-like CO_2-Philic Agent for Highly Efficient CO_2 Capture. *Nat. Commun.* **2017**, *8*, 2107. [CrossRef] [PubMed]
60. McKeown, N.B. Polymers of Intrinsic Microporosity (PIMs). *Polymer* **2020**, *202*, 122736. [CrossRef]
61. Comesaña-Gándara, B.; Chen, J.; Bezzu, C.G.; Carta, M.; Rose, I.; Ferrari, M.C.; Esposito, E.; Fuoco, A.; Jansen, J.C.; McKeown, N.B. Redefining the Robeson Upper Bounds for CO_2/CH_4 and CO_2/N_2 Separations Using a Series of Ultrapermeable Benzotriptycene-Based Polymers of Intrinsic Microporosity. *Energy Environ. Sci.* **2019**, *12*, 2733–2740. [CrossRef]
62. Lai, H.W.H.; Benedetti, F.M.; Ahn, J.M.; Robinson, A.M.; Wang, Y.; Pinnau, I.; Smith, Z.P.; Xia, Y. Hydrocarbon Ladder Polymers with Ultrahigh Permselectivity for Membrane Gas Separations. *Science* **2022**, *375*, 1390–1392. [CrossRef] [PubMed]
63. Robeson, L.M. The Upper Bound Revisited. *J. Memb. Sci.* **2008**, *320*, 390–400. [CrossRef]
64. Scholes, C.A.; Kanehashi, S. Polymer of Intrinsic Microporosity (PIM-1) Membranes Treated with Supercritical CO_2. *Membranes* **2019**, *9*, 41. [CrossRef]
65. Tan, X.; Robijns, S.; Thür, R.; Ke, Q.; De Witte, N.; Lamaire, A.; Li, Y.; Aslam, I.; Van Havere, D.; Donckels, T.; et al. Truly Combining the Advantages of Polymeric and Zeolite Membranes for Gas Separations. *Science* **2022**, *378*, 1189–1194. [CrossRef]
66. Torre-Celeizabal, A.; Casado-Coterillo, C.; Garea, A. Biopolymer-Based Mixed Matrix Membranes (MMMs) for CO_2/CH_4 Separation: Experimental and Modeling Evaluation. *Membranes* **2022**, *12*, 561. [CrossRef]
67. Casado-Coterillo, C.; Fernández-Barquín, A.; Zornoza, B.; Téllez, C.; Coronas, J.; Irabien, Á. Synthesis and Characterisation of MOF/Ionic Liquid/Chitosan Mixed Matrix Membranes for CO_2/N_2 Separation. *RSC Adv.* **2015**, *5*, 102350–102361. [CrossRef]
68. Borgohain, R.; Jain, N.; Prasad, B.; Mandal, B.; Su, B. Carboxymethyl Chitosan/Carbon Nanotubes Mixed Matrix Membranes for CO_2 Separation. *React. Funct. Polym.* **2019**, *143*, 104331. [CrossRef]

69. Casado-Coterillo, C.; Fernández-Barquín, A.; Irabien, A. Effect of Humidity on CO_2/N_2 and CO_2/CH_4 Separation Using Novel Robust Mixed Matrix Composite Hollow Fiber Membranes: Experimental and Model Evaluation. *Membranes* **2019**, *10*, 6. [CrossRef] [PubMed]
70. Shen, J.; Nada, A.A.; Abou-Zeid, N.Y.; Hudson, S.M. Synthesis of Chitosan Iodoacetamides via Carbodiimide Coupling Reaction: Effect of Degree of Substitution on the Hemostatic Properties. *Carbohydr. Polym.* **2020**, *229*, 115522. [CrossRef]
71. Nada, A.A.; Ali, E.A.; Soliman, A.A.F.; Shen, J.; Abou-Zeid, N.Y.; Hudson, S.M. Multi-Layer Dressing Made of Laminated Electrospun Nanowebs and Cellulose-Based Adhesive for Comprehensive Wound Care. *Int. J. Biol. Macromol.* **2020**, *162*, 629–644. [CrossRef]
72. Nada, A.A.; Abdellatif, F.H.H.; Soliman, A.A.F.; Shen, J.; Hudson, S.M.; Abou-Zeid, N.Y. Fabrication and Bioevaluation of a Medicated Electrospun Mat Based on Azido-Cellulose Acetate via Click Chemistry. *Cellulose* **2019**, *26*, 9721–9736. [CrossRef]
73. Ji, Y.; Zhang, M.; Guan, K.; Zhao, J.; Liu, G.; Jin, W. High-Performance CO_2 Capture through Polymer-Based Ultrathin Membranes. *Adv. Funct. Mater.* **2019**, *29*, 1–9. [CrossRef]
74. Sandru, M.; Sandru, E.M.; Ingram, W.F.; Deng, J.; Stenstad, P.M.; Deng, L.; Spontak, R.J. An Integrated Materials Approach to Ultrapermeable and Ultraselective CO_2 Polymer Membranes. *Science* **2022**, *376*, 90–94. [CrossRef]
75. Zhang, Y.; Wang, H.; Liu, J.; Hou, J.; Zhang, Y. Enzyme-Embedded Metal-Organic Framework Membranes on Polymeric Substrates for Efficient CO_2 Capture. *J. Mater. Chem. A* **2017**, *5*, 19954–19962. [CrossRef]
76. Saeed, M.; Deng, L. CO_2 Facilitated Transport Membrane Promoted by Mimic Enzyme. *J. Memb. Sci.* **2015**, *494*, 196–204. [CrossRef]
77. Gilassi, S.; Taghavi, S.M.; Rodrigue, D.; Kaliaguine, S. Techno-Economic Evaluation of Membrane and Enzymatic-Absorption Processes for CO_2 Capture from Flue-Gas. *Sep. Purif. Technol.* **2020**, *248*, 116941. [CrossRef]
78. Bao, L.; Trachtenberg, M.C. Facilitated Transport of CO_2 across a Liquid Membrane: Comparing Enzyme, Amine, and Alkaline. *J. Memb. Sci.* **2006**, *280*, 330–334. [CrossRef]
79. Cheng, L.H.; Zhang, L.; Chen, H.L.; Gao, C.J. Hollow Fiber Contained Hydrogel-CA Membrane Contactor for Carbon Dioxide Removal from the Enclosed Spaces. *J. Memb. Sci.* **2008**, *324*, 33–43. [CrossRef]
80. Bednár, A.; Nemestóthy, N.; Bakonyi, P.; Fülöp, L.; Zhen, G.; Lu, X.; Kobayashi, T.; Kumar, G.; Xu, K.; Bélafi-Bakó, K. Enzymatically-Boosted Ionic Liquid Gas Separation Membranes Using Carbonic Anhydrase of Biomass Origin. *Chem. Eng. J.* **2016**, *303*, 621–626. [CrossRef]
81. De Castro, A.M.; Prasavath, D.; Bevilaqua, J.V.; Portugal, C.A.M.; Neves, L.A.; Crespo, J.G. Role of Water on Deep Eutectic Solvents (DES) Properties and Gas Transport Performance in Biocatalytic Supported DES Membranes. *Sep. Purif. Technol.* **2021**, *255*, 117763. [CrossRef]
82. Nemestóthy, N.; Bakonyi, P.; Németh, Z.; Bélafi-Bakó, K. Evaluation of Pectin-Reinforced Supported Liquid Membranes Containing Carbonic Anhydrase: The Role of Ionic Liquid on Enzyme Stability and CO_2 Separation Performance. *J. CO_2 Util.* **2018**, *24*, 59–63. [CrossRef]
83. Rivero, J.R.; Panagakos, G.; Lieber, A.; Hornbostel, K. Hollow Fiber Membrane Contactors for Post-Combustion Carbon Capture: A Review of Modeling Approaches. *Membranes* **2020**, *10*, 382. [CrossRef]
84. Vadillo, J.M.; Gómez-Coma, L.; Garea, A.; Irabien, A. Hollow Fiber Membrane Contactors in CO_2 Desorption: A Review. *Energy Fuels* **2021**, *35*, 111–136. [CrossRef]
85. Porcheron, F.; Ferré, D.; Favre, E.; Nguyen, P.T.; Lorain, O.; Mercier, R.; Rougeau, L. Hollow Fiber Membrane Contactors for CO_2 Capture: From Lab-Scale Screening to Pilot-Plant Module Conception. *Energy Procedia* **2011**, *4*, 763–770. [CrossRef]
86. Ibrahim, M.H.; El-Naas, M.H.; Zhang, Z.; Van Der Bruggen, B. CO_2 Capture Using Hollow Fiber Membranes: A Review of Membrane Wetting. *Energy Fuels* **2018**, *32*, 963–978. [CrossRef]
87. Zare, A.; Perna, L.; Nogalska, A.; Ambrogi, V.; Cerruti, P.; Tylkowski, B.; García-Valls, R.; Giamberini, M. Polymer Blends for Improved CO_2 Capture Membranes. *Polymers* **2019**, *11*, 1662. [CrossRef]
88. Yong, J.K.J.; Cui, J.; Cho, K.L.; Stevens, G.W.; Caruso, F.; Kentish, S.E. Surface Engineering of Polypropylene Membranes with Carbonic Anhydrase-Loaded Mesoporous Silica Nanoparticles for Improved Carbon Dioxide Hydration. *Langmuir* **2015**, *31*, 6211–6219. [CrossRef] [PubMed]
89. Kim, T.J.; Lang, A.; Chikukwa, A.; Sheridan, E.; Dahl, P.I.; Leimbrink, M.; Skiborowski, M.; Roubroeks, J. Enzyme Carbonic Anhydrase Accelerated CO_2 Absorption in Membrane Contactor. *Energy Procedia* **2017**, *114*, 17–24. [CrossRef]
90. RASOULI, H.; ILIUTA, I.; BOUGIE, F.; GARNIER, A.; ILIUTA, M.C. Hybrid Enzymatic CO_2 Capture Process in Intensified Flat Sheet Membrane Contactors with Immobilized Carbonic Anhydrase. *Sep. Purif. Technol.* **2022**, *287*, 120505. [CrossRef]
91. Nguyen, K.; Iliuta, I.; Bougie, F.; Pasquier, L.C.; Iliuta, M.C. Techno-Economic Assessment of Enzymatic CO_2 Capture in Hollow Fiber Membrane Contactors with Immobilized Carbonic Anhydrase. *Sep. Purif. Technol.* **2023**, *307*, 122702. [CrossRef]
92. Kurayama, F.; Matsuyama, T.; Yamamoto, H. Kinetic Study of a New Photosynthesis Bioreactor Design Using TiO_2 Particles Combined with Enzymes. *Adv. Powder Technol.* **2005**, *16*, 517–533. [CrossRef]
93. Gundersen, M.T.; Gladis, A.; Fosbøl, P.L.; Von Solms, N.; Woodley, J.M. Operating Considerations of Ultrafiltration in Enzyme Enhanced Carbon Capture. *Energy Procedia* **2017**, *114*, 735–743. [CrossRef]
94. Guo, M.; Gu, F.; Meng, L.; Liao, Q.; Meng, Z.; Liu, W. Synthesis of Formaldehyde from CO_2 Catalyzed by the Coupled Photo-Enzyme System. *Sep. Purif. Technol.* **2022**, *286*, 120480. [CrossRef]

95. Wang, Y.Z.; Zhao, Z.P.; Li, M.F.; Chen, Y.Z.; Liu, W. fang Development of a Hollow Fiber Membrane Micro-Reactor for Biocatalytic Production of Formate from CO_2. *J. Memb. Sci.* **2016**, *514*, 44–52. [CrossRef]
96. Zhang, Y.; Liu, J. Bioinspired Photocatalytic NADH Regeneration by Covalently Metalated Carbon Nitride for Enhanced CO_2 Reduction. *Chem.–Eur. J.* **2022**, *28*, e202201430. [CrossRef]
97. Nilouyal, S.; Karahan, H.E.; Isfahani, A.P.; Yamaguchi, D.; Gibbons, A.H.; Ito, M.M.M.; Sivaniah, E.; Ghalei, B. Carbonic Anhydrase-Mimicking Supramolecular Nanoassemblies for Developing Carbon Capture Membranes. *ACS Appl. Mater. Interfaces* **2022**, *14*, 37595–37607. [CrossRef]
98. Wang, Z.; Zhang, Y.; Wang, J.; Zhang, Y. Bioinspired Porous Organic Polymer-Functionalized Membranes for Efficient CO_2 Capture. *Sustain. Energy Fuels* **2020**, *4*, 1191–1198. [CrossRef]
99. Jahan, Z.; Niazi, M.B.K.; Gul, S.; Sher, F.; Kakar, S.J.; Hägg, M.B.; Gregersen, Ø.W. Mimic Enzyme Based Cellulose Nanocrystals/PVA Nanocomposite Membranes for Enrichment of Biogas as a Natural Gas Substitute. *J. Polym. Environ.* **2021**, *29*, 2598–2608. [CrossRef]
100. Saeed, M.; Deng, L. Carbon Nanotube Enhanced PVA-Mimic Enzyme Membrane for Post-Combustion CO_2 Capture. *Int. J. Greenh. Gas Control* **2016**, *53*, 254–262. [CrossRef]
101. Duan, S.; Kai, T.; Nakao, S.I. Effect of Carbonic Anhydrase on CO_2 Separation Performance of Thin Poly(Amidoamine) Dendrimer/Poly(Ethylene Glycol) Hybrid Membranes. *Membranes* **2019**, *9*, 167. [CrossRef] [PubMed]
102. Fan, L.H.; Liu, N.; Yu, M.R.; Yang, S.T.; Chen, H.L. Cell Surface Display of Carbonic Anhydrase on Escherichia Coli Using Ice Nucleation Protein for CO_2 Sequestration. *Biotechnol. Bioeng.* **2011**, *108*, 2853–2864. [CrossRef]
103. Favre, N.; Pierre, A.C. Synthesis and Behaviour of Hybrid Polymer-Silica Membranes Made by Sol Gel Process with Adsorbed Carbonic Anhydrase Enzyme, in the Capture of CO_2. *J. Sol-Gel Sci. Technol.* **2011**, *60*, 177–188. [CrossRef]
104. Neves, L.A.; Afonso, C.; Coelhoso, I.M.; Crespo, J.G. Integrated CO_2 Capture and Enzymatic Bioconversion in Supported Ionic Liquid Membranes. *Sep. Purif. Technol.* **2012**, *97*, 34–41. [CrossRef]
105. Fu, Y.; Jiang, Y.B.; Dunphy, D.; Xiong, H.; Coker, E.; Chou, S.; Zhang, H.; Vanegas, J.M.; Croissant, J.G.; Cecchi, J.L.; et al. Ultra-Thin Enzymatic Liquid Membrane for CO_2 Separation and Capture. *Nat. Commun.* **2018**, *9*, 1–12. [CrossRef]
106. Craveiro, R.; Neves, L.A.; Duarte, A.R.C.; Paiva, A. Supported Liquid Membranes Based on Deep Eutectic Solvents for Gas Separation Processes. *Sep. Purif. Technol.* **2021**, *254*, 117593. [CrossRef]
107. Mondal, S.; Alke, B.; de Castro, A.M.; Ortiz-Albo, P.; Syed, U.T.; Crespo, J.G.; Brazinha, C. Design of Enzyme Loaded W/O Emulsions by Direct Membrane Emulsification for CO_2 Capture. *Membranes* **2022**, *12*, 797. [CrossRef]
108. De Castro, A.M.; Neves, L.A.; Corvo, M.C.; Cabrita, E.J.; Crespo, J.G. Effect of Carbonic Anhydrase on CO_2 Absorption Promoted by Choline Hydroxide Using Supported Liquid Membranes. *Sep. Purif. Technol.* **2022**, *280*, 119921. [CrossRef]
109. Enns, T. Facilitation by Carbonic Anhydrase of Carbon Dioxide Transport. *Science.* **1967**, *155*, 44–47. [CrossRef] [PubMed]
110. Abdelrahim, M.Y.M.; Martins, C.F.; Neves, L.A.; Capasso, C.; Supuran, C.T.; Coelhoso, I.M.; Crespo, J.G.; Barboiu, M. Supported Ionic Liquid Membranes Immobilized with Carbonic Anhydrases for CO_2 Transport at High Temperatures. *J. Memb. Sci.* **2017**, *528*, 225–230. [CrossRef]
111. Zhang, Y.-T.; Zhang, L.; Chen, H.-L.; Zhang, H.-M. Selective Separation of Low Concentration CO_2 Using Hydrogel Immobilized CA Enzyme Based Hollow Fiber Membrane Reactors. *Chem. Eng. Sci.* **2010**, *65*, 3199–3207. [CrossRef]
112. COWAN, R.M.; GE, J.-J.; QIN, Y.-J.; McGREGOR, M.L.; TRACHTENBERG, M.C. CO_2 Capture by Means of an Enzyme-Based Reactor. *Ann. N. Y. Acad. Sci.* **2003**, *984*, 453–469. [CrossRef]
113. Amann, J.M.G.; Bouallou, C. CO_2 Capture from Power Stations Running with Natural Gas (NGCC) and Pulverized Coal (PC): Assessment of a New Chemical Solvent Based on Aqueous Solutions of N-Methyldiethanolamine + Triethylene Tetramine. *Energy Procedia* **2009**, *1*, 909–916. [CrossRef]
114. Liu, S.; Gao, H.; He, C.; Liang, Z. Experimental Evaluation of Highly Efficient Primary and Secondary Amines with Lower Energy by a Novel Method for Post-Combustion CO_2 Capture. *Appl. Energy* **2019**, *233–234*, 443–452. [CrossRef]
115. Vega, F.; Baena-Moreno, F.M.; Gallego Fernández, L.M.; Portillo, E.; Navarrete, B.; Zhang, Z. Current Status of CO_2 Chemical Absorption Research Applied to CCS: Towards Full Deployment at Industrial Scale. *Appl. Energy* **2020**, *260*, 114313. [CrossRef]
116. N.Borhani, T.; Wang, M. Role of Solvents in CO_2 Capture Processes: The Review of Selection and Design Methods. *Renew. Sustain. Energy Rev.* **2019**, *114*, 109299. [CrossRef]
117. Zhang, S.; Lu, Y. Kinetic Performance of CO_2 Absorption into a Potassium Carbonate Solution Promoted with the Enzyme Carbonic Anhydrase: Comparison with a Monoethanolamine Solution. *Chem. Eng. J.* **2015**, *279*, 335–343. [CrossRef]
118. Leimbrink, M.; Neumann, K.; Kupitz, K.; Górak, A.; Skiborowski, M. Enzyme Accelerated Carbon Capture in Different Contacting Equipment—A Comparative Study. *Energy Procedia* **2017**, *114*, 795–812. [CrossRef]
119. Zhao, S.; Feron, P.H.M.; Deng, L.; Favre, E.; Chabanon, E.; Yan, S.; Hou, J.; Chen, V.; Qi, H. Status and Progress of Membrane Contactors in Post-Combustion Carbon Capture: A State-of-the-Art Review of New Developments. *J. Memb. Sci.* **2016**, *511*, 180–206. [CrossRef]
120. Arazawa, D.T.; Oh, H.I.; Ye, S.H.; Johnson, C.A.; Woolley, J.R.; Wagner, W.R.; Federspiel, W.J. Immobilized Carbonic Anhydrase on Hollow Fiber Membranes Accelerates CO_2 Removal from Blood. *J. Memb. Sci.* **2012**, *403–404*, 25–31. [CrossRef]
121. Kimmel, J.D.; Arazawa, D.T.; Ye, S.H.; Shankarraman, V.; Wagner, W.R.; Federspiel, W.J. Carbonic Anhydrase Immobilized on Hollow Fiber Membranes Using Glutaraldehyde Activated Chitosan for Artificial Lung Applications. *J. Mater. Sci. Mater. Med.* **2013**, *24*, 2611–2621. [CrossRef]

122. Arazawa, D.T.; Kimmel, J.D.; Finn, M.C.; Federspiel, W.J. Acidic Sweep Gas with Carbonic Anhydrase Coated Hollow Fiber Membranes Synergistically Accelerates CO_2 Removal from Blood. *Acta Biomater.* **2015**, *25*, 143–149. [CrossRef]
123. Arazawa, D.T.; Kimmel, J.D.; Federspiel, W.J. Kinetics of CO_2 Exchange with Carbonic Anhydrase Immobilized on Fiber Membranes in Artificial Lungs. *J. Mater. Sci. Mater. Med.* **2015**, *26*, 1–8. [CrossRef] [PubMed]
124. Koonaphapdeelert, S.; Wu, Z.; Li, K. Carbon Dioxide Stripping in Ceramic Hollow Fibre Membrane Contactors. *Chem. Eng. Sci.* **2009**, *64*, 1–8. [CrossRef]
125. Kosaraju, P.; Kovvali, A.S.; Korikov, A.; Sirkar, K.K. Hollow Fiber Membrane Contactor Based CO_2 Absorption-Stripping Using Novel Solvents and Membranes. *Ind. Eng. Chem. Res.* **2005**, *44*, 1250–1258. [CrossRef]
126. Saunders, P.; Salmon, S.; Borchert, M.; Lessard, L.P. Modular Membrane Reactor and Process for Carbon Dioxide Extraction. US Patent 20110223650 A1, 15 September 2011.
127. Hou, J.; Dong, G.; Xiao, B.; Malassigne, C.; Chen, V. Preparation of Titania Based Biocatalytic Nanoparticles and Membranes for CO_2 Conversion. *J. Mater. Chem. A* **2015**, *3*, 3332–3342. [CrossRef]
128. Hou, J.; Ji, C.; Dong, G.; Xiao, B.; Ye, Y.; Chen, V. Biocatalytic Janus Membranes for CO_2 Removal Utilizing Carbonic Anhydrase. *J. Mater. Chem. A* **2015**, *3*, 17032–17041. [CrossRef]
129. Hou, J.; Zulkifli, M.Y.; Mohammad, M.; Zhang, Y.; Razmjou, A.; Chen, V. Biocatalytic Gas-Liquid Membrane Contactors for CO_2 Hydration with Immobilized Carbonic Anhydrase. *J. Memb. Sci.* **2016**, *520*, 303–313. [CrossRef]
130. Saeed, M.; Deng, L. Post-Combustion CO_2 Membrane Absorption Promoted by Mimic Enzyme. *J. Memb. Sci.* **2016**, *499*, 36–46. [CrossRef]
131. Yong, J.K.J.; Stevens, G.W.; Caruso, F.; Kentish, S.E. In Situ Layer-by-Layer Assembled Carbonic Anhydrase-Coated Hollow Fiber Membrane Contactor for Rapid CO_2 Absorption. *J. Memb. Sci.* **2016**, *514*, 556–565. [CrossRef]
132. Yong, J.K.J.; Stevens, G.W.; Caruso, F.; Kentish, S.E. The Resilience of Carbonic Anhydrase Enzyme for Membrane-Based Carbon Capture Applications. *Int. J. Greenh. Gas Control* **2017**, *62*, 122–129. [CrossRef]
133. Malankowska, M.; Martins, C.F.; Rho, H.S.; Neves, L.A.; Tiggelaar, R.M.; Crespo, J.G.; Pina, M.P.; Mallada, R.; Gardeniers, H.; Coelhoso, I.M. Microfluidic Devices as Gas—Ionic Liquid Membrane Contactors for CO_2 Removal from Anaesthesia Gases. *J. Memb. Sci.* **2018**, *545*, 107–115. [CrossRef]
134. Xu, Y.; Lin, Y.; Chew, N.G.P.; Malde, C.; Wang, R. Biocatalytic PVDF Composite Hollow Fiber Membranes for CO_2 Removal in Gas-Liquid Membrane Contactor. *J. Memb. Sci.* **2019**, *572*, 532–544. [CrossRef]
135. Kim, H.S.; Hong, S.G.; Yang, J.; Ju, Y.; Ok, J.; Kwon, S.J.; Yeon, K.M.; Dordick, J.S.; Kim, J. 3D-Printed Interfacial Devices for Biocatalytic CO_2 Conversion at Gas-Liquid Interface. *J. CO_2 Util.* **2020**, *38*, 291–298. [CrossRef]
136. Liu, Q.; Bai, X.; Pham, H.; Hu, J.; Dinu, C.Z. Active Nanointerfaces Based on Enzyme Carbonic Anhydrase and Metal–Organic Framework for Carbon Dioxide Reduction. *Nanomaterials* **2021**, *11*, 1008. [CrossRef]
137. Rasouli, H.; Iliuta, I.; Bougie, F.; Garnier, A.; Iliuta, M.C. Enzyme-Immobilized Flat-Sheet Membrane Contactor for Green Carbon Capture. *Chem. Eng. J.* **2021**, *421*, 129587. [CrossRef]
138. Iliuta, I.; Iliuta, M.C. Investigation of CO_2 Removal by Immobilized Carbonic Anhydrase Enzyme in a Hollow-Fiber Membrane Bioreactor. *AIChE J.* **2017**, *63*, 2996–3007. [CrossRef]
139. Wang, X.; Zhou, D.; Sinniah, K.; Clarke, C.; Birch, L.; Li, H.; Rayment, T.; Abell, C. Electrostatic Orientation of Enzymes on Surfaces for Ligand Screening Probed by Force Spectroscopy. *Langmuir* **2006**, *22*, 887–892. [CrossRef] [PubMed]
140. Zeuner, B.; Ma, N.; Berendt, K.; Meyer, A.S.; Andric, P.; Jørgensen, J.H.; Pinelo, M. Immobilization of Alcohol Dehydrogenase on Ceramic Silicon Carbide Membranes for Enzymatic CH_3OH Production. *J. Chem. Technol. Biotechnol.* **2018**, *93*, 2952–2961. [CrossRef]
141. Wang, Y.; Chen, Y.; Wang, C.; Sun, J.; Zhao, Z.; Liu, W. Polyethylenimine-Modified Membranes for CO_2 Capture and in Situ Hydrogenation. *ACS Appl. Mater. Interfaces* **2018**, *10*, 29003–29009. [CrossRef] [PubMed]
142. Barin, R.; Biria, D.; Rashid-Nadimi, S.; Asadollahi, M.A. Enzymatic CO_2 Reduction to Formate by Formate Dehydrogenase from Candida Boidinii Coupling with Direct Electrochemical Regeneration of NADH. *J. CO_2 Util.* **2018**, *28*, 117–125. [CrossRef]
143. Zhu, D.; Ao, S.; Deng, H.; Wang, M.; Qin, C.; Zhang, J.; Jia, Y.; Ye, P.; Ni, H. Ordered Coimmobilization of a Multienzyme Cascade System with a Metal Organic Framework in a Membrane: Reduction of CO_2 to Methanol. *ACS Appl. Mater. Interfaces* **2019**, *11*, 33581–33588. [CrossRef] [PubMed]
144. Gu, F.J.; Wang, Y.Z.; Meng, Z.H.; Liu, W.F.; Qiu, L.Y. A Coupled Photocatalytic/Enzymatic System for Sustainable Conversion of CO_2 to Formate. *Catal. Commun.* **2020**, *136*, 105903. [CrossRef]
145. Guo, M.Y.; Zhai, T.T.; Wang, C.H.; Meng, Z.H.; Liu, W.F. Immobilization of Formate Dehydrogenase on Polyethyleneimine Modified Carriers for the Enhancement of Catalytic Performance. *Catal. Commun.* **2021**, *149*, 106259. [CrossRef]
146. Chai, M.; Razmjou, A.; Chen, V. Metal-Organic-Framework Protected Multi-Enzyme Thin-Film for the Cascade Reduction of CO_2 in a Gas-Liquid Membrane Contactor. *J. Memb. Sci.* **2021**, *623*, 118986. [CrossRef]
147. Lin, G.; Zhang, Y.; Hua, Y.; Zhang, C.; Jia, C.; Ju, D.; Yu, C.; Li, P.; Liu, J. Bioinspired Metalation of the Metal-Organic Framework MIL-125-NH_2 for Photocatalytic NADH Regeneration and Gas-Liquid-Solid Three-Phase Enzymatic CO_2 Reduction. *Angew. Chemie Int. Ed.* **2022**, *61*, e202206283. [CrossRef]
148. Tian, Y.; Zhou, Y.; Zong, Y.; Li, J.; Yang, N.; Zhang, M.; Guo, Z.; Song, H. Construction of Functionally Compartmental Inorganic Photocatalyst-Enzyme System via Imitating Chloroplast for Efficient Photoreduction of CO_2 to Formic Acid. *ACS Appl. Mater. Interfaces* **2020**, *12*, 34795–34805. [CrossRef]

149. Hollmann, F.; Witholt, B.; Schmid, A. $[Cp^*Rh(Bpy)(H_2O)]^{2+}$: A Versatile Tool for Efficient and Non-Enzymatic Regeneration of Nicotinamide and Flavin Coenzymes. *J. Mol. Catal. B Enzym.* **2002**, *19–20*, 167–176. [CrossRef]
150. Yuan, M.; Sahin, S.; Cai, R.; Abdellaoui, S.; Hickey, D.P.; Minteer, S.D.; Milton, R.D. Creating a Low-Potential Redox Polymer for Efficient Electroenzymatic CO_2 Reduction. *Angew. Chem. -Int. Ed.* **2018**, *57*, 6582–6586. [CrossRef] [PubMed]
151. Sahin, S.; Cai, R.; Milton, R.D.; Abdellaoui, S.; Macazo, F.C.; Minteer, S.D. Molybdenum-Dependent Formate Dehydrogenase for Formate Bioelectrocatalysis in a Formate/O_2 Enzymatic Fuel Cell. *J. Electrochem. Soc.* **2018**, *165*, H109–H113. [CrossRef]
152. Minteer, S.D.; Liaw, B.Y.; Cooney, M.J. Enzyme-Based Biofuel Cells. *Curr. Opin. Biotechnol.* **2007**, *18*, 228–234. [CrossRef] [PubMed]
153. Moore, C.M.; Akers, N.L.; Hill, A.D.; Johnson, Z.C.; Minteer, S.D. Improving the Environment for Immobilized Dehydrogenase Enzymes by Modifying Nafion with Tetraalkylammonium Bromides. *Biomacromolecules* **2004**, *5*, 1241–1247. [CrossRef]
154. Moehlenbrock, M.J.; Minteer, S.D. Extended Lifetime Biofuel Cells. *Chem. Soc. Rev.* **2008**, *37*, 1188–1196. [CrossRef]
155. Arechederra, R.L.; Treu, B.L.; Minteer, S.D. Development of Glycerol/O_2 Biofuel Cell. *J. Power Sources* **2007**, *173*, 156–161. [CrossRef]
156. Rubenwolf, S.; Sané, S.; Hussein, L.; Kestel, J.; Von Stetten, F.; Urban, G.; Krueger, M.; Zengerle, R.; Kerzenmacher, S. Prolongation of Electrode Lifetime in Biofuel Cells by Periodic Enzyme Renewal. *Appl. Microbiol. Biotechnol.* **2012**, *96*, 841–849. [CrossRef]
157. Herkendell, K.; Stemmer, A.; Tel-Vered, R. Extending the Operational Lifetimes of All-Direct Electron Transfer Enzymatic Biofuel Cells by Magnetically Assembling and Exchanging the Active Biocatalyst Layers on Stationary Electrodes. *Nano Res.* **2019**, *12*, 767–775. [CrossRef]

membranes

MDPI

Article

Study on Organo-Silica-Derived Membranes Using a Robeson-like Plot

Lucas Bünger [1,*], Tim van Gestel [2], Tim Kurtz [1], Krassimir Garbev [1], Peter Stemmermann [1], Wilhelm A. Meulenberg [2], Olivier Guillon [2] and Dieter Stapf [1]

[1] Institute for Technical Chemistry, Karlsruhe Institute of Technology, 76344 Karlsruhe, Germany; tim.kurtz@kit.edu (T.K.); krassimir.garbev@kit.edu (K.G.); dieter.stapf@kit.edu (D.S.)

[2] Institute of Energy Materials and Devices-IMD-2, Forschungszentrum Jülich GmbH, 52425 Jülich, Germany; t.van.gestel@fz-juelich.de (T.v.G.); w.a.meulenberg@fz-juelich.de (W.A.M.); o.guillon@fz-juelich.de (O.G.)

* Correspondence: lucas.buenger@kit.edu; Tel.: +49-721-6052-4405

Abstract: For industrial CO_2 utilization, the supply of concentrated CO_2 within a continuous, high-volume stream at high temperatures remains a substantial requirement. Membrane processes offer a simple and efficient method to provide CO_2 in this form. While several organo-silica-based membranes have been developed for CO_2/N_2 separation under these conditions, there is no standardized framework guiding comparability and optimization. Therefore, we present these membranes in a Robeson-like plot across various temperatures. Utilizing a standard 1,2-bis(triethoxysilyl)-ethane (BTESE) precursor and a simplified sol–gel method, we prepared a microporous membrane layer and characterized it for an exemplary comparison. This characterization includes key parameters for mixed-gas applications: (1) temperature-dependent single- and mixed-gas permeances to observe interactions, (2) the impact of the driving forces in mixtures (vacuum and concentration) to distinguish between permselectivity and the separation factor clearly, and (3) influence of the support structure to enable permeability calculations at elevated temperatures. Furthermore, a quick interpretation method for assessing the membrane's microstructure is presented. A qualitative microstructure assessment can be achieved by analyzing the temperature dependencies of the three major diffusion mechanisms that simultaneously occur—Knudsen, surface, and activated diffusion.

Keywords: microporous membrane; BTESE; CO_2 separation; binary mixtures; Robeson-like plot

1. Introduction

An important way to compensate for remaining fossil carbon emissions is to capture carbon from unavoidable industrial emissions and reutilize it as a carbon source in chemical processes. According to Favre et al. [1], processes in the industry generate continuous exhaust streams with CO_2 concentrations between 5 and 30%. Carbon utilization processes, such as the reverse water–gas shift reaction (rWGS), require temperatures above 250 °C, making CO_2 capture at these temperatures more economical [2]. The state-of-the-art separation methods based on washing and adsorption typically yield CO_2 discontinuously at low temperatures, which is less efficient for subsequent utilization. Although it is possible to implement intelligent process interconnections to achieve a pseudo-continuous process, like in pressure swing adsorption (PSA), this would require advanced process control and high investment costs [3]. A membrane separation process with a CO_2-affine membrane appears as an attractive, simple, and straightforward alternative because it

https://doi.org/10.3390/membranes15030083

enables removing CO_2 continuously from an industrial emission stream and constantly delivering a concentrated CO_2 stream for further utilization [4]. Notably, a membrane can be regarded as a continuous PSA application by spatially separating adsorption and desorption through diffusion across the membrane [5].

In recent years, major progress has been made in gas-separation polymer membranes. New classes of CO_2-philic polymers have been developed with improved performance for CO_2 separation and capture [6]. Many polymeric membranes investigated so far are summarized in the Robeson plot in Figure 1, with its upper bound defined by the intrinsic mass transport properties of CO_2 and N_2 in polymers [7]. However, for industrial applications, the operating conditions in the industry (e.g., high temperature) present a major challenge for polymer membrane development [8]. There are no Robeson plots for membranes above standard temperature due to the changing membrane properties at high temperatures that render CO_2 separation impossible.

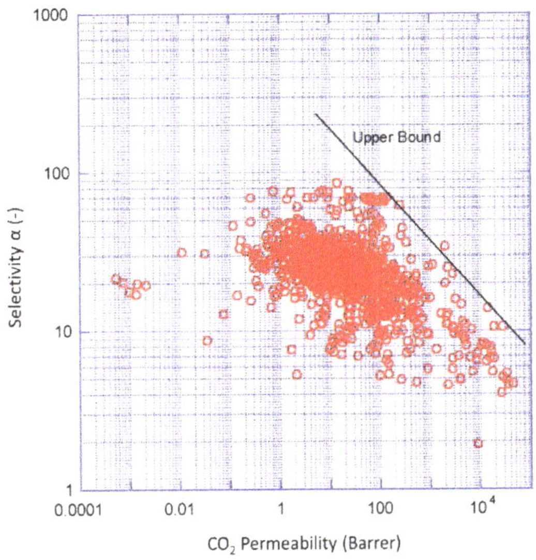

Figure 1. Robeson plot upper bound correlation for CO_2/N_2 separation around standard temperature for various polymeric membranes. Reprinted from Robeson, L.M. The upper bound revisited. *J. Membr. Sci.* 2008, 320, 390–400 with permission from Elsevier [7].

Merkel et al. [9] displayed a variety of porous membranes in a Robeson-like plot, with permeance rather than permeability as the membrane kinetics (Figure 2). Considering the economic suitability, the authors included an area of optimal membrane properties as a target for future developments. Membranes with optimal properties have at least a permeance for CO_2 above 1000 GPU (GPU = 10^{-6} cm^3 (STP)/(cm^2 s cmHg)) and a single-gas selectivity higher than 20. Optimizing membranes to reach the upper boundary may lead to high selectivities but low permeabilities. These membranes have a very low permeability for the unselective component, causing relatively high partial pressures of the selective component in the permeate. As a result, this leads to low fluxes of the selective compound caused by the small partial pressure difference and requires larger membrane areas to separate high-feed streams. Therefore, the (Robeson) upper bound is not the technically relevant limitation, as membranes with low fluxes and high selectivity have no technical relevance in separating gas mixtures [9].

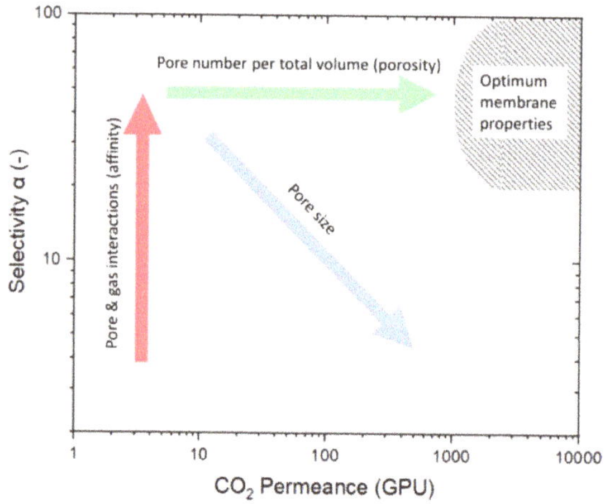

Figure 2. Robeson-like plot for porous materials. This schematic illustrates the ways to influence the membrane design criteria to reach the optimum membrane properties as described by Merkel et al. [9]. The increase in selectivity is predominantly determined by the membrane material (red arrow), whereas the permeance is based on pore size (blue arrow) and pore number (green arrow).

Several design criteria must be met to reach the envisioned optimal membrane properties. Most importantly, the membrane must be made of selective material for the preferred component to permeate with higher kinetics. Differences in the affinity between gas and membrane determine the selectivity (marked by red arrow). For this affinity to become relevant, pores must be below 2 nm, as only these micropores have a favorable surface-to-volume ratio where the flow along the pore surface exceeds the volume flow. However, as the pores become larger, the permeance increases, but the selectivity decreases (marked by the blue arrow in Figure 2). Increasing the pore number is the only way to shift a microstructure with small pores and high selectivity into the envisioned area. In other words, the ratio of the number of micropores to membrane volume (porosity) needs to be high (marked by the green arrow) [10,11].

In designing a membrane to separate CO_2 from N_2, it is necessary to make use of the right property difference. The kinetic diameters of CO_2 (0.33 nm) and N_2 (0.365 nm) are too close, and a precise pore size would be necessary. The weight of N_2 (28 g mol^{-1}) is smaller than that of CO_2 (44 g mol^{-1}). Therefore, Knudsen-based separation would prefer N_2 permeation with low selectivity. Thus, only a separation based on their affinity with surfaces is applicable. CO_2 has a quadrupole that enables adsorption on basic (oxygen) sites [12]. This property enables surface diffusion, which can lead to high selectivities if the microstructure is set appropriately. At first glance, a membrane with ultramicropores would be preferable for CO_2 separation and purification as high selectivities are possible. However, a small pore size down to the range where molecular sieving dominates is also accompanied by very low flow rates for the non-selective component, demanding very high membrane areas to be technically relevant [9]. Therefore, we opted for a membrane with slightly larger micropores consisting of a CO_2-philic material. In the literature, various CO_2 affine materials with a network of connected micropores of approximately 0.5 nm diameter have been reported, including zeolites [13], metal–organic frameworks (MOFs) [14], carbon-based membranes [12], and sol–gel-derived silica [15]. Silica membranes, in particular, have been intensively investigated due to their relatively simple formation by the sol–gel coating process. They are characterized by relatively good selectivity for CO_2 while maintaining adequate permeability. However, interest in this material has diminished considerably due to concerns about its stability in water (vapor) [16]. An extension of the work on silica membranes is the introduction of hybrid carbon-modified silicas. They are produced similarly to the silica membranes but, with alkane-bridged silicon precursors. The precursor used in this work is the 1,2-bis(triethoxysilyl) ethane (BTESE). This compound is a precursor for silica-bridged aerogels with versatile applications, i.e., in environmental applications in oil/water separation [17] or even in antibiotic adsorption from water [18]. Notably, the organo-silica aerogels showed excellent behavior in gas separation applications [16,19–22].

Focusing on the temperature-dependent transport behavior can provide insights into the pore structure, extending the work of Lee et al. [23]. A comparison with the reported organo-silica-derived membranes exhibits permeances ranging from as low as 30 GPU [24] to as high as 9500 GPU [25]. Although all the membranes are fabricated similarly, no comprehensive analysis for this wide range of reported kinetics has yet been provided.

As all technical applications aim to operate in mixtures, the mixed-gas permeances are of paramount relevance, and there is no operational reliability that the gases behave in the same manner in both mixtures and single-gas experiments. To our knowledge, only three reports currently investigate mixed-gas behavior in such membranes. A comparison between binary CO_2 and single-gas permeances shows very different results within the limited data available. Whereas Yang et al. [26] report a decreasing permeance for binary mixtures, Rubner et al. [24] observed an increasing one and van Gestel. et al. [27] found no difference between mixtures and single-gas experiments. For this reason, we add further data and aim to explain the observed variations.

To our knowledge, no permeation data are available for organo-silica membranes operating in vacuum and atmospheric modes, making it difficult to compare the two. Only in other microporous systems were the vacuum permeation data entirely given [28]. It is assumed that low permeate pressures enhance the CO_2 desorption kinetics and thus increase the permeance [29].

The present study aims to re-evaluate the permeation and separation behavior in organo-silica membranes, focusing on general and specific cases, using an organo-silica membrane fabricated in our lab. The investigation examines the CO_2 and N_2 permeance of the fabricated membrane in a single-gas operation. It evaluates the mass transport properties of each layer (macroporous support, mesoporous intermediate layer, and selective cover layer). We aim to correlate the permeance levels with the respective mass transport mechanism by examining the temperature trajectories. As these membranes operate to provide CO_2 for processes like rWGS, we present the available data for various published organo-silica-based membranes in the Robeson-like plot for the respective temperatures.

After the single-gas examination, we conduct experiments on mixed-gas transport phenomena by testing the fabricated membrane in binary CO_2/N_2 mixtures and providing mixed-gas permeances as one of the first for this membrane. Finally, we also study how the mode of operation affects mass transport kinetics. This involves varying factors such as the CO_2/N_2 composition of the feed mixtures and the driving force (atmospheric or vacuum permeate pressure). We compare operation modes with permeation data to show why separation selectivity and separation factors are often confused, leading to misleading conclusions.

2. Materials and Methods

2.1. Membrane Formation

The support for membrane deposition was a polished macroporous α-alumina disk with a diameter of 39 mm, a thickness of 2.2 mm, and a pore size of 80 nm, purchased from Pervatech (PT) BV (the Netherlands). On top of this, a mesoporous γ-Al_2O_3 layer was coated by an aqueous sol–gel method using a boehmite (γ-AlO(OH)) sol. The synthesis has been described in the literature based on the original work of Yoldas [30]. The sol was mixed with a PVA solution (3.5 g PVA/100 mL water) in a ratio of 3:2 and then poured into a 50 mL Petri dish. We used a specially designed dip-coater, which moved clockwise through the Petri dish so that only the polished side of the support was immersed (holding time was 15 s). For this, the support was attached to a vacuum suction cup mounted on a rotating pendulum, and the dish with the sol/PVA mixture was positioned at the lowest point of the pendulum (see Figure S1). After coating, it was isothermally treated at 500 °C for 3 h with a pre-heating and post-cooling ramp of 1 °C/min.

On top of the obtained mesoporous γ-Al_2O_3 layer, a microporous organo-silica (Org-Sil) top layer was deposited by an alcoholic sol–gel method using the same pendulum dip-coater. Our synthesis and coating process was mainly based on previous work by Castricum et al. [20] and van Gestel et al. [27]. In summary, 16.66 mL of the membrane-forming precursor (1,2-bis(triethoxysilyl)-ethane (BTESE)) was mixed with 28.14 mL of ethanol (both thermo scientific chemicals), 0.63 mL of nitric acid (65 wt.%), and 4.57 mL of water at room temperature ([water]:[hydrolyzable ethoxy groups] = 1). The resulting mixture was heated to 60 °C under reflux conditions for 90 min, followed by natural cooling to room temperature. Then, the sol was diluted with 50 mL of ethanol and stored in a refrigerator (stock sol). Before the coating step, this stock sol was further diluted 20 times with ethanol to obtain defect-free films reliably. Finally, after dip-coating, the membrane sample was dried for ~1 h and thermally treated at 300 °C for 3 h under a N_2 or air atmosphere to create different microporous structures [20,31–33]. A drawing of the resulting microstructure can be found elsewhere [5], and photographs of the membrane are shown in Figure S2.

2.2. Structural and Chemical Characterization

High-resolution scanning electron microscope images (SEM) were made using a Zeiss Supra 55 VP microscope with an acceleration voltage of 18 kV. The membrane disk was shock-frozen in liquid N_2 for a few minutes to obtain high-quality images. Then, the membrane was cut into small 1 cm^2 pieces using a side cutter, and the obtained piece was sputter-coated with a thin layer of gold to ensure electrical conductivity. Chemical characterization by ATR-FTIR and Raman spectroscopy, as well as X-ray diffraction (XRD) and thermo gravimetric (TG) analysis, was conducted in detail and are reported elsewhere [34].

2.3. Gas Permeation and Separation Experiments

Single-gas experiments were performed using a custom-made test cell designed to hold 39 mm disks, with the feed gas inlet and the retentate outlet on the upper side and the permeate outlet at the bottom side of the cell. A small opening at the top side allowed the application of a thermocouple non-invasive close to the membrane. The feed gas, CO_2 or N_2, was supplied by a gas bottle with a mounted pressure reducer, allowing measurements at pressures up to 6 bar. A Bronkhorst El Press pressure controller valve was installed in the retentate line. Pressure was measured in the feed p_f and the permeate p_p line. All the measurements were conducted with the same transmembrane pressure difference of 3.2 bar. Experiments with atmospheric pressure on the permeate side held a feed pressure of 4.2 bar; for experiments with a vacuum (70 mbar) on the permeate side, the feed pressure was lowered to 3.27 bar. No sweep gas was used. To quantify the mass flow of the permeating stream Ji, three Bronkhorst EL-Flow Prestige mass flow meters with different measurement ranges (0–1 g/h, 0–10 g/h, and 200 g/h) to reduce the measurement error were installed in the permeate line (error: ±0.5% of the measured value, ±0.1% of endpoint value). The membrane cell was placed in an oven to vary the temperature. To apply a lower pressure, a membrane pump (KNF LABORPORT N938.50) was installed in the permeate line. Before the permeation experiments, the membrane surface was sealed with a Viton O-Ring (i.d. 32 mm), resulting in an effective permeation area A_m of 804 mm^2. Due to the application of Viton, the testing temperature was limited to 200 °C; with the vacuum pump, a pressure of 70 mbar (absolute pressure) was obtained.

$$Q_i = \frac{J_i}{A_m \cdot (p_{i,f} - p_{i,p})} = \frac{P_i}{d_i} \tag{1}$$

The permeance Q was calculated from the measured values according to Equation (1). Accordingly, the permeability P_i as the intrinsic membrane kinetic was calculated by multiplying the permeance with the respective layer thickness d_i.

$$S_{CO_2/N_2} = \frac{Q_{CO_2}}{Q_{N_2}} \tag{2}$$

The permselectivity S was calculated as the ratio of the CO_2 to N_2 permeance as shown in Equation (2).

$$\beta_{CO_2/N_2} = \frac{\left(x_{CO_2}/x_{N_2}\right)_{Permeate}}{\left(x_{CO_2}/x_{N_2}\right)_{Feed}} \tag{3}$$

In binary gas permeation experiments with CO_2/N_2 mixtures, Bronkhorst EL-Flow mass flow controllers were applied to create mixtures with 10, 25, and 50 vol % CO_2. The composition of the permeate was analyzed with a micro gas chromatograph (μGC) (Agilent 490, Manchester, UK). The separation efficiency was evaluated using the separation factor β, defined in Equation (3) as the ratio of the mole fraction x of CO_2 to N_2 in the permeate compared to the same ratio in the feed:

$$Q_{composite} = \left(\frac{d_{\alpha-Al_2O_3}}{P_{\alpha-Al_2O_3}} + \frac{d_{\gamma-Al_2O_3}}{P_{\gamma-Al_2O_3}} + \frac{d_{org-sil}}{P_{org-sil}}\right)^{-1} \tag{4}$$

A resistance-in-series model was employed to calculate the consecutive mass transport resistances in the composite membrane [35].

First, the substrate's permeance was measured (Equation (1)) as a stand-alone membrane, and with the respective membrane thickness d_i, obtained from SEM images, the substrate permeability was calculated by the relevant summand of Equation (4). In the second step, the same substrate was coated with an γ-Al_2O_3 layer and combined with the information about the substrate and the γ-Al_2O_3 layer thickness; the permeability of this layer was calculated by Equation (4). This was repeated for the organo-silica layer.

3. Results

Figure 3 displays the SEM images of a membrane sample prepared as described in the experimental section. The left panel (A) shows the defect-free surface, uniformly covered by the organo-silica layer. An edge view is presented in panel (B), where distinct macro-, meso-, and microporous parts are visible. Layer 3 represents the macroporous α-Al_2O_3 substrate, followed by a γ-Al_2O_3 layer with a thickness of 4 μm (layer 2), which is relatively large compared with the other publications in this field, possibly due to the higher amount of PVA in the coating liquid used here. However, as can be seen, the roughness of the α-Al_2O_3 substrate is, in this way, very effectively smoothened. In the insert of Figure 3B, the BTESE-derived functional organo-silica top layer 1 is also clearly visible. The estimated layer thickness is around 100 nm, which is comparable with values for similar membranes in the literature [20,27,36], and the surface looks very clean and defect-free.

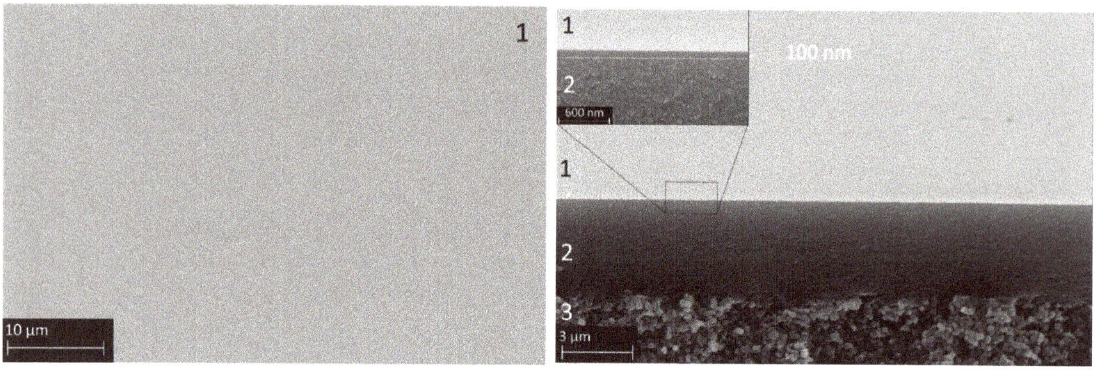

(A) **(B)**

Figure 3. SEM secondary electron images of a composite membrane comprising the α-Al_2O_3 macroporous support (3), the mesoporous γ-Al_2O_3 layer (2), and the microporous organo-silica top layer fired in N_2 (1). Panel (**A**) shows the surface view, and Panel (**B**) shows a cross-section with distinctive layer separation. The insert shows the membrane at a higher magnification to better visualize the organo-silica layer thickness.

Figure 4A shows the single-gas permeation results for CO_2 and N_2. The measurements include the stand-alone α-Al_2O_3 substrate (3), (3) + γ-Al_2O_3 layer (2), and (3 + 2) + organo-silica top layer, both the thermally treated version in N_2 and air. The support (3) shows a lower permeance for CO_2 (3.25×10^{-7} mol m^{-2} s^{-1} Pa^{-1}) than for N_2 (3.6×10^{-7} mol m^{-2} s^{-1} Pa^{-1}) at ambient temperature. The permeances for both components decrease with increasing temperature according to the Knudsen theory. For all the temperatures, the observed permselectivity is around 0.8, which equals the theoretical Knudsen selectivity. The same behavior is also observed after adding the γ-Al_2O_3 layer. However, the permeance is lower (CO_2: 2.87×10^{-7} mol m^{-2} s^{-1} Pa^{-1}; N_2: 3.6×10^{-7} mol m^{-2} s^{-1} Pa^{-1}) as an additional resistance is introduced. Such pores could enhance the surface diffusion mechanism for CO_2 as described in reference [37].

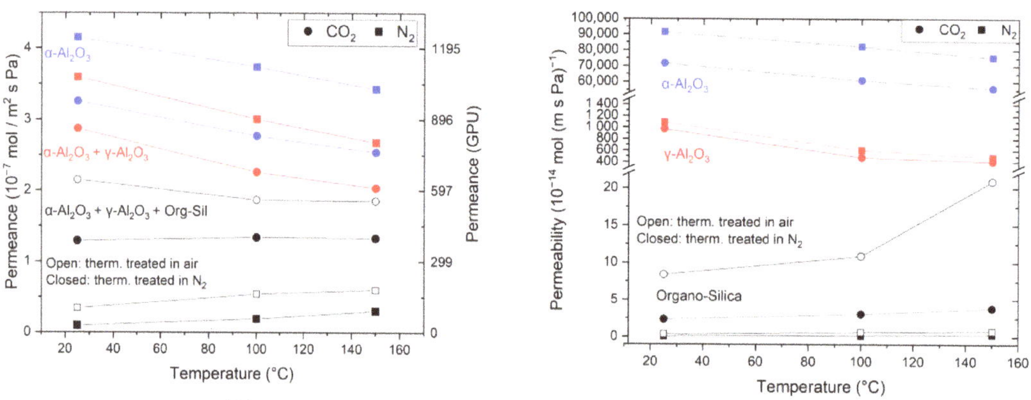

(A) **(B)**

Figure 4. Temperature-dependent single-gas permeation through the α-Al_2O_3 support (blue), the support + γ-Al_2O_3 (red), and the selective cover layer made of organo-silica on top of the others (black). Panel (**A**) shows the measured permeance, and (**B**) the derived permeability.

Coating with the organo-silica layer leads to essential changes in the gas permeation behavior. As can be seen, the CO_2 permeance exceeds the N_2 permeance, albeit at a lower level. The permeance values are reduced to 1.3×10^{-7} mol m^{-2} s^{-1} Pa^{-1} for CO_2 and 0.1×10^{-7} mol m^{-2} s^{-1} Pa^{-1} for N_2 for the organo-silica thermally treated in N_2. Again, the same trend is observed in the air-treated one, but the permeance levels are higher for both components. Notably, the permeance of N_2 is now much lower than that of CO_2, demonstrating that the membrane became CO_2-affine and suitable for CO_2 separation applications. We assume that N_2 molecules can only diffuse through the free pore volume in the membrane layer. In contrast, the CO_2 molecules can use an additional diffusion path along the inner pore surface, making surface diffusion the prominent mass transport mechanism. The obtained CO_2/N_2 permselectivity of 13 for organo-silica treated in N_2 is around 15 times higher than that for a non-selective Knudsen membrane. When treated in air the CO_2/N_2 permselectivity of 6.3 is around 8 times higher than for a non-selective Knudsen membrane.

Equally interesting are also the trends observed upon increasing the testing temperature. For N_2, a slight but measurable increase is obtained going from room temperature to 100 °C and further to 150 °C. This suggests an activated gas transport mechanism typical for relatively narrow microporous structures (pore diameter < 1 nm) and the absence of a substantial Knudsen transport contribution [38]. Remarkably, the CO_2 permeance remains constant, which is unexpected considering the lower adsorptive affinity for CO_2 but already observed elsewhere [26,27,29,39]. In these tests, activated transport and surface diffusion contribute equally to the overall CO_2 transport through the membrane and neutralize each other. From these results, it is also clear that the permselectivity of the membrane is negatively affected by these temperature-dependent trends. An increase in the temperature to 100 °C and further to 150 °C leads to a lowering of the permselectivity to 7 and 4.3 for the membrane treated in N_2, and for the membrane treated in air, the selectivity decreases from 3.4 to 3. Both cases are still larger than the Knudsen value of 0.8.

In a multi-layered system consisting of a macroporous support, mesoporous layers, and a microporous selective top layer as described here, the measured permeance and permselectivity values reflect the mass transfer resistance in the permeation through the membrane, which is expressed in terms of flux normalized per unit of pressure and area (mol/m^2 s Pa). To facilitate comparison, the permeation data in Figure 4A were then normalized per unit of thickness for each part of the membrane, resulting in the intrinsic permeability values, see Equation (4). As expected, the results summarized in Figure 4B show that the macroporous α-Al_2O_3 support (thickness 2.2 mm) exhibits by far the highest permeability and, thus, the lowest resistance in the membrane structure. When we consider the mesoporous γ-Al_2O_3 layer as a stand-alone membrane (thickness 4 µm), the permeability is much lower, and a CO_2/N_2 permselectivity of 0.9 is obtained. We can see further a similar trend for the stand-alone organo-silica layer thermally treated in N_2, which exhibits a permselectivity of ca. 23 at 20 °C, decreasing to 15 and 11 at 100 and 150 °C, respectively. Notably, the selectivity increases for the in-air treated membrane from 23 at 20 °C to 16 at 100 °C to 26 at 150 °C. The differences between the theoretically calculated and measured values suggest faster CO_2 permeating kinetics in the organo-silica layer of the membrane structure. However, this is negatively compensated by the dominating Knudsen diffusion contribution of the support structure, where N_2 shows faster kinetics. Remarkably, in the single organo-silica layer, increased permeance of CO_2 is seen for the membrane treated in air. This phenomenon is not seen in the composites as the slower CO_2 kinetics in the support structure overlay and dominate the overall kinetics. To evaluate the quality and separation performance of the membrane's top layer, the transport kinetics of the underlying layers must also be analyzed, as the entire multi-layer membrane has a lower selectivity than the top layer.

In the next series of experiments, CO_2 and N_2 are applied as a mixture to the membrane treated in N_2. A transmembrane pressure of 3.2 bar was applied. To quantify the influence of CO_2 desorption on the permeation behavior, the pressure on the permeate side was set to 70 mbar. Figure 5 shows the observed CO_2 and N_2 permeances for different feed mixtures with 10 vol%, 25 vol%, and 50 vol% CO_2 at both pressure levels as a function of the testing temperature. It can be seen that the N_2 permeance increases from room temperature to 100 °C and further to 150 °C for each feed composition by analogy with the single-gas permeation tests (100% N_2). This confirms that N_2 exhibits activated diffusion in the membrane's micropores in a mixed-gas situation. In contrast, a constant CO_2 permeation with increasing temperature in the single-gas permeation tests was observed in all the mixed-gas experiments regardless of the composition of the feed. This indicates once again that activated permeation and surface diffusion balance each other out or that the influence of the support structure overlays the temperature increase.

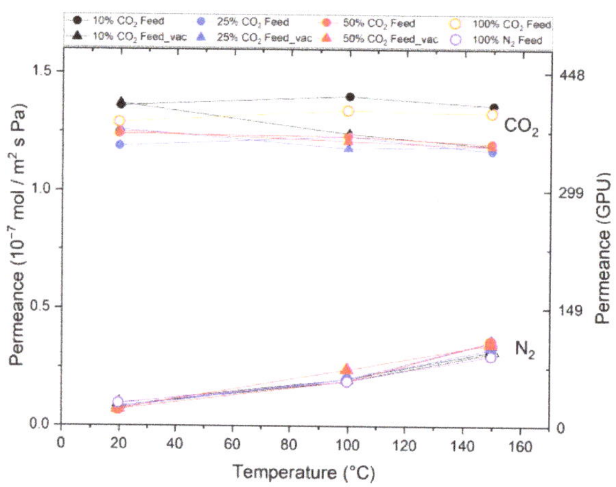

Figure 5. Mixed-gas permeation through Org-Sil membrane treated in N_2. Permeances of CO_2 and N_2 for different gas compositions (10%, 25%, and 50% CO_2) and permeate pressures (atmospheric (○) and 70 mbar (△)) at different temperatures. Single-gas measurements (100% CO_2 and 100% N_2 as reference). The transmembrane pressure is kept constant at Δp = 3.2 bar.

The trends observed in the mixed-gas results are consistent with those from the single-gas experiments, including CO_2/N_2 selectivity, which shows agreement with the permselectivity values calculated from the single-gas tests. For example, an identical selectivity of 13.5 is achieved at room temperature, which decreases to 3–4 for the test at 150 °C. This indicates that the widely accepted mechanism of pore blocking by CO_2 and consequent reduction in N_2 permeation does not occur. If such a pore-blocking mechanism were to happen when both gases are applied as a mixture to a microporous membrane, the N_2 permeance would decrease in the mixed-gas situation.

Figure 5 also shows the results of alternative permeation measurements for the case where the pressure on the permeate side is reduced to 70 mbar. We do not observe differences from the previous pressure-driven experiments with atmospheric pressure on the permeate side, which shows that the CO_2 and N_2 permeation is not dependent on the pressure on the permeate side. Furthermore, as for all the measuring points, a different CO_2 partial pressure on the feed side was applied, and the same permeance was observed. A pressure dependence on the permeance can be ruled out.

In studying the permeation and separation properties of the top layer, the effective separation factor is an important parameter. However, descriptions of permeation and separation in microporous membranes covering this aspect are hard to find. The separation factor is a characteristic parameter that describes the separation efficiency of a binary mixture. It measures the enrichment of a gas component (e.g., CO_2) after it has passed the membrane. Thus, it is, together with the permeance, the main factor determining the practical applicability of the membrane. Figure 6A shows the effect of feed composition, pressure, and temperature on the separation factor and the related CO_2 enrichment from the feed gas. As can be seen, the higher the CO_2/N_2 ratio in the feed, the higher the separation efficiency. When comparing atmospheric pressure with reduced pressure at the permeate side, the separation factor shifts to higher values when a vacuum is applied on the permeate side. As shown in Figure 6B, this effect is pronounced for the room temperature results with, e.g., a separation factor increase from 10 to 16 for a 50/50% CO_2/N_2 feed mixture. Calculated in terms of CO_2 purity in the permeate flow, this results in very high values of 90–95%. Following the single-gas permselectivity, the separation factor decreases continuously with increasing temperature, converging at 150 °C to values of 2 to 4, which the increased N_2 kinetics and a temperature constant CO_2 permeance can explain.

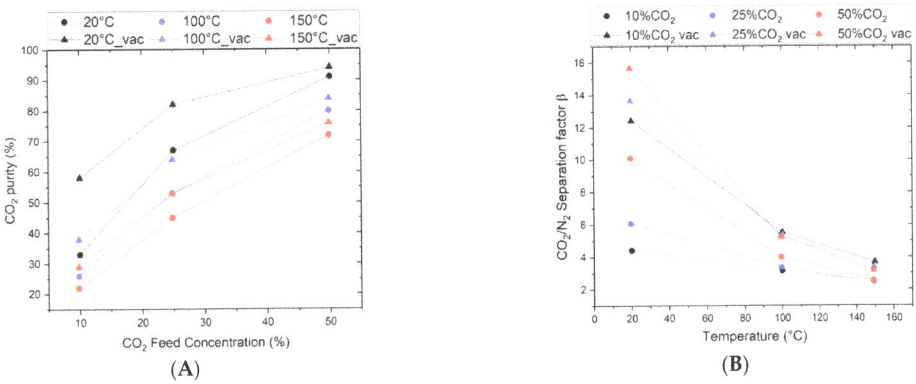

Figure 6. CO_2/N_2 gas mixture separation through Org-Sil membrane. (**A**) CO_2 purity in permeate as a function of the CO_2 Feed Composition, (**B**) CO_2/N_2 separation factor as a function of temperature for different feed compositions (10, 25, and 50 vol% CO_2, Δ denotes permeate vacuum). The transmembrane pressure is kept constant at Δp = 3.2 bar.

4. Discussion

Figure 7 displays a Robeson-like plot of various organo-silica and primarily BTESE-derived membranes. As the future interest lies in the application at higher temperatures, the available data for temperatures up to >200 °C are also shown. Although no data at 50 °C are presented, this temperature panel is included as the membrane reported by Guo et al. [25] at 50 °C is the only one that reaches the envisioned area of optimum membrane properties. As the temperature increases, even the best-performing membrane leaves this area.

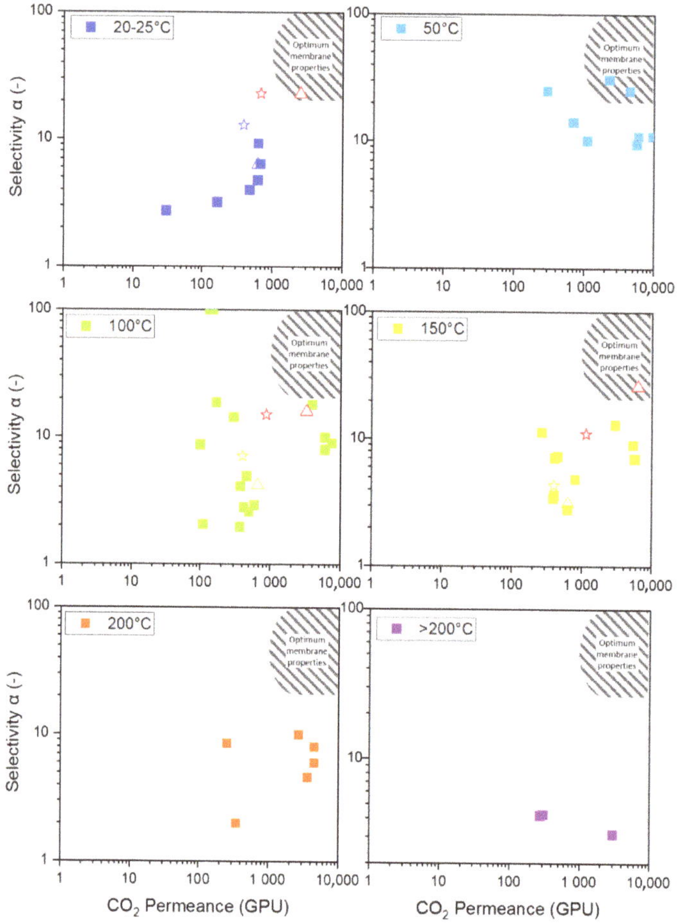

Figure 7. Display of available BTESE-derived permeance and selectivity in a Robeson-like plot for temperatures from 20 °C to over 200 °C. Each square □ represents a membrane from the references. The star ☆ indicates the membrane presented in this study, treated in N_2, while the triangle △ represents the membrane treated in air. The red stars and triangles correspond to theoretical stand-alone membranes. Data from references [24–27,29,31,36,39–42].

The membrane discussed in this article is indicated by the star (in N_2) and the triangle (in air). The membrane treated in N_2 is located at the arrow's starting point, and the arrowhead points to the membrane treated in air. The initial approach was to modify the pore structure to increase the number of pores while keeping the pore size constant, as illustrated in Figure 2 by the green arrow. However, this approach did not produce the intended result, suggesting increased pore size. This behavior was also observed by Ibrahim et al., who describe the increased permeance by an increase in pore size. Thermal treatment in air leads to a higher degree of decomposition of organic ethoxide groups [33].

Comparing the membranes at different temperatures is difficult for two major reasons. Firstly, all membranes have different properties, as shown in Figure 2. The microstructure of the membranes has (i) different porosity for CO_2 (green arrow), (ii) different pore sizes (blue arrow), and (iii) a varying degree of active sites that participate in surface diffusion (red arrow). However, the pore structure and number are often undisclosed along the permeation data. This is either because of laborious investigation procedures or influences on the pore structure, which are various and usually not easy to control. Membranes fabricated similarly often vary in their permeation data because of varying influences. The discrepancies can be attributed to the specific fabrication processes, including the water/silica and acid/silica ratio, reaction time, reaction temperature, stirring, sol aging (dilution and particle growth), drying conditions (humidity, duration, temperature, and defects), sintering (temperature, time, atmosphere, and defects), coating technique (dilution, coating times, and airborne particles), and membrane aging (measurement immediately after fabrication or after a storage period).

Secondly, the practice of reporting the permeation data as the permeance of the composite and not as the permeability of the respective layer complicates the comparability. In Figure 7, organo-silica-derived membranes on different support structures are shown, making it difficult to compare the resulting permeation data. In Figure 4B, we presented permeability data for our membranes and noticed a different temperature trajectory for the composite than for the single layers. This is possibly due to differences in the mass transport phenomena of the mesoporous support structure, which affects the behavior of the separation layer. We calculated the permeance of a theoretical single layer and placed it in the Robeson-like plot to illustrate the impact of the support. The red squares in the panels for 20 °C, 100 °C, and 150 °C represent the theoretical single-layer permeances for the fabricated membrane treated in N_2 and air. This demonstrates that the air-treated membrane now meets the necessary permeation properties for economic application. We understand that some support structure is needed, and we believe that with material optimization, some of the membranes currently not in the optimum area will reach that point. We want to highlight how the mesoporous support becomes more influential at higher temperatures because the dominating Knudsen diffusion decreases its flux. Therefore, high flux supports can help make the application economically viable at high temperatures [43].

Nevertheless, we tried to explain the different permeation and selectivity levels in the published data and how apparent contradictions can be resolved. Examining the temperature trajectories of different gas permeances offers insights into the pore structure. Three different cases are observed. Firstly, Castricum et al. and Kanezashi et al. report decreasing permeances for CO_2 with increasing temperature [31,41]. Secondly, Grekou et al. [24,29] and Rubner et al. demonstrate the opposite trend, showing an increase in CO_2 permeance with rising temperature [24,29]. Thirdly, Van Gestel et al. and Yang et al., as well as our work, observe no changes in permeances with increasing temperatures [26,27]. Three distinct mass transport mechanisms—Knudsen diffusion, surface diffusion, and activated permeation—coexist depending on the number and size of pores, participating in mass transport in varying ratios. Higher proportions of Knudsen diffusion and surface diffusion result in higher permeances but at a decreasing trend with increasing temperature. Therefore, in the first case, it can be concluded that Castricum et al. [41] and Kanezashi et al. [31] produced membranes with many micro- and mesopores. Membranes with activated diffusion as the predominating mass transport mechanism give low permeances but show an increasing trend with increasing temperature. Therefore, in the second case, the membranes reported by Grekou et al. [29] and Rubner et al. [24] indicate a microstructure with a high proportion of ultramicropores. In the third case, membranes with a pore structure in between can have mass transport mechanisms that offset each other, resulting in constant permeances with increasing temperature. The membranes studied by van Gestel et al. [27], Yang et al. [26], and those investigated in this work are likely to have structures with micro- and ultramicropores. Therefore, it is concluded that for a membrane design, only microstructures with pores below 0.7 nm are suitable, as only these show large permeances with good selectivity. The problem with decreasing selectivity at higher temperatures is assumed to be overcome by providing active sites for surface diffusion at higher temperatures.

Using permeances obtained from single-gas measurements for mixed-gas applications can cause problems since the interactions of mixed gases can alter their behavior. Membranes with a large portion of ultramicropores can enable interactions between the slow and fast permeating compounds by a pore-blocking or condensation mechanism [44]. Yang et al. showed a decrease in CO_2 in the mixture compared to the single gases and accounted for some hindrance as its cause, as reported in zeolites [13,26]. Rubner et al. even observed increased CO_2 permeances in the mixture compared to the single-gas results, indicating that the non-adsorbing gas enhances the mass transport of the adsorbing gas. Our results fall within the range of reported data, showing no deviations from the single-gas results, indicating that no interactions occur in micropores.

Since most publications do not disclose mixed-gas permeances, they assess their mixed-gas results based on the separation factor and compare them to the single-gas permselectivity. If the separation factor is lower than the permselectivity, some hindrance or blockage is attributed to the decreased selectivity. This approach can lead to incorrect conclusions since the separation factor depends on kinetics (permeance) and the driving force (partial pressure difference) governing mass transport.

As illustrated in Figure 6B, the same membrane exhibits variations in its separation factor from 5 to 10 by merely changing the feed composition (10% to 50%). Additionally, for the same composition (10% CO_2), applying a vacuum on the permeate side causes the separation factor to change from 4 to 13. However, the permselectivity remains constant throughout (see Figure 5). For a proper comparison, mixed-gas permeances are necessary.

Operating mixed-gas separations on the permeate side under vacuum conditions yields higher separation factors. Contrary to the proposition by Grekou et al. [29] that the higher separation factors are caused by the enhanced desorption kinetics of the adsorbing component under lower pressure, our observations indicate otherwise. If the adsorption were truly enhanced, the permeances for different operation modes (atmospheric vs. vacuum) would show variation. Operating the membrane with a vacuum increases the driving force (partial pressure difference) for the faster-permeating component, leading to higher fluxes and higher permeate concentrations at low pressures [45].

5. Conclusions

In this study, we investigated ceramic microporous CO_2-philic membranes. We analyzed the transport behavior of CO_2 and N_2 focusing on the possible future implementation of such a membrane in a CO_2 separation process at elevated temperatures.

The investigated membrane shows a constant CO_2 permeance with increasing temperature and an even increasing permeability for the membrane thermally treated in air. This shows that focusing on permeances overlooks the actual transport behavior of the fabricated organo-silica layer, as the influence of the support structure is non-neglectable. A single-gas separation efficiency derived from permeances at 150 °C yields a selectivity of 4.3. In contrast, the same membrane has a selectivity of 11 when single-layer permeabilities are used as mass transport kinetics. It is important to note that the Knudsen behavior of the support structure often masks the properties of the selective cover layer. Only conclusions based on permeabilities are meaningful when evaluating different organo-silica fabrication methods.

As all membranes are aimed to perform in mixed-gas applications, single-gas kinetics must be compared to mixed-gas behavior to derive the relevant kinetics and separation performance. The presented membrane shows the same kinetics and selectivities as in single-gas experiments. The elsewhere reported separation performance increases based on a pore-blocking mechanism can not be confirmed as the anticipated decrease in N_2 permeance was not observed.

However, the separation factor based on operation conditions showed that the membrane is pressure ratio limited, and the separation factor increases with increasing pressure ratios (Figure 6A,B). This is achieved by increasing the partial pressure in the feed or by lowering the partial pressure in the permeate via a vacuum. An enhanced CO_2 desorption kinetic caused by permeate vacuums was also not observed, as the CO_2 permeance stayed constant for all the operation modes.

This study introduces an extension of the Robeson-like plot for different temperatures, including membrane data reported in the literature and data from the present study. Only one membrane at 50 °C reaches the optimum membrane properties across all the available data. Until now, no organo-silica-based membrane has reached the properties for economic application above 50 °C.

From temperature trajectories introduced into the Robeson plot, we conclude that only membranes with micropores in the range for surface diffusion can reach the envisioned area of optimum membrane properties. This is especially important for high-temperature applications where the surface affinity is declining. Combining micropores with surfaces that offer adsorption sites that enable surface diffusion at higher temperatures is very promising [5,34].

Supplementary Materials: The following supporting information can be downloaded at: https://www.mdpi.com/article/10.3390/membranes15030083/s1, Figure S1: Rotary dip coating setup. The top left picture shows the pendulum moved by a motor in a rotary motion. At the end of the pendulum is a substrate mounted by suction. This is moved through a petri dish filled with the coating sol. The right panel shows a schematic drawing of the laboratory setup. The bottom left picture shows a magnification of the critical point where the coated substrate leaves the sol and a drop forms at the rim; Figure S2: Photograph of the BTESE-coated Substrate in the Laboratory on the left. The same BTESE membrane inside the permeation cell on the right.

Author Contributions: Conceptualization, L.B.; methodology, L.B., T.v.G., T.K. and K.G.; validation, L.B.; formal analysis, L.B.; investigation, L.B.; data curation, L.B.; writing—original draft L.B. and T.v.G.; writing—review and editing, T.v.G., T.K., K.G., P.S. and D.S.; visualization; L.B. and T.K.; supervision, K.G., P.S., W.A.M., O.G. and D.S.; funding acquisition, D.S. All authors have read and agreed to the published version of the manuscript.

Funding: Program-oriented funding for this research was provided through the German Helmholtz Association in the frame of the "Materials and Technologies for the Energy Transition" program.

Institutional Review Board Statement: Not applicable.

Data Availability Statement: The data that support the findings of this study are available from the corresponding author upon request.

Conflicts of Interest: Tim van Gestel, Wilhelm A. Meulenberg and Olivier Guillon were employed by the company Forschungszentrum Jülich GmbH. The authors declare that the research was conducted in the absence of any commercial or financial relationships that could be construed as a potential conflict of interest.

References

1. Favre, E. Membrane Separation Processes and Post-Combustion Carbon Capture: State of the Art and Prospects. *Membranes* **2022**, *12*, 884. [CrossRef] [PubMed]
2. Zhu, M.; Ge, Q.; Zhu, X. Catalytic Reduction of CO_2 to CO via Reverse Water Gas Shift Reaction: Recent Advances in the Design of Active and Selective Supported Metal Catalysts. *Trans. Tianjin Univ.* **2020**, *26*, 172–187. [CrossRef]
3. Sircar, S. Pressure Swing Adsorption Technology. In *Adsorption: Science and Technology*; Rodrigues, A.E., LeVan, M.D., Tondeur, D., Eds.; Springer: Dordrecht, The Netherlands, 1989; pp. 285–321.

4. Kenarsari, S.D.; Yang, D.; Jiang, G.; Zhang, S.; Wang, J.; Russell, A.G.; Wei, Q.; Fan, M. Review of recent advances in carbon dioxide separation and capture. *RSC Adv.* **2013**, *3*, 22739. [CrossRef]
5. Bünger, L.; Kurtz, T.; Garbev, K.; Stemmermann, P.; Stapf, D. Mixed-Matrix Organo–Silica–Hydrotalcite Membrane for CO_2 Separation Part 2: Permeation and Selectivity Study. *Membranes* **2024**, *14*, 156. [CrossRef]
6. Amooghin, A.E.; Sanaeepur, H.; Pedram, M.Z.; Omidkhah, M.; Kargari, A. New advances in polymeric membranes for CO_2 separation. In *Polymer Science: Research Advances, Practical Applications and Educational Aspects*; Formatex Research Center: Norristown, PA, USA, 2016; pp. 354–368.
7. Robeson, L.M. The upper bound revisited. *J. Membr. Sci.* **2008**, *320*, 390–400. [CrossRef]
8. Rezakazemi, M.; Sadrzadeh, M.; Matsuura, T. Thermally stable polymers for advanced high-performance gas separation membranes. *Prog. Energy. Combust. Sci.* **2018**, *66*, 1–41. [CrossRef]
9. Merkel, T.C.; Lin, H.; Wei, X.; Baker, R. Power plant post-combustion carbon dioxide capture: An opportunity for membranes. *J. Membr. Sci.* **2010**, *359*, 126–139. [CrossRef]
10. Burggraaf, A.J. Transport and separation properties of membranes with gases and vapours. In *Fundamentals of Inorganic Membrane Science and Technology*; Burggraaf, A.J., Cot, L., Eds.; Elsevier: Amsterdam, The Netherlands, 1996; pp. 331–434.
11. Baker, R.W. Membrane Transport Theory. In *Membrane Technology and Applications*; Baker, R.W., Ed.; Wiley: Chichester, UK, 2023; pp. 15–87.
12. Richter, H.; Voss, H.; Kaltenborn, N.; Kämnitz, S.; Wollbrink, A.; Feldhoff, A.; Caro, J.; Roitsch, S.; Voigt, I. High-Flux Carbon Molecular Sieve Membranes for Gas Separation. *Angew. Chem. Int. Ed Engl.* **2017**, *56*, 7760–7763. [CrossRef]
13. Bernal, M.P.; Coronas, J.; Menéndez, M.; Santamaría, J. Separation of CO_2/N_2 mixtures using MFI-type zeolite membranes. *AIChE J.* **2004**, *50*, 127–135. [CrossRef]
14. Liu, Y.; Wang, N.; Diestel, L.; Steinbach, F.; Caro, J. MOF membrane synthesis in the confined space of a vertically aligned LDH network. *Chem. Commun.* **2014**, *50*, 4225–4227. [CrossRef]
15. de Vos, R.M.; Verweij, H. High-selectivity, high-flux silica membranes for gas separation. *Science* **1998**, *279*, 1710–1711. [CrossRef] [PubMed]
16. ten Elshof, J.E. Hybrid Materials for Molecular Sieves. In *Handbook of Sol-Gel Science and Technology*; Klein, L., Aparicio, M., Jitianu, A., Eds.; Springer International Publishing: Cham, Switzerland, 2018; pp. 2973–2999.
17. Ben Rejeb, Z.; Abidli, A.; Zaoui, A.; Fashandi, M.; Selka, A.; Naguib, H.E.; Park, C.B. One-pot synthesis of rationally-designed flexible, robust, and hydrophobic ambient-dried molecularly-bridged silica aerogels with efficient and versatile oil/water separation applications. *Adv Compos Hybrid Mater* **2024**, *7*, 188. [CrossRef]
18. Çok, S.S.; Koç, F.; Len, A.; Kriechbaum, M.; Dudás, Z. Antibiotic adsorption from aqueous media by using silica aerogels derived from amine-bridged silsesquioxane networks. *J. Water Process Eng.* **2024**, *68*, 106538. [CrossRef]
19. Castricum, H.L.; Kreiter, R.; van Veen, H.M.; Blank, D.H.; Vente, J.F.; ten Elshof, J.E. High-performance hybrid pervaporation membranes with superior hydrothermal and acid stability. *J. Membr. Sci.* **2008**, *324*, 111–118. [CrossRef]
20. Castricum, H.L.; Sah, A.; Kreiter, R.; Blank, D.H.A.; Vente, J.F.; ten Elshof, J.E. Hydrothermally stable molecular separation membranes from organically linked silica. *J. Mater. Chem.* **2008**, *18*, 2150. [CrossRef]
21. Castricum, H.L.; Sah, A.; Geenevasen, J.A.J.; Kreiter, R.; Blank, D.H.A.; Vente, J.F.; ten Elshof, J.E. Structure of hybrid organic–inorganic sols for the preparation of hydrothermally stable membranes. *J. Sol. Gel. Sci. Technol.* **2008**, *48*, 11–17. [CrossRef]
22. Kreiter, R.; Rietkerk, M.D.A.; Castricum, H.L.; van Veen, H.M.; ten Elshof, J.E.; Vente, J.F. Stable hybrid silica nanosieve membranes for the dehydration of lower alcohols. *ChemSusChem* **2009**, *2*, 158–160. [CrossRef]
23. Lee, H.R.; Kanezashi, M.; Shimomura, Y.; Yoshioka, T.; Tsuru, T. Evaluation and fabrication of pore-size-tuned silica membranes with tetraethoxydimethyl disiloxane for gas separation. *AIChE J.* **2011**, *57*, 2755–2765. [CrossRef]
24. Rubner, J.; Skribbe, S.; Roth, H.; Kleines, L.; Dahlmann, R.; Wessling, M. On the Mixed Gas Behavior of Organosilica Membranes Fabricated by Plasma-Enhanced Chemical Vapor Deposition (PECVD). *Membranes* **2022**, *12*, 994. [CrossRef]
25. Guo, M.; Zhang, Y.; Xu, R.; Ren, X.; Huang, W.; Zhong, J.; Tsuru, T.; Kanezashi, M. Ultrahigh permeation of CO_2 capture using composite organosilica membranes. *Sep. Purif. Technol.* **2022**, *282*, 120061. [CrossRef]
26. Yang, X.; Du, H.; Lin, Y.; Song, L.; Zhang, Y.; Gao, X.; Kong, C.; Chen, L. Hybrid organosilica membrane with high CO_2 permselectivity fabricated by a two-step hot coating method. *J. Membr. Sci.* **2016**, *506*, 31–37. [CrossRef]
27. van Gestel, T.; Velterop, F.; Meulenberg, W.A. Zirconia-supported hybrid organosilica microporous membranes for CO_2 separation and pervaporation. *Sep. Purif. Technol.* **2021**, *259*, 118114. [CrossRef]
28. Li, S.; Fan, C.Q. High-Flux SAPO-34 Membrane for CO_2/N_2 Separation. *Ind. Eng. Chem. Res.* **2010**, *49*, 4399–4404. [CrossRef]
29. Grekou, T.K.; Koutsonikolas, D.E.; Karagiannakis, G.; Kikkinides, E.S. Tailor-Made Modification of Commercial Ceramic Membranes for Environmental and Energy-Oriented Gas Separation Applications. *Membranes* **2022**, *12*, 307. [CrossRef]
30. Yoldas, B.E. Alumina Sol Preparation from Alkoxides. *Amer. Ceram. Soc. Bull.* **1975**, *54*, 289–290.
31. Kanezashi, M.; Yada, K.; Yoshioka, T.; Tsuru, T. Organic–inorganic hybrid silica membranes with controlled silica network size: Preparation and gas permeation characteristics. *J. Membr. Sci.* **2010**, *348*, 310–318. [CrossRef]

32. Yu, X.; Nagasawa, H.; Kanezashi, M.; Tsuru, T. Improved thermal and oxidation stability of bis(triethoxysilyl)ethane (BTESE)-derived membranes, and their gas-permeation properties. *J. Mater. Chem. A* **2018**, *6*, 23378–23387. [CrossRef]
33. Ibrahim, S.M.; Nagasawa, H.; Kanezashi, M.; Tsuru, T. Organosilica bis(triethoxysilyl)ethane (BTESE) membranes for gas permeation (GS) and reverse osmosis (RO): The effect of preparation conditions on structure, and the correlation between gas and liquid permeation properties. *J. Membr. Sci.* **2017**, *526*, 242–251. [CrossRef]
34. Bünger, L.; Garbev, K.; Ullrich, A.; Stemmermann, P.; Stapf, D. Mixed-Matrix Organo-Silica–Hydrotalcite Membrane for CO_2 Separation Part 1: Synthesis and Analytical Description. *Membranes* **2024**, *14*, 170. [CrossRef]
35. Kim, T.W.; Sahimi, M.; Tsotsis, T.T. Preparation of Hydrotalcite Thin Films Using an Electrophoretic Technique. *Ind. Eng. Chem. Res.* **2008**, *47*, 9127–9132. [CrossRef]
36. Castricum, H.L.; Paradis, G.G.; Mittelmeijer-Hazeleger, M.C.; Kreiter, R.; Vente, J.F.; Ten Elshof, J.E. Tailoring the Separation Behavior of Hybrid Organosilica Membranes by Adjusting the Structure of the Organic Bridging Group. *Adv. Funct. Mater.* **2011**, *21*, 2319–2329. [CrossRef]
37. Keizer, K.; Uhlhorn, R.; van Vuren, R.J.; Burggraaf, A.J. Gas separation mechanisms in microporous modified γ-Al_2O_3 membranes. *J. Membr. Sci.* **1988**, *39*, 285–300. [CrossRef]
38. de Lange, R.; Keizer, K.; Burggraaf, A.J. Analysis and theory of gas transport in microporous sol-gel derived ceramic membranes. *J. Membr. Sci.* **1995**, *104*, 81–100. [CrossRef]
39. Yu, X.; Meng, L.; Niimi, T.; Nagasawa, H.; Kanezashi, M.; Yoshioka, T.; Tsuru, T. Network engineering of a BTESE membrane for improved gas performance via a novel pH-swing method. *J. Membr. Sci.* **2016**, *511*, 219–227. [CrossRef]
40. Castricum, H.L.; Qureshi, H.F.; Nijmeijer, A.; Winnubst, L. Hybrid silica membranes with enhanced hydrogen and CO_2 separation properties. *J. Membr. Sci.* **2015**, *488*, 121–128. [CrossRef]
41. Castricum, H.L.; Paradis, G.G.; Mittelmeijer-Hazeleger, M.C.; Bras, W.; Eeckhaut, G.; Vente, J.F.; Rothenberg, G.; ten Elshof, J.E. Tuning the nanopore structure and separation behavior of hybrid organosilica membranes. *Microporous Microporous Mater.* **2014**, *185*, 224–234. [CrossRef]
42. Ren, X.; Nishimoto, K.; Kanezashi, M.; Nagasawa, H.; Yoshioka, T.; Tsuru, T. CO_2 Permeation through Hybrid Organosilica Membranes in the Presence of Water Vapor. *Ind. Eng. Chem. Res.* **2014**, *53*, 6113–6120. [CrossRef]
43. Park, H.B.; Kamcev, J.; Robeson, L.M.; Elimelech, M.; Freeman, B.D. Maximizing the right stuff: The trade-off between membrane permeability and selectivity. *Science* **2017**, *356*, eaab0530. [CrossRef]
44. Zhou, C.; Yuan, C.; Zhu, Y.; Caro, J.; Huang, A. Facile synthesis of zeolite FAU molecular sieve membranes on bio-adhesive polydopamine modified Al_2O_3 tubes. *J. Membr. Sci.* **2015**, *494*, 174–181. [CrossRef]
45. Huang, Y.; Merkel, T.C.; Baker, R.W. Pressure ratio and its impact on membrane gas separation processes. *J. Membr. Sci.* **2014**, *463*, 33–40. [CrossRef]

membranes

MDPI

Article

Simultaneously Enhanced Permeability and Selectivity of Pebax-1074-Based Mixed-Matrix Membrane for CO_2 Separation

Rujing Hou, Junwei Xie, Yawei Gu, Lei Wang and Yichang Pan *

State Key Laboratory of Materials-Oriented Chemical Engineering, College of Chemical Engineering, Nanjing Tech University, Nanjing 210009, China; rujing.hou@njtech.edu.cn (R.H.); 202261104011@njtech.edu.cn (J.X.); 202362042052@njtech.edu.cn (Y.G.); 202362042066@njtech.edu.cn (L.W.)

* Correspondence: panyc@njtech.edu.cn

Abstract: Membrane technology is a promising methodology for carbon dioxide separation due to its benefit of a small carbon footprint. However, the trade-off relationship between gas permeability and selectivity is one obstacle to limiting its application. Herein, branched polyethyleneimine (BPEI) containing a rich amino group was successfully grafted on the surface of the metal–organic framework (MOF) of AIFFIVE-1-Ni (KAUST-8) through coordination between N in BPEI and open metal sites in the MOF and with the resultant maintained BET surface area and pore volume. Both the strengthened CO_2 solubility coefficients coming from the additional CO_2 adsorption sites of amino groups in BPEI and the reinforced CO_2 diffusivity coefficients originating from the fast transport channels created by KAUST-8 led to the promising CO_2 separation performance for KAUST-8@BPEI/Pebax-1074 MMM. With 5 wt.% KAUST-8@BPEI loading, the MMM showed the CO_2 permeability of 156.5 Barrer and CO_2/N_2 selectivity of 16.1, while the KAUST-8-incorporated MMM (5 wt.% loading) only exhibited the CO_2 permeability of 86.9 Barrer and CO_2/N_2 selectivity of 13.0. Such enhancement is superior to most of the reported Pebax-1074-based MMMs for CO_2 separation indicating a wide application for the coordination method for MOF fillers with open metal sites.

Keywords: mixed-matrix membrane; metal–organic framework; carbon capture; grafting

1. Introduction

Urgent action is required to limit warming to 2 °C by the year 2100 according to Intergovernmental Panel on Climate Change (IPCC) [1,2]. One of the main greenhouse gases is carbon dioxide. Various techniques have been developed for carbon capture including distillation, sorption, and membrane separation. The mature market-dominated technology of distillation and sorption, such as liquid amine-scrubbing and sorbents adsorption, which have a large carbon footprint and equipment footprint for CO_2 capture, due to the phase change involved in the distillation and the absorbent regeneration step in sorption [3,4]. Comparatively, membrane technology stands out due to its high energy efficiency, small physical footprint, and low carbon footprint [5,6]. Baker's study demonstrated the potential of membranes for carbon capture by showing a reduction of 44% energy cost relative to the amine absorption method when increasing the CO_2 feed content from 10% to 30% with CO_2 purity more than 99%, which meets the requirement set by the Department of Energy [7].

With intensive studies of membrane separation, there is yet industrial application for flue gas treatment. For polymeric membranes, the main obstacle is the trade-off relationship between gas permeability and selectivity that the increased gas permeability is generally

https://doi.org/10.3390/membranes15010026

coupled with the reduced selectivity, and vice versa [8–10]. For inorganic membranes, the inevitable intercrystallite defects limit its further scale-up [5]. Correspondingly, the mixed-matrix membranes (MMMs) fabricated by incorporating nanoparticle filler as the dispersed phase into the continuous phase of polymer matrix have gained increasing attention in recent years because they combine the advantages of solvent-processability from polymers and precise pore-regulation ability from nanoparticles, which is helpful to break the trade-off issue [8]. Diverse particles have been investigated as fillers in the MMMs including inorganic metal oxide, organic porous aromatic framework (PAF) [11], and the organic–inorganic hybrid of mental-organic framework (MOF) [12].

Among these, metal–organic framework (MOF) is one of the most popular and efficient fillers due to its highly controllable pore size, shape, and functionality arising from the diversity of metal ions and ligands that can accommodate the specific application scenario [13,14]. For example, the 5 wt.% loading of micron-sized hollow ZIF-8 in the MMM made the CO_2 permeability 224.6% and 63.5% higher than the neat PEO membrane and nanosized hollow ZIF-8/PEO MMM due to the accumulated transport highway constructed by the micron-sized hollow ZIF-8 [15]. The 5 wt.% loading of Cu@Al(bpydc) in Pebax accelerated the diffusion and solubility of CO_2 in the MMMs and led to the simultaneous improvement of CO_2 permeability and CO_2/CH_4 selectivity by 92.90% and 27.10%, respectively [16]. A similar concurrent enhancement for CO_2 permeability and CO_2/N_2 selectivity was also reported in the 10 wt.% PHZ-2/Pebax MMMs owing to the improved free volume and low gas transfer resistance caused by the hollow structure of filler [17].

As a pillar-layered metal–organic framework, AIFFIVE-1-Ni (KAUST-8) with periodic square-grid layer established by linking Ni(II) ion with pyrazine ligand and further pillared by the $[AlF_5(H_2O)]^{2-}$ anion in the third dimension (primitive cubic (pcu) topology) was applied as fillers in this study (Figure 1a) [18–21]. The pillar was demonstrated as an octahedral with an Al^{3+} metal center accommodated with five F atoms and one H_2O molecule with three F atoms and one H_2O molecule facing the 1D channel (Figure 1b, left). Upon dehydration, with one coordinated H_2O molecule lost away from the Al^{3+} site, the pillar geometry changed from a square to a trigonal bipyramidal-like structure with a reported pore size of around 5 Å (Figure 1b). The reported structure demonstrated double loaded site for CO_2 with both electropositive interaction between C with F atoms in pillars ($F \cdots C_{CO2}$ = ~2.8 Å) and electronegative O with H in pyrazine ligand ($H \cdots O_{CO2}$ = ~2.6 Å, Figure 1c) [13,21,22]. The open metal site of Al^{3+} was declared as a barrier for more CO_2 loading as it blocks its simultaneous interaction with the four F atoms [14,21]. Meanwhile, the open metal site was well known for anchoring the desired functional moiety to boost the targeted application [23–25]. To utilize the CO_2 affinity sites and further enhance their CO_2 separation ability, branched polyethyleneimine (BPEI)-grafted KAUST-8 was constructed and applied as fillers in Pebax 1074. Pebax 1074 was chosen as the polymer matrix because it consists of 55% soft phase polyethylene oxide and 45% hard phase polyamide 12, which can correspondingly supply the CO_2 affinity property and the mechanical strength [26]. With 5 wt.% SAPO zeolite, the Pebax-1074-based MMM exhibited the CO_2 permeability of 98.2 Barrer and CO_2/N_2 selectivity of 72 [27]; The 8 wt.% loading of ZnO in Pebax 1074 accelerated the CO_2 permeability and CO_2/N_2 selectivity of the MMM from 110.67 to 152.27 Barrer and 50.08 to 62.15, separately [28]. The function of the grafted BPEI in KAUST-8 and its mechanism for CO_2 separation were thoroughly investigated in this work.

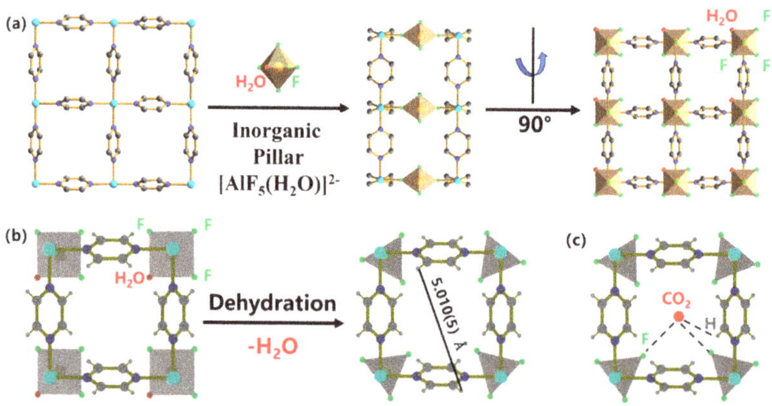

Figure 1. (**a**) Schematic diagram of KAUST-8 formation, (**b**) Dehydration process of KAUST-8, and (**c**) CO_2 interaction sites with KAUST-8. (Ni, cyan; N, blue; C, grey; O, red; F, green; Al, orange).

2. Experimental Section

2.1. Materials

Nickel (II) acetate ($Ni(CH_3COO)_2$, 99%), pyrazine ($C_4H_4N_2$, 99%), hydrofluoric acid (HF, 40 wt.% in H_2O), and n-butanol ($C_4H_{10}O$, 99%) were purchased from Shanghai Aladdin Biochemical Technology Co., Ltd., Shanghai, China. Aluminum (III) hydroxide ($Al(OH)_3$, AR), methanol (CH_4O, 99.9%), and ethanol (C_2H_6O, 99.9%) were obtained from Sinopharm Chemical Reagent Co., Ltd., Shanghai, China. Branched polyethyleneimine (BPEI, 99%, M_n = 600 g·mol^{-1}) was purchased from Merck Ltd., Beijing, China. Pebax 1074 polymer pellet was provided by Arkema Inc., Beijing, China. Gases of N_2, CO_2, and He were supplied by Nanjing Chuangda Group Co., Ltd., Nanjing, China. All purchased materials were used as received without further purification.

2.2. Preparation of KAUST-8

The KAUST-8 was synthesized according to our previous work [29]. Pyrazine (1.80 g) and nickel acetate (2.20 g) were separately dissolved in ethanol (6 mL) and deionized water (3 mL), followed by mixing and sonicating for 3 min. Aluminum hydroxide (0.7 g) was dissolved in a mixed solvent of HF and deionized water (7 mL, *v*:*v* = 6:1). The above two solutions were then mixed and sonicated for 3 min. The mixture was poured into a 50 mL Teflon liner and heated to 80 °C for 6 h. After cooling down to room temperature, the sky-blue product (Figure S1a) was collected by centrifugation and washed with deionized water three times and ethanol once. Finally, the product was dried at 60 °C for 24 h.

2.3. Preparation of KAUST-8@BPEI

The BPEI-grafted KAUST-8 was synthesized according to the reported work (abbreviated as KAUST-8@BPEI) [25]. BPEI (0.3 g) was dissolved in methanol (20 mL) and sonicated for 10 min. KAUST-8 (2.0 g) was added to the above solution and followed by stirring at 25 °C for 24 h. Then, KAUST-8@BPEI (Figure S1b) was collected by washing with methanol three times. Finally, the product was dried at 70 °C for 24 h.

2.4. Membranes Fabrication

The membranes were synthesized according to our previous work [30]. Pebax 1074 was dried in a vacuum oven at 80 °C for 6 h before use to remove any adsorbed moisture. The ratio of polymer/(polymer + solvent) is controlled at 2.5 wt.%. For the neat membrane fabrication, Pebax 1074 was dissolved in n-butanol at 90 °C under a water bath for 3 h.

After dissolution, the solution was poured rapidly into the Teflon dish followed by placing in an oven at 60 °C for 14 h to obtain the membrane. Finally, the membrane was carefully removed from the Teflon dish and dried in a vacuum oven at 60 °C for 12 h to remove any residual solvent before the gas permeation test. The fabrication of the MMMs was similar to the neat membrane. The quantitative MOF (KAUST-8 and KAUST-8@BPEI) was dispersed in n-butanol and followed by stirring for 3 h. At the same time, Pebax 1074 was dissolved in n-butanol and heated for 3 h. Then, the Pebax 1074 solution was gradually added to the MOF-dispersed solution in 2 steps (1/10 for the first time and 9/10 for the second time) and each time it was further stirred at 90 °C for 2 h. The following procedures were conducted in the same way as the neat membrane. For easy note, the KAUST-8 and KAUST-8@BPEI-based MMMs were separately labeled as KAUST-8/Pebax-X and KAUST-8@BPEI/Pebax-X, where X means the weight percentage of the loading (X = 2.5 wt.%, 5 wt.%, and 10 wt.%). Other loadings of more than 10 wt.% led to membrane cracks or serious agglomeration for the incorporated fillers. The thickness of all membranes was measured with a digital micrometer (Mitutoyo, Kanagawa, Japan) and around 60 ± 10 μm. The gas permeability and CO_2/N_2 selectivity of KAUST-8@BPEI/Pebax-5 with different thickness are shown in Table S1. It could be seen that membrane thickness has a slight effect on membrane separation performance.

2.5. Gas Permeation Measurements

The membrane gas permeation measurements were carried out by using a home-made Wicke–Kallenbach technique shown in Figure 2. All gas permeation tests were carried out under the condition that the temperature was maintained at 25 °C and the feed pressure was controlled from 1 to 6 bar. To avoid gas leakage, the membrane sample with an effective membrane area (3.8 cm^2) was placed in the sample holder and sealed with a rubber O-ring. The CO_2/N_2 feed gas mixture (v:v = 15:85) was fed to the membrane module. As shown in Table S2, the flow rate of 20 mL/min is the optimal experimental condition. Hence, the total flow rate of the feed side was controlled at 20 mL/min. The permeated gas was introduced into the gas chromatography (Agilent 7890, Santa Clara, CA, USA) and swept by Helium at a flow rate of 20 mL/min. After the test system was stable, the gas permeation data were tested three times and averaged. The gas permeability was calculated according to Equation (1).

$$P_i = \frac{LN_i}{A\Delta P_i} \tag{1}$$

where P_i is the gas permeability of the gas i (1 Barrer = 10^{-10} cm^3(STP)·cm·cm^{-2}·s^{-1}·$cmHg^{-1}$), L is the membrane thickness (cm), N_i is the volumetric flux through the membrane of the gas i (cm^3/s), A is the membrane effective area (cm^2), and ΔP_i is the pressure difference across the membrane (cmHg).

The selectivity ($\alpha_{i/j}$) of membrane was calculated according to Equation (2).

$$\alpha_{i/j} = \frac{P_i}{P_j} \tag{2}$$

where P_i and P_j are the gas permeability of the gas i and j, respectively.

Based on the results obtained from the gas (CO_2 and N_2) adsorption, the solution coefficient (S) can be calculated according to Equation (3).

$$S = \frac{c}{f} \tag{3}$$

where S is the solution coefficient ($cm^3(STP) \cdot cm^{-3} \cdot cmHg^{-1}$), c is the concentration for adsorbed CO_2 of the sample ($cm^3(STP) \cdot cm^{-3}$), and f is the corresponding upstream fugacity driving force (cmHg).

The diffusion coefficient (D) can be obtained indirectly based on the Equation (4).

$$D = \frac{P}{S} \quad (4)$$

where D is the diffusion coefficient ($cm^2 \cdot s^{-1}$).

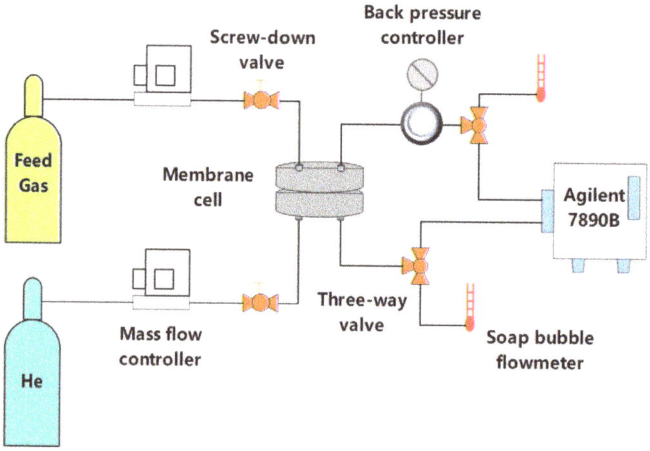

Figure 2. Schematic diagram for gas permeation test.

2.6. Characterizations

X-ray diffraction (XRD) patterns of KAUST-8, KAUST-8@BPEI, and the corresponding MMMs were collected from the Rigaku Smartlab TM 9 KW equipment, Tokyo, Japan (Cu Kα radiation, 40 kV, 40 mA, and λ = 1.542 Å). Thermogravimetric analysis (TGA) of KAUST-8, KAUST-8@BPEI, and the corresponding MMMs was obtained from the thermal analysis system (NETZSCH STA 449, Selby, Germany) under N_2 atmosphere from 30 °C to 800 °C with a heating rate of 10 °C/min. The morphologies of KAUST-8, KAUST-8@BPEI, and the corresponding MMMs were collected from field emission scanning electron microscopes (FESEM, S4800, Hitachi, Tokyo, Japan). X-ray photoelectron spectroscopy (XPS) of KAUST-8 and KAUST-8@BPEI was conducted on Kratos Amicus. The XPS spectra were conducted for C, N, Al, and Ni atoms and referenced to the C 1s level to correct the peak shift. The CO_2 and N_2 adsorption isotherms of the samples were measured by BSD-660M at 298 K and 77 K, respectively. Based on the N_2 adsorption data, the multipoint BET (Brunauer–Emmett–Teller, Beijing, China) was used to determine the specific surface areas and pore widths of KAUST-8 and KAUST-8@BPEI. Fourier transform infrared (FTIR) spectroscopy analysis of the samples was performed on the Nicolet IS10 equipment (Waltham, MA, USA) in the wavenumber range of 400–4000 cm^{-1}.

3. Results and Discussion

3.1. Characterization of KAUST-8@BPEI

The crystallinity of the synthesized MOF was examined by XRD patterns in Figure 3a. Both KAUST-8 and KAUST-8@BPEI exhibited the identical pattern relative to the simulated pattern and the prior reported work [21,29] with the main peaks of 110, 004, and 220 at 2θ of 12.96°, 23.34° and 25.74°, respectively. This indicates the right product and the

BPEI modification did not affect the crystallinity of KAUST-8. The weaker intensity of the KAUST-8@BPEI relative to the pristine KAUST-8 might be due to the interaction between KAUST-8 and BPEI, which was consistent with the trend of the BPEI-modified UiO-66 [31]. The decomposition temperature (Figure S2) was up to 350 °C aligning with the previous work, which proved the favorable thermal stability and the potential application in industrial scale of KAUST-8 and KAUST-8@BPEI [32]. The weight loss before 200 °C was due to the loss of the coordinated water in the framework of KAUST-8 as shown in Figure 1b [21]. Compared to KAUST-8, the more reduced weight loss (~5%) after 800 °C for KAUST-8@BPEI was due to the grafted BPEI functional groups, which supported the existence of BPEI on KAUST-8@BPEI surface. Similar thermal weight loss phenomena were found for BPIE-grafted ZIF-8 and carbon nanotubes in the previous works [25,33]. The grating of BPEI on KAUST-8 was proved by the FTIR spectra (Figure 3b). Newly emerged peaks at wavenumber of 1672 cm^{-1} assigned to the bending vibration of NH and 990 cm^{-1} and 646 cm^{-1} assigned to the out-of-plan wagging vibration of NH were detected for KAUST-8@BPEI compared to the pristine KAUST-8, which confirmed the successful grafting of the BPEI on KAUST-8 particles [33] and was echoed by the weaker intensity in the XRD pattern (Figure 3a). The minor reduced BET surface area (Figure 3c, Table S3, 489.9 m^2/g vs. 433.1 m^2/g), pore volume (7.6% drop, Table S3), and the slightly narrowed pore size (Figure 3d, 4.73 Å vs. 4.66 Å) certified that one end of the BPEI was grated into KAUST-8 framework. The CO_2 uptake of KAUST-8 was 2.83 mmol/g, which was consistent with Eddaoudi's work [21]. Meanwhile, the enhanced CO_2 adsorption property (3.3% up, Figure 3e) revealed that the other end of BPEI functionalized as CO_2 adsorption site. This was further confirmed by the XPS results, which showed a slight left shift to the lower binding energy for Ni 2p from 855.2 eV to 855.1 eV (Figure 3h) after KAUST-8 was grafted with BPEI, suggesting an enhancement of electron density in Ni [31]. The confirmation for the covalent attachment of BPEI with the KAUST-8 framework was supported by the detected increased ratio of N/Ni^{2+} (Table S4).

3.2. Characterization of MMMs

The morphology of the neat membrane and MMMs was investigated by the SEM cross-sectional images (Figure 4a–c). The neat membrane demonstrated a smooth cross-sectional morphology (Figure 4a), which was consistent with our prior work [30]. With the addition of KAUST-8 and KAUST-8@BPEI at a loading lower than 10 wt.%, the MMMs showed a similar cross-sectional morphology to that of the neat membrane without obvious interface voids or even a visible interface between the polymer and the additive (Figures 4b and S3a–c), reflecting the compatible interface and defect-free membrane structure. Such behavior matched our previous work [34]. The OH-PDCX/PIM-1 MMMs exhibited similar morphology to that of the neat PIM-1 membrane without obviously distinguishing between additive and polymer under the compatible solvent, leading to a more compatible polymer–additive interface and therefore to a more enhanced membrane performance relative to the obvious polymer–additive interface membrane sample (CO_2/N_2 enhancement: 22% vs. 4.4%). When additive loading up to 10 wt.%, no matter with KAUST-8 or KAUST-8@BPEI, an obvious additive agglomeration was observed (highlighted in the red circle in Figures 4c and S3d). This is a general phenomenon that a high additive loading of more than 10 wt.% has the tendency to join together for stabilization due to the high surface energy of the particle [29,30]. The FTIR spectra of the neat membrane showed peaks at 981 cm^{-1}, 1513 cm^{-1}, 1608 cm^{-1}, 2792 cm^{-1}, and 3170 cm^{-1}, which are assigned to the typical C–O–C, H–N–C=O, O–C=O, C–H and N–H groups in Pebax 1074 polymer (Figure 3i) [35–37]. With the incorporation of KAUST-8 or KAUST-8@BPEI from 2.5 wt.% to 5 wt.%, and 10 wt.%, typical peaks were retained, indicating there was no detectable chemical change through the FTIR examination and the existence of additive did not affect the crystallinity of the polymer matrix. For CO_2 adsorption (Figure 3f), the KAUST-8 additive-incorporated KAUST-8/Pebax MMMs showed slightly higher CO_2 adsorption property relative to the controlled sample (4.08 cm^3/g vs. 3.98 cm^3/g). This was caused by the CO_2 affinity sites in the framework of KAUST-8 (CO_2 with both electropositive interaction between C with F atoms in pillars and electronegative O with H in pyrazine ligand, Figure 1c) [13,21,22]. The enhancement degree was small due to the low loading of KAUST-8 (5 wt.%) in the polymer matrix. With BPEI further grafted on KAUST-8, an obvious improvement of CO_2 adsorption to 4.57 cm^3/g was detected, revealing the successful grating of BPEI in KAUST-8 framework and consistency to the result in XPS (Figure 3g,h), along with the reported enhanced CO_2 adsorption for BPEI, which was grated into MOF as noted in Zheng's work [25].

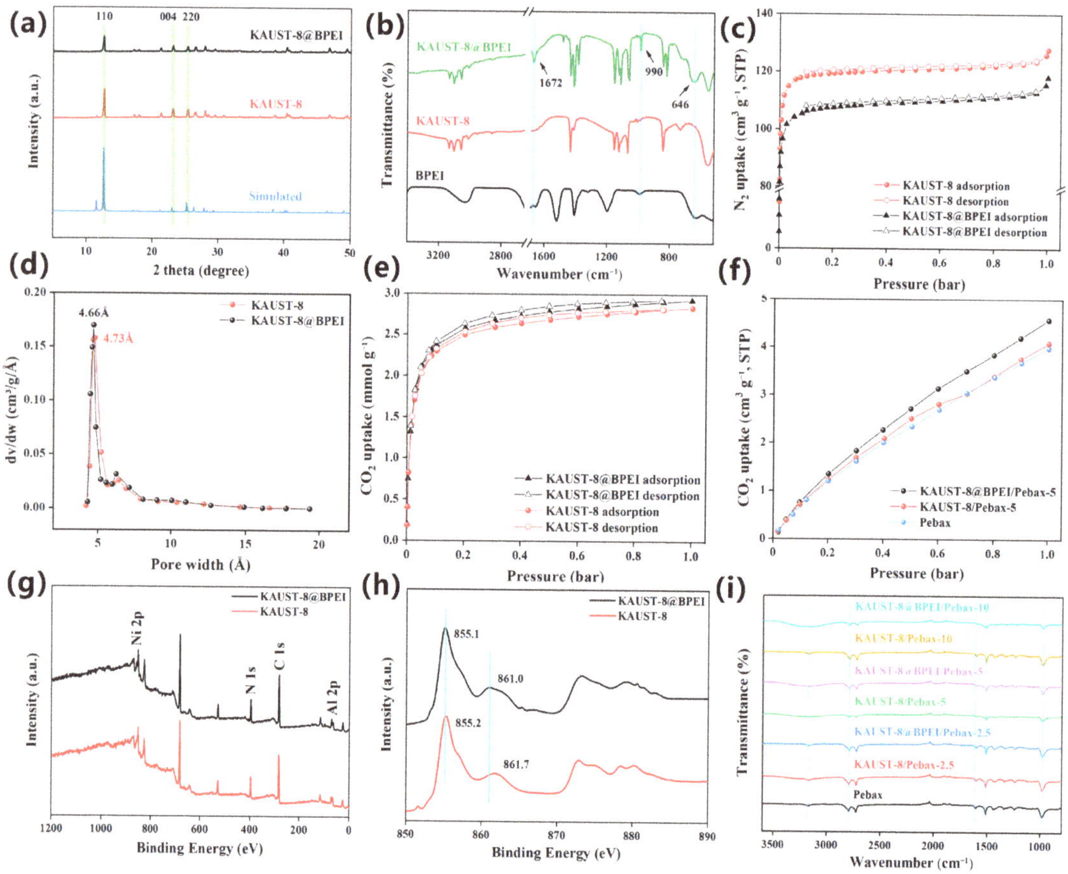

Figure 3. (**a**) XRD for KAUST-8 and KAUST-8@BPEI, (**b**) FTIR spectra for KAUST-8, KAUST-8@BPEI and BPEI, (**c**) N_2 adsorption and desorption isotherms for KAUST-8 and KAUST-8@BPEI at 77 K, (**d**) Pore size distribution of KAUST-8 and KAUST-8@BPEI calculated from 77 K N_2 adsorption isotherm based on the Horvath–Kawazoe model, (**e**) CO_2 sorption isotherms for KAUST-8 and KAUST-8@BPEI at 298 K, (**f**) CO_2 sorption isotherms for neat membrane, KAUST-8/Pebax-5 and KAUST-8@BPEI/Pebax-5 at 298 K, (**g**) XPS spectra of KAUST-8 and KAUST-8@BPEI, (**h**) Ni XPS spectra of KAUST-8 and KAUST-8@BPEI, and (**i**) FTIR spectra for membranes. All green and blue vertical lines are drawn to guide the eye.

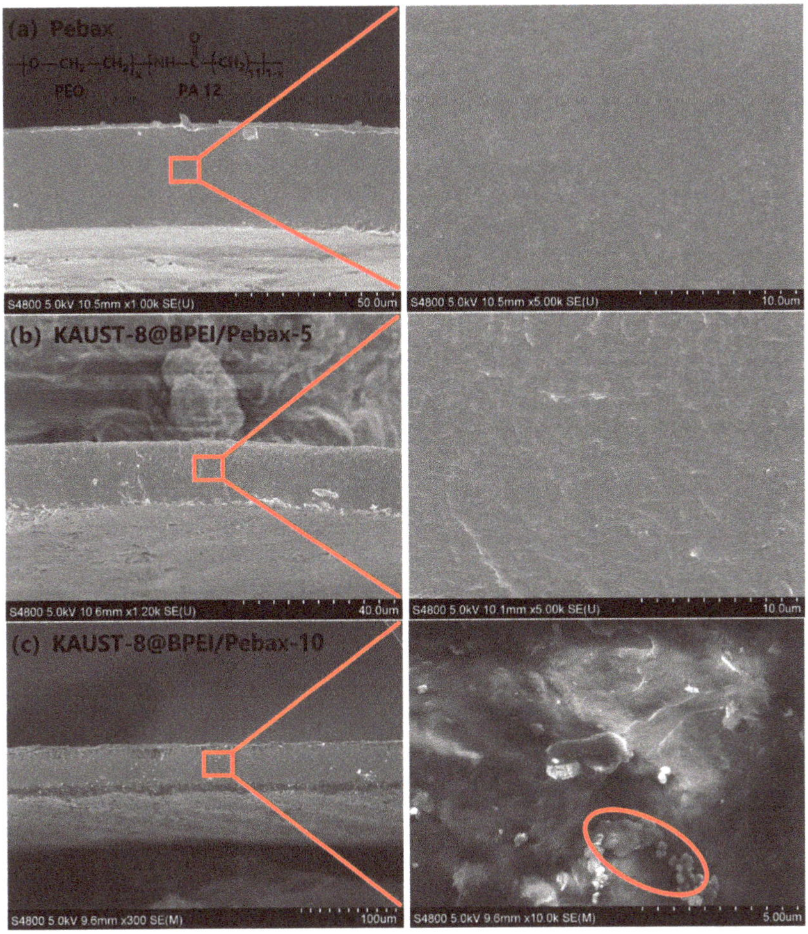

Figure 4. Cross-sectional SEM images for (**a**) neat membrane, (**b**) KAUST-8@BPEI/Pebax-5, and (**c**) KAUST-8@BPEI/Pebax-10, Left: low magnification, Right: high magnification, filler agglomeration highlighted in the red circle in the inset in Figure 4c.

3.3. Separation Performance

3.3.1. Loading Effect

The loading effect was evaluated by increasing loading from 0 to 2.5 wt.%, 5.0 wt.%, and 10 wt.% in Figure 5. The neat membrane demonstrated CO_2 permeability of 71.2 Barrer, N_2 permeability of 6 Barrer, and ideal CO_2/N_2 selectivity of 11.9, which were within the range of the reported work since the CO_2 permeability and CO_2/N_2 selectivity of the Pebax-1074-based MMMs, were influenced by several factors, such as Pebax-1074 from different batches, different kinds of solvent for membrane preparation, and so on. As a result, the membrane performance differences between the different Pebax-1074-based MMMs were significant [26,27,38,39]. With KAUST-8 loading (Figure 5a), CO_2 permeability and CO_2/N_2 selectivity demonstrated a concomitant increasing trend till the optimal loading of 5 wt.% with an enhancement of 22% and 9%, respectively. Such an enhancement mainly came from the CO_2 adsorption property of the embedded KAUST-8 additive (63 cm^3/g in Figure 3e) as both F atoms in the pillar of $[AlF_5(H_2O)]^{2-}$ and the H atoms in the pyrazine ligand can act as CO_2 affinity sites. With further KAUST-8 loading to 10 wt.%, both CO_2 permeability and

CO_2/N_2 selectivity dropped, although they were still higher than the controlled sample. This was caused by the additive agglomeration at a higher loading, which reduced the ability of KAUST-8. A similar behavior was reported in the Y-abtc@Pebax MMMs for CO_2/N_2 separation [40]. By comparison, the KAUST-8@BPEI/Pebax MMMs demonstrated a similar gas permeation trend to that of the KAUST-8/Pebax MMMs under the function of various loading with the same optimal loading at 5 wt.%. However, higher loading also led to serious particle accumulation (Figure 4c). The difference for gas permeation change behavior under the function of KAUST-8@BPEI as the additive was explored in detail in Figure 6.

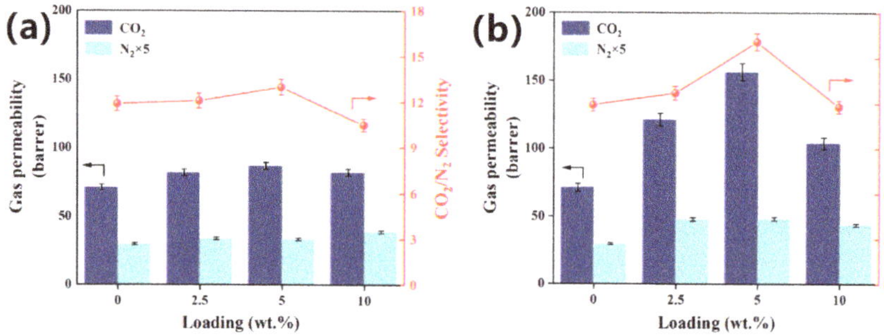

Figure 5. Loading effect on the change of gas permeability and CO_2/N_2 selectivity of (**a**) KAUST-8/Pebax MMMs and (**b**) KAUST-8@BPEI/Pebax MMMs. Arrows are drawn to guide the eye.

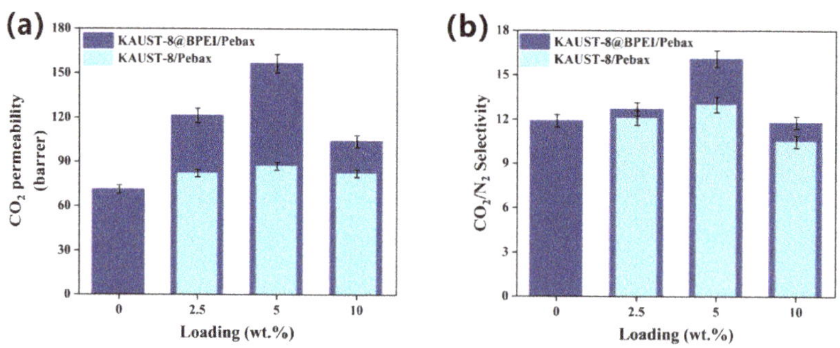

Figure 6. BPEI effect on the change of (**a**) CO_2 permeability and (**b**) CO_2/N_2 selectivity of KAUST-8/Pebax MMMs and KAUST-8@BPEI/Pebax MMMs.

3.3.2. BPEI Effect

In Figure 6, the KAUST-8@BPEI/Pebax MMMs showed a clear strengthened enhancement both for CO_2 permeability (Figure 6a) and CO_2/N_2 selectivity (Figure 6b) relative to the KAUST-8/Pebax MMMs throughout various loading from 2.5 wt.% to 5 wt.%, and 10 wt.%. For example, at the optimal loading of 5 wt.%, the KAUST-8@BPEI/Pebax MMM showed a 120% increment for CO_2 permeance and a 35.3% increment for CO_2/N_2 selectivity relative to the neat membrane, which were higher than most of the other Pebax-based MMM samples. Hence, the effect of the BPEI functionalization is effective and we intend to further enhance the role of BPEI in our future work. Generally, as an amine-functionalized moiety, BPEI possessed the ability for more CO_2 adsorption. This was confirmed by the CO_2 adsorption property both from the MOF particle (Figure 3e) and the corresponding membrane samples (Figure 3f), and it was consistent with prior work [25,31]. Consequently, a more CO_2-selective and faster CO_2 transport membrane structure was constructed. The reason for the specific function of BPEI for CO_2 separation and the separation mechanism is discussed in the following section.

3.3.3. Pressure, Temperature Effect and Long-Term Stability

Pressure effect on membrane separation performance was examined under both the single gas and mixed gas conditions (Figure 7). Generally, high pressure can lead to the following effect on membrane performance. Firstly, it can compact the membrane structure with the generated condensed fractional free volume, which corresponds to the reduced gas permeability; secondly, it can provide a more driving force across the membrane for the gas transportation and can be coupled with enhanced gas permeability; thirdly, the membrane can adsorb more condensable gas (such as CO_2) and be plasticized with the resultant enlarged fractional free volume and the enhanced permeability. For CO_2 permeability in the studied three samples (Figure 7a), they all increased with increasing the feed pressure under both single and mixed gas conditions and with a lower value under the mixture gas condition than that in the single gas condition. The increasing CO_2 permeability with increasing feed pressure was caused by the plasticization effect. This was revealed by the opposite N_2 permeability change trend under the single and mixed gas conditions (Figure 7b). Under the single gas condition, N_2 permeability reduced gradually with the increasing feed pressure due to the polymer condensability effect under high pressure. However, under mixed gas conditions, N_2 permeability gradually increased with increasing feed pressure, indicating that the membrane samples were already plasticized under the CO_2/N_2 mixture gases. Moreover, the presence of N_2 in the mixed gas could prevent further adsorption and condensation of CO_2 and reduce its solubility, which called for the competition sorption and was a general trend for mixed gas tests [26,28,41,42]. Hence, the increase rate for CO_2 permeability under the mixed gas condition was lower than that of the single gas condition. Both enhanced CO_2 and N_2 permeability in the mixed gas condition also led to the fairly constant CO_2/N_2 selectivity relative to the single gas test (Figure 7c). Although it exhibited a plasticization effect in the MMM samples, the highest and constant CO_2/N_2 selectivity for the KAUST-8@BPEI/Pebax MMM in this study relative to the KAUST-8/Pebax MMM and the neat Pebax membrane (15.7 vs. 10.3 vs. 9.4) revealed the benefits of the synergistic function of KAUST-8 and BPEI grating for carbon capture.

Figure S4 showed the temperature effect on the change of gas permeability and CO_2/N_2 selectivity of KAUST-8@BPEI-5/Pebax. As the temperature rose, both CO_2 and N_2 permeability of KAUST-8@BPEI-5/Pebax increased because it facilitated the gas diffusion through the membrane. Similar to our previous work, due to the higher permeation activation energy, the N_2 permeability was more susceptible to temperature, resulting in the larger degree of N_2 permeability enhancement compared to the CO_2 permeability and hence the reduced CO_2/N_2 selectivity [29,38]. Hence, a lower temperature is more suitable for separating CO_2 and N_2 with excellent selectivity. In addition, the three studied membrane samples demonstrated a similarly stable separation performance (Figure 7d) within the 240 h test under mixed gas condition, which demonstrated the highest CO_2 permeability of 150.8 Barrer and the highest CO_2/N_2 selectivity of 16.5 for the KAUST-8@BPEI/Pebax MMM compared to the KAUST-8/Pebax MMM and the neat Pebax membrane. As the testing time reached 1000 h, the CO_2 permeability of all test samples decreased, accompanied by an increase in CO_2/N_2 selectivity. This phenomenon was due to the decreased fractional free volume caused by the physical aging in the membranes [42]. Generally, the permeability of larger gas molecules declines more rapidly than that of smaller ones. Since the molecule size of N_2 is larger than CO_2, the degree of decrease in N_2 permeability is severer than that of CO_2, resulting in the slight increase in CO_2/N_2 selectivity.

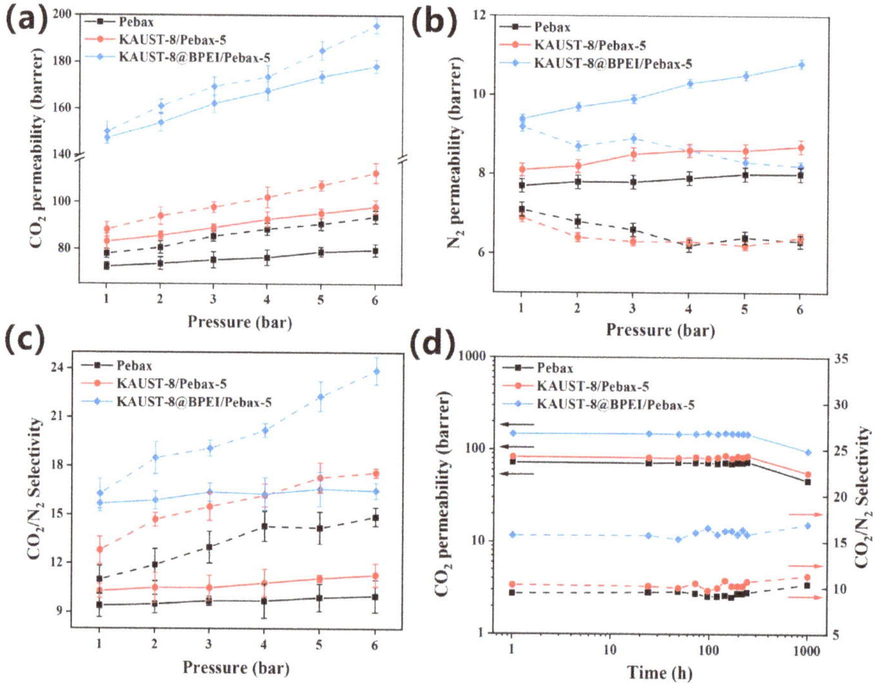

Figure 7. Pressure effect on the change of (**a**) CO_2 permeability, (**b**) N_2 permeability, and (**c**) CO_2/N_2 selectivity of neat membrane, KAUST-8/Pebax-5, and KAUST-8@BPEI/Pebax-5, dashed line: single gas condition, solid line: mixed gas condition. (**d**) Long-term gas permeation test of neat membrane, KAUST-8/Pebax-5, and KAUST-8@BPEI/Pebax-5 according to feed pressures of 1 bar, operated under 25 °C for mixed gas with a volume ratio of 15/85. Arrows are drawn to guide the eye.

3.4. Separation Mechanism

To clarify the function of KAUST-8 and BPEI for CO_2 separation performance, the solubility and diffusivity coefficient deduced from the membrane gas sorption results (Figure 3f) are grouped in Table 1, and the XRD patterns for the d-spacing change are shown in Figure 8. Three membrane samples including neat membrane and with the optimal 5 wt.% loading MMMs with KAUST-8 and KAUST-8@BPEI were selected to clearly illustrate the effect of KAUST-8 and BPEI. As can be seen in Table 1, the neat Pebax membrane showed a CO_2 solubility of 4.56×10^{-2} $cm^3(STP)\cdot cm^{-3}\cdot cmHg^{-1}$ and CO_2 diffusivity of 15.63×10^{-8} $cm^2\cdot s^{-1}$. With the addition of KAUST-8, the CO_2 solubility was nearly retained with a slight enhancement of 2% but with an obvious CO_2 diffusivity coefficient enhancement of 19%. The retained solubility coefficient was aligned to the nearly unchanged CO_2 adsorption property by adding 5 wt.% into the polymer matrix, as illustrated in Figure 3f, and the obvious enhanced diffusivity coefficient was mainly from the porosity of the KAUST-8 framework with the reported pore size (~5 Å, Figure 1b). Similarly, the maintained N_2 solubility was due to the lack of extra N_2 affinity sites from the embedded KAUST-8 additives, and the minor increased N_2 diffusivity coefficient was caused by the small addition of the porous structure of KAUST-8. Consequently, the obvious increment of CO_2 diffusivity coefficient and the minor enhancement of N_2 diffusivity coefficient led to the more boosted CO_2 permeability than N_2 permeability (Figure 5). With BPEI grafting, the amine group in BPEI contributed to the clearly strengthened CO_2 solubility (15% up, Table 1) due to its CO_2 affinity (Figure 3e,f) and slightly reduced N_2

solubility due to the occupied surface area and pore volume in KAUST-8 through grating of BPEI (Figure 3c). Meanwhile, the branched structure of BPEI contributed to the further intensified CO_2 and N_2 diffusivity (Table 1) due to the fact that its branches can disrupt the polymer chain packing during the membrane formation process [25]. As a result, both CO_2 and N_2 permeability were further enhanced (Figure 6a). The d-spacing change in these three membrane samples (Figure 8) correlated to the diffusivity change in Table 1 and the corresponding membrane gas separation performance in Figures 5 and 6. As seen in Figure 8, the XRD pattern for the neat Pebax membrane demonstrated a strong and typical diffraction peak at about 2θ of 22° aligned to the (001) form of the crystal from the PA diffraction [39]. The d-spacing for the neat membrane was 0.397 nm, which is similar to the reported work [39]. With the addition of KAUST-8, the typical peak at 2θ of 22° showed a slight left to 0.398 nm due to the disrupted polymer chain through the addition of KAUST-8 particle. With further BPEI grafting, the branched characteristic in BPEI intensified the polymer chain disruption and led to a reinforced left shift to 0.401 nm. This correlated to the enlarged d-spacing and extra gas transport channels for the continuously increased CO_2 and N_2 permeability (Figures 5 and 6) and explained the enhanced gas diffusivity coefficient demonstrated in Table 1. Conclusively, the CO_2 separation mechanism for the simultaneously enhanced gas permeability, and the selectivity is illustrated in Figure 9. In brief, the additional CO_2 adsorption sites of -NH_2 (adsorption site 1) and -NH- (adsorption site 2) from the grafted BPEI along with the H from pyridine (ligand, adsorption site 3) and F from pillar ($[AlF_5(H_2O)]^{2-}$, adsorption site 4) contributed to the enhanced CO_2 solubility. The porosity and branched structure of BPEI disrupted the polymer chain packing behavior and were responsible for the created fast gas transport channels as well as the increased CO_2 and N_2 diffusivity. The overall balanced function between the enhanced diffusivity coefficient for CO_2 and N_2 as well as the reduced N_2 solubility coefficient led to the simultaneously enhanced CO_2 permeability and CO_2/N_2 selectivity (Figure 6).

Table 1. Solubilities and diffusivities of CO_2 and N_2 for neat Pebax membrane, KAUST-8/Pebax-5, and KAUST-8@BPEI/Pebax-5 MMMs.

Membrane	CO_2 Solubility [10^{-2} $cm^3(STP)\cdot cm^{-3}\cdot cmHg^{-1}$]	CO_2 Diffusivity [10^{-8} $cm^2\cdot s^{-1}$]	N_2 Solubility [10^{-2} $cm^3(STP)\cdot cm^{-3}\cdot cmHg^{-1}$]	N_2 Diffusivity [10^{-8} $cm^2\cdot s^{-1}$]
Pebax	4.56	15.63	2.12	2.83
KAUST-8/Pebax-5	4.67	18.61	2.15	3.11
KAUST-8@BPEI/Pebax-5	5.23	29.92	1.58	6.14

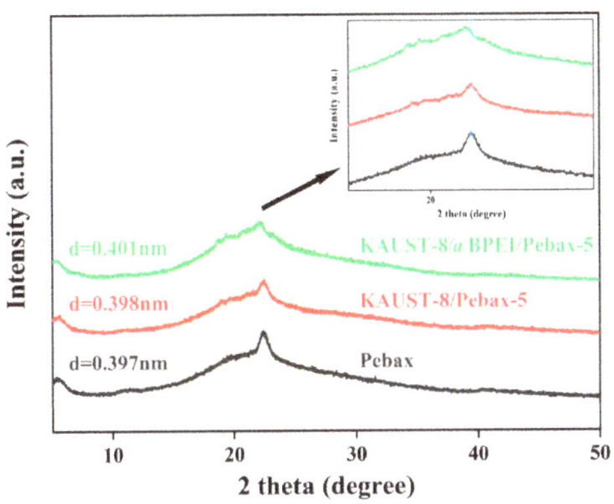

Figure 8. XRD for neat Pebax membrane, KAUST-8/Pebax-5, and KAUST-8@BPEI/Pebax-5 MMMs. The blue line is drawn to guide the eye.

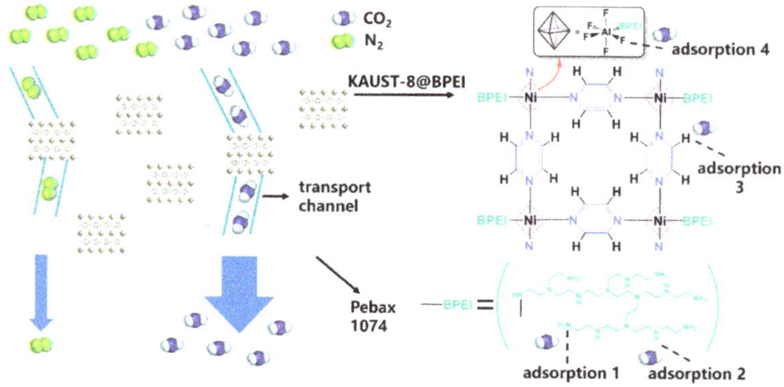

Figure 9. The gas transport behavior in KAUST@BPEI/Pebax MMM.

3.5. Comparison with Other Membrane Materials

The CO_2/N_2 separation for the BPEI-grafted KAUST-8 MMMs was compared to other the Pebax-based MMMs by normalizing the enhancement of CO_2 permeability and CO_2/N_2 selectivity to clearly demonstrate the function of the fillers. As shown in Figure 10a, even the gas separation performance of the representative MMM (KAUST-8@BPEI/Pebax-5) failed to surpass 2008 Robeson upper bound. It could be seen in Figure 10b that the proportion of improvement was relatively high [29,43–45]. The KAUST-8@BPEI/Pebax MMM showed 120% increment for CO_2 permeance (left column) relative to the neat membrane, which was higher than most of the other MMM samples with the only exception of SUM-9, NH_2-MIL-53, and MgO in the Pebax-based MMMs and the KAUST-8/XLPEO MMMs. Among these exceptions, all membrane samples showed a lower enhancement ratio for CO_2/N_2 selectivity (right column), with only one named NH_2-MIL-53/Pebax exhibiting the simultaneously slightly higher CO_2/N_2 selectivity (39.4% vs. 35.3%) [45]. Such results confirmed the synergistic function of CO_2-philic from BPEI and porous framework of KAUST-8 in Pebax for CO_2/N_2 separation. As a result, fillers in the MMMs specific for targeting in simultaneously increasing solubility and diffusivity coefficient for the targeted gases are a promising way to largely enhance the separation performance. In the long run, the grafting of BPEI can be applied to other kinds of MOFs, such as ZIF-8, UiO-66 and so on. Meanwhile, 6FDA-DAM or PIM-1 can be chosen as the polymer matrix to prepare the MMMs and achieve better performance.

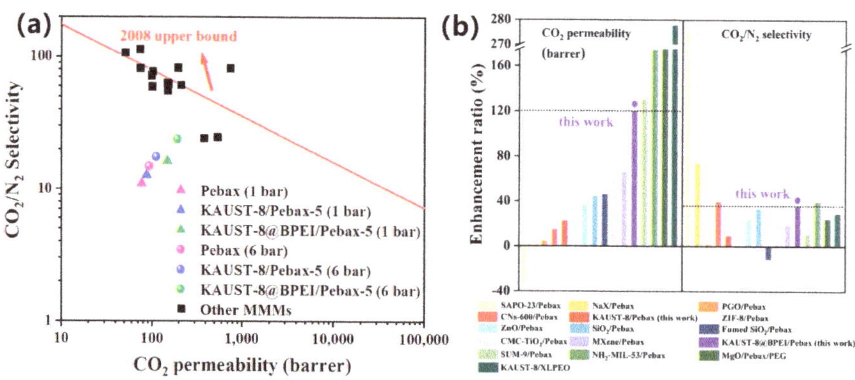

Figure 10. Membrane separation performance comparison with (**a**) 2008 Robeson upper bound and (**b**) enhancement ratio in terms of CO_2 permeability and CO_2/N_2 selectivity (individual data refer to Table S5).

4. Conclusions

In this work, branched polyethyleneimine (BPEI) was successfully grafted on the surface of KAUST-8 (AlFFIVE-1-Ni) through the coordination between the N element in BPEI and the open metal sites of Ni^{2+} in the framework of KAUST-8. The extra added CO_2 adsorption sites from the amino group in the BPEI-grafted KAUST-8 combining the intrinsic CO_2 adsorption sites of F from the inorganic pillar and H from the ligand led to the enhanced CO_2 solubility coefficients in the KAUST-8@BPEI/Pebax MMMs. The additional fast transport channels arising from the porous structure of KAUST-8 and the enlarged d-spacing from the disrupted polymer chains caused by the branched structure of BPEI contributed to the overall enhanced gas diffusivity coefficient in the KAUST-8@BPEI/Pebax MMMs. The synergic strengthened CO_2 solubility and diffusivity coefficients are responsible for the simultaneously enhanced CO_2 permeability and CO_2/N_2 selectivity by the increment of 120% and 35%, respectively. The resultant MMMs exhibited the superior enhancement ratio for CO_2/N_2 separation and H_2O/N_2 separation performance to most of the reported membranes by demonstrating either higher gas permeability or selectivity or both higher than other membranes. The constant CO_2/N_2 selectivity in mixed gas under high pressure (6 bar) and stable long-term stability test running at 240 h for the KAUST-8@BPEI/Pebax MMMs relative to the counterpart samples highlights the efficiency of open metal sites engineering of MOF fillers in the MMMs for carbon capture. Moreover, the low cost of Pebax and filler provides the advantage for the industrial application of the MMMs.

Supplementary Materials: The following supporting information can be downloaded at: https://www.mdpi.com/article/10.3390/membranes15010026/s1, Figure S1: SEM images for (a) KAUST-8, Inset: KAUST-8 dispersion after soaking in n-butanol for 24 h and (b) KAUST-8@BPEI, Inset: KAUST-8@BPEI dispersion after soaking in n-butanol for 24 h; Figure S2: TGA for KAUST-8, KAUST-8@BPEI, neat membrane, KAUST-8/Pebax-5, and KAUST-8@BPEI/Pebax-5; Figure S3: Cross-sectional SEM images for (a) KAUST-8/Pebax-2.5, (b) KAUST-8@BPEI/Pebax-2.5, (c) KAUST-8/Pebax-5, and (d) KAUST-8/Pebax-10. Left: low magnification, Right: high magnification, filler agglomeration is highlighted in the red circle in the inset in Figure S3d; Figure S4: Temperature effect on the change of (a) CO_2 and N_2 permeability and (b) CO_2/N_2 selectivity of KAUST-8@BPEI-5/Pebax; Table S1: Membrane thickness effect on membrane separation performance of CO_2 and N_2 permeability and CO_2/N_2 selectivity; Table S2: Flow rates effect on membrane separation performance of CO_2 and N_2 permeability and CO_2/N_2 selectivity; Table S3: BET surface area and pore volume of KAUST-8 and KAUST-8@BPEI; Table S4: The atomic concentration of Al, C, N, and Ni for KAUST-8 and KAUST-8@BPEI from XPS, respectively; Table S5: Membrane separation performance comparison between this work and other published works. References [46–53] are cited in the Supplementary Materials.

Author Contributions: Conceptualization, R.H. and Y.P.; methodology, J.X.; software, J.X.; validation, Y.G. and L.W.; formal analysis, J.X.; investigation, J.X. and R.H.; resources, R.H. and Y.P.; data curation, J.X.; writing—original draft preparation, J.X. and R.H.; writing—review and editing, Y.G. and L.W.; visualization, Y.G.; supervision, Y.P.; project administration, R.H.; funding acquisition, R.H. and Y.P. All authors have read and agreed to the published version of the manuscript.

Funding: This research was funded by the National Key Research and Development Program of China (2021YFB3801200), the Natural Science Foundation of Jiangsu Province Grant No. BK20240556), Jiangsu Provincial Double-Innovation Doctor Program (Grant No. JSSCBS2023115), and the State Key Laboratory of Materials-Oriented Chemical Engineering (Grant No. SKL-MCE-23A14).

Institutional Review Board Statement: Not Applicable.

Data Availability Statement: The original contributions presented in this study are included in the article/Supplementary Materials. Further inquiries can be directed to the corresponding author(s).

Acknowledgments: The authors are grateful to the characterization equipment including SEM, XPS, FTIR, XRD, and TGA provided by the State Key Laboratory of Materials-Oriented Chemical Engineering.

Conflicts of Interest: The authors declare no conflict of interest.

Abbreviations

The list of symbols and acronyms:

BPEI	Branched polyethyleneimine
D	Diffusion coefficient
FESEM	Field emission scanning electron microscopes
FTIR	Fourier transform infrared
HCP	Hyper-cross-linked polymer
HF	Hydrofluoric acid
IPCC	Intergovernmental Panel on Climate Change
KAUST-8	AIFFIVE-1-Ni
MMMs	Mixed-matrix membranes
MOF	Metal–organic framework
P	Permeability
PAF	Porous aromatic framework
PEO	Polyethylene oxide
PTMSP	Poly(1-trimethylsilyl1-propyne)
S	Solution coefficient
TGA	Thermogravimetric analysis
XLPEO	Crosslinked poly(ethylene oxide)
XPS	X-ray photoelectron spectroscopy
XRD	X-ray diffraction

References

1. Climate Change 2022: Mitigation of Climate Change. Working Group III Contribution to the IPCC Sixth Assessment Report. 2022. Available online: https://www.ipcc.ch/report/ar6/wg3/chapter/chapter-1/ (accessed on 2 June 2024).
2. Vitillo, J.G.; Eisaman, M.D.; Aradóttir, E.S.P.; Passarini, F.; Wang, T.; Sheehan, S.W. The Role of Carbon Capture, Utilization, and Storage for Economic Pathways That Limit Global Warming to below 1.5 °C. *iScience* **2022**, *25*, 104237. [CrossRef]
3. Raganati, F.; Ammendola, P. CO_2 Post-Combustion Capture: A Critical Review of Current Technologies and Future Directions. *Energy Fuels* **2024**, *38*, 13858–13905. [CrossRef]
4. Hou, R.; Yu, J.; Xie, J.; Gu, Y.; Wang, L.; Zhao, B.; Pan, Y. Membrane Technology for Carbon Capture Sequestration and Direct Air Capture—Current Status and Perspective. *Sep. Purif. Technol.* **2025**, *360*, 131077. [CrossRef]
5. Cheng, Y.; Datta, S.J.; Zhou, S.; Jia, J.; Shekhah, O.; Eddaoudi, M. Advances in Metal-Organic Framework-Based Membranes. *Chem. Soc. Rev.* **2022**, *51*, 8300–8350. [CrossRef] [PubMed]
6. Hou, R.; Fong, C.; Freeman, B.D.; Hill, M.R.; Xie, Z. Current Status and Advances in Membrane Technology for Carbon Capture. *Sep. Purif. Technol.* **2022**, *300*, 121863. [CrossRef]
7. Baker, R.W.; Freeman, B.; Kniep, J.; Huang, Y.I.; Merkel, T.C. CO_2 Capture from Cement Plants and Steel Mills Using Membranes. *Ind. Eng. Chem. Res.* **2018**, *57*, 15963–15970. [CrossRef]
8. Park, H.B.; Kamcev, J.; Robeson, L.M.; Elimelech, M.; Freeman, B.D. Maximizing the Right Stuff: The Trade-off between Membrane Permeability and Selectivity. *Science* **2017**, *356*, eaab0530. [CrossRef]
9. Robeson, L.M. The Upper Bound Revisited. *J. Membr. Sci.* **2008**, *320*, 390–400. [CrossRef]
10. Robeson, L.M.; Liu, Q.; Freeman, B.D.; Paul, D.R. Comparison of Transport Properties of Rubbery and Glassy Polymers and the Relevance to the Upper Bound Relationship. *J. Membr. Sci.* **2015**, *476*, 421–431. [CrossRef]
11. Hou, R.; Smith, S.J.D.; Konstas, K.; Doherty, C.M.; Easton, C.D.; Park, J.; Yoon, H.; Wang, H.; Freeman, B.D.; Hill, M.R. Synergistically Improved PIM-1 Membrane Gas Separation Performance by PAF-1 Incorporation and UV Irradiation. *J. Mater. Chem. A* **2022**, *10*, 10107–10119. [CrossRef]
12. Kamble, A.R.; Patel, C.M.; Murthy, Z.V.P. A Review on the Recent Advances in Mixed Matrix Membranes for Gas Separation Processes. *Renew. Sustain. Energy Rev.* **2021**, *145*, 111062. [CrossRef]

13. Ozcan, A.; Fan, D.; Datta, S.J.; Diaz-Marquez, A.; Semino, R.; Cheng, Y.; Joarder, B.; Eddaoudi, M.; Maurin, G. Tuning MOF/Polymer Interfacial Pore Geometry in Mixed Matrix Membrane for Upgrading CO_2 Separation Performance. *Sci. Adv.* **2024**, *10*, eadk5846. [CrossRef] [PubMed]
14. Belmabkhout, Y.; Zhang, Z.; Adil, K.; Bhatt, P.M.; Cadiau, A.; Solovyeva, V.; Xing, H.; Eddaoudi, M. Hydrocarbon Recovery Using Ultra-Microporous Fluorinated MOF Platform with and without Uncoordinated Metal Sites: I- Structure Properties Relationships for C_2H_2/C_2H_4 and CO_2/C_2H_2 Separation. *Chem. Eng. J.* **2019**, *359*, 32–36. [CrossRef]
15. Zheng, W.; Ding, R.; Li, Z.; Ruan, X.; Dai, Y.; Yu, M.; Li, X.; Yan, X.; Jiang, X.; Zhang, X.; et al. Highspeed Transport Pathway Dominated by Continuous Arrangement of Micron-Sized Hollow MOFs in MMMs to Accelerate CO_2 Permeation. *Chem. Eng. J.* **2024**, *490*, 151639. [CrossRef]
16. Gou, M.; Guo, R.; Cao, H.; Zhu, W.; Liu, F.; Wei, Z. An MOF-Tailored Hierarchical Porous Microenvironment for CO_2 as an Efficient Filler for Mixed Matrix Membranes. *Chem. Eng. J.* **2022**, *438*, 135651. [CrossRef]
17. Ding, R.; Dai, Y.; Zheng, W.; Li, X.; Yan, X.; Liu, Y.; Ruan, X.; Li, S.; Yang, X.; Yang, K.; et al. Vesicles-Shaped MOF-Based Mixed Matrix Membranes with Intensified Interfacial Affinity and CO_2 Transport Freeway. *Chem. Eng. J.* **2021**, *414*, 128807. [CrossRef]
18. Liu, G.; Cadiau, A.; Liu, Y.; Adil, K.; Chernikova, V.; Carja, I.-D.; Belmabkhout, Y.; Karunakaran, M.; Shekhah, O.; Zhang, C.; et al. Enabling Fluorinated MOF-Based Membranes for Simultaneous Removal of H_2S and CO_2 from Natural Gas. *Angew. Chem. Int. Ed. Engl.* **2018**, *57*, 14811–14816. [CrossRef]
19. Tchalala, M.R.; Bhatt, P.M.; Chappanda, K.N.; Tavares, S.R.; Adil, K.; Belmabkhout, Y.; Shkurenko, A.; Cadiau, A.; Heymans, N.; De Weireld, G.; et al. Fluorinated MOF Platform for Selective Removal and Sensing of SO_2 from Flue Gas and Air. *Nat. Commun.* **2019**, *10*, 1328. [CrossRef] [PubMed]
20. Datta, S.J.; Mayoral, A.; Murthy Srivatsa Bettahalli, N.; Bhatt, P.M.; Karunakaran, M.; Carja, I.D.; Fan, D.; Graziane M. Mileo, P.; Semino, R.; Maurin, G.; et al. Rational Design of Mixed-Matrix Metal-Organic Framework Membranes for Molecular Separations. *Science* **2022**, *376*, 1080–1087. [CrossRef]
21. Cadiau, A.; Belmabkhout, Y.; Adil, K.; Bhatt, P.M.; Pillai, R.S.; Shkurenko, A.; Martineau-Corcos, C.; Maurin, G.; Eddaoudi, M. Hydrolytically Stable Fluorinated Metal-Organic Frameworks for Energy-Efficient Dehydration. *Science* **2017**, *356*, 731–735. [CrossRef]
22. Belmabkhout, Y.; Bhatt, P.M.; Adil, K.; Pillai, R.S.; Cadiau, A.; Shkurenko, A.; Maurin, G.; Liu, G.; Koros, W.J.; Eddaoudi, M. Natural Gas Upgrading Using a Fluorinated MOF with Tuned H_2S and CO_2 Adsorption Selectivity. *Nat. Energy* **2018**, *3*, 1059–1066. [CrossRef]
23. Lin, Y.; Yan, Q.; Kong, C.; Chen, L. Polyethyleneimine Incorporated Metal-Organic Frameworks Adsorbent for Highly Selective CO_2 Capture. *Sci. Rep.* **2013**, *3*, 1859. [CrossRef] [PubMed]
24. Knebel, A.; Bavykina, A.; Datta, S.J.; Sundermann, L.; Garzon-Tovar, L.; Lebedev, Y.; Durini, S.; Ahmad, R.; Kozlov, S.M.; Shterk, G.; et al. Solution Processable Metal-Organic Frameworks for Mixed Matrix Membranes Using Porous Liquids. *Nat. Mater.* **2020**, *19*, 1346–1353. [CrossRef]
25. Li, X.; Yao, D.; Wang, D.; He, Z.; Tian, X.; Xin, Y.; Su, F.; Wang, H.; Zhang, J.; Li, X.; et al. Amino-Functionalized ZIFs-Based Porous Liquids with Low Viscosity for Efficient Low-Pressure CO_2 Capture and CO_2/N_2 Separation. *Chem. Eng. J.* **2022**, *429*, 132296. [CrossRef]
26. Wang, Y.; Li, H.; Dong, G.; Scholes, C.; Chen, V. Effect of Fabrication and Operation Conditions on CO_2 Separation Performance of PEO-PA Block Copolymer Membranes. *Ind. Eng. Chem. Res.* **2015**, *54*, 7273–7283. [CrossRef]
27. Zhang, S.; Zheng, Y.; Wu, Y.; Zhang, B. Fabrication of Pebax/SAPO Mixed Matrix Membranes for CO_2/N_2 Separation. *J. Appl. Polym. Sci.* **2021**, *138*, 51336. [CrossRef]
28. Azizi, N.; Mohammadi, T.; Behbahani, R.M. Synthesis of a PEBAX-1074/ZnO Nanocomposite Membrane with Improved CO_2 Separation Performance. *J. Energy. Chem.* **2017**, *26*, 454–465. [CrossRef]
29. Hou, R.; Wang, S.; Wang, L.; Li, C.; Wang, H.; Xu, Y.; Wang, C.; Pan, Y.; Xing, W. Enhanced CO_2 Separation Performance by Incorporating KAUST-8 Nanosheets into Crosslinked Poly(Ethylene Oxide) Membrane. *Sep. Purif. Technol.* **2023**, *309*, 123057. [CrossRef]
30. Xu, Y.; Xie, J.; Lin, R.; Zhu, Y.; Wu, J.; Hou, R.; Niu, R.; Pan, Y. Building Extra Transport Channels in Mixed-Matrix Membrane for Air Dehumidification. *J. Environ. Chem. Eng.* **2024**, *12*, 112969. [CrossRef]
31. Liu, B.; Li, D.; Yao, J.; Sun, H. Enhanced CO_2 Selectivity of Polyimide Membranes through Dispersion of Polyethyleneimine Decorated UiO-66 Particles. *J. Appl. Polym. Sci.* **2020**, *137*, 49068. [CrossRef]
32. Panta, J.; Rider, A.N.; Wang, J.; Yang, R.; Brack, N.; Zhang, Y.X. Grafting of Branched Polyethyleneimine onto Carbon Nanotubes to Efficiently Enhance the Lap Shear Strength of Epoxy Adhesives. *Appl. Surf. Sci.* **2023**, *634*, 157691. [CrossRef]
33. Abd El-Mongy, S.; Seoudi, R.; Hussein, A.M. The Influence of Branched Polyethyleneimine Concentration on the Spectroscopic and Morphology of BPEI/AuNPs for Photocatalytic Reduction of 4-Nitrophenol. *J. Mater. Sci. Mater. Electron.* **2024**, *35*, 1440. [CrossRef]

34. Hou, R.; Smith, S.J.D.; Wood, C.D.; Mulder, R.J.; Lau, C.H.; Wang, H.; Hill, M.R. Solvation Effects on the Permeation and Aging Performance of PIM-1-Based MMMs for Gas Separation. *ACS Appl. Mater. Interfaces.* **2019**, *11*, 6502–6511. [CrossRef] [PubMed]
35. Atash Jameh, A.; Mohammadi, T.; Bakhtiari, O. Preparation of PEBAX-1074/Modified ZIF-8 Nanoparticles Mixed Matrix Membranes for CO_2 Removal from Natural Gas. *Sep. Purif. Technol.* **2020**, *231*, 115900. [CrossRef]
36. Azizi, S.; Azizi, N.; Homayoon, R. Experimental Study of CO_2 and CH_4 Permeability Values Through Pebax®-1074/Silica Mixed Matrix Membranes. *Silicon* **2019**, *11*, 2045–2057. [CrossRef]
37. Jiang, Y.; Zhang, B.; Zheng, Y.; Wu, Y. Highly Permselective Pebax/MWCNTs Mixed Matrix Membranes for CO_2/N_2 Separation. *Polym. Bull.* **2024**, *81*, 9699–9719. [CrossRef]
38. Ortiz-Albo, P.; Alves, V.D.; Kumakiri, I.; Crespo, J.; Neves, L.A. A Greener Route to Prepare PEBAX®1074 Membranes for Gas Separation Processes. *J. Membr. Sci.* **2024**, *693*, 122346. [CrossRef]
39. Ariazadeh, M.; Farashi, Z.; Azizi, N.; Khajouei, M. Influence of functionalized SiO_2 nanoparticles on the morphology and CO_2/CH_4 separation efficiency of Pebax-based mixed-matrix membranes. *Korean J. Chem. Eng.* **2020**, *37*, 295–306. [CrossRef]
40. Wu, L.; Sun, J.; Zhu, W.; Wang, C.; Zhang, L.; Niu, R.; Hou, R.; Pan, Y. Constructing Molecular Sieve Channels in Mixed Matrix Membranes for Efficient CO_2 Separation. *Ind. Eng. Chem. Res.* **2024**, *63*, 7760–7768. [CrossRef]
41. Azizi, N.; Mohammadi, T.; Mosayebi Behbahani, R. Comparison of Permeability Performance of PEBAX-1074/TiO_2, PEBAX-1074/SiO_2 and PEBAX-1074/Al_2O_3 Nanocomposite Membranes for CO_2/CH_4 Separation. *Chem. Eng. Res. Des.* **2017**, *117*, 177–189. [CrossRef]
42. Fan, S.-T.; Tan, M.; Liu, W.-T.; Li, B.-J.; Zhang, S. MOF-Layer Composite Polyurethane Membrane Increasing Both Selectivity and Permeability: Pushing Commercial Rubbery Polymer Membranes to Be Attractive for CO_2 Separation. *Sep. Purif. Technol.* **2022**, *297*, 121452. [CrossRef]
43. Feng, X.; Qin, Z.; Lai, Q.; Zhang, Z.; Shao, Z.-W.; Tang, W.; Wu, W.; Dai, Z.; Liu, C. Mixed-Matrix Membranes Based on Novel Hydroxamate Metal-Organic Frameworks with Two-Dimensional Layers for CO_2/N_2 Separation. *Sep. Purif. Technol.* **2023**, *305*, 122476. [CrossRef]
44. Meshkat, S.; Kaliaguine, S.; Rodrigue, D. Mixed Matrix Membranes Based on Amine and Non-Amine MIL-53(Al) in Pebax® MH-1657 for CO_2 Separation. *Sep. Purif. Technol.* **2018**, *200*, 177–190. [CrossRef]
45. Azizi, N.; Jahanmahin, O.; Homayoon, R.; Khajouei, M. A New Ternary Mixed-Matrix Membrane (PEBAX/PEG/MgO) to Enhance CO_2/CH_4 and CO_2/N_2 Separation Efficiency. *Korean J. Chem. Eng.* **2023**, *40*, 1457–1473. [CrossRef]
46. Chen, H.; Zhou, Y.; Sun, J.; Liu, Y.; Zhong, Y.; Du, W.; Lan, J. An Experimental Study of Membranes for Capturing Water Vapor from Flue Gas. *J. Energy. Inst.* **2018**, *91*, 339–348. [CrossRef]
47. Casadei, R.; Giacinti Baschetti, M.; Yoo, M.J.; Park, H.B.; Giorgini, L. Pebax® 2533/Graphene Oxide Nanocomposite Membranes for Carbon Capture. *Membranes* **2020**, *10*, 188. [CrossRef] [PubMed]
48. Zheng, W.; Ding, R.; Yang, K.; Dai, Y.; Yan, X.; He, G. ZIF-8 Nanoparticles with Tunable Size for Enhanced CO_2 Capture of Pebax Based MMMs. *Sep. Purif. Technol.* **2019**, *214*, 111–119. [CrossRef]
49. Aghaei, Z.; Naji, L.; Hadadi Asl, V.; Khanbabaei, G.; Dezhagah, F. The Influence of Fumed Silica Content and Particle Size in Poly (Amide 6-b-Ethylene Oxide) Mixed Matrix Membranes for Gas Separation. *Sep. Purif. Technol.* **2018**, *199*, 47–56. [CrossRef]
50. Shamsabadi, A.A.; Seidi, F.; Salehi, E.; Nozari, M.; Rahimpour, A.; Soroush, M. Efficient CO_2-Removal Using Novel Mixed-Matrix Membranes with Modified TiO_2 Nanoparticles. *J. Mater. Chem. A* **2017**, *5*, 4011–4025. [CrossRef]
51. Wang, H.; Zheng, W.; Yang, X.; Ning, M.; Li, X.; Xi, Y.; Yan, X.; Zhang, X.; Dai, Y.; Liu, H.; et al. Pebax-Based Mixed Matrix Membranes Derived from Microporous Carbon Nanospheres for Permeable and Selective CO_2 Separation. *Sep. Purif. Technol.* **2021**, *274*, 119015. [CrossRef]
52. Shi, F.; Sun, J.; Wang, J.; Liu, M.; Yan, Z.; Zhu, B.; Li, Y.; Cao, X. MXene versus Graphene Oxide: Investigation on the Effects of 2D Nanosheets in Mixed Matrix Membranes for CO_2 Separation. *J. Membr. Sci.* **2021**, *620*, 118850. [CrossRef]
53. Salehi Maleh, M.; Raisi, A. Comparison of Porous and Nonporous Filler Effect on Performance of Poly (Ether-Block-Amide) Mixed Matrix Membranes for Gas Separation Applications. *Chem. Eng. Res. Des.* **2019**, *147*, 545–560. [CrossRef]

membranes

MDPI

Article

Syngas Production via Oxidative Reforming of Propane Using a CO_2- and O_2-Permeating Membrane

José A. Fabián-Anguiano [1], Lorena G. Cuéllar-Herrera [1], José A. Romero-Serrano [1], Issis C. Romero-Ibarra [2], Antonieta García-Murillo [3], Felipe Carrillo-Romo [3] and José Ortiz-Landeros [1,*]

1 Instituto Politécnico Nacional, Escuela Superior de Ingeniería Química e Industrias Extractivas, Departamento de Ingeniería en Metalurgia y Materiales, UPALM-Zacatenco, IPN Avenue, Mexico City 07738, Mexico; jfabiana1000@alumno.ipn.mx (J.A.F.-A.); lcuellarh1000@alumno.ipn.mx (L.G.C.-H.); aromeros@ipn.mx (J.A.R.-S.)

2 Instituto Politécnico Nacional, Unidad Profesional Interdisciplinaria en Ingeniería y Tecnologías Avanzadas, IPN Avenue, Mexico City 07340, Mexico; iromero@ipn.mx

3 Instituto Politécnico Nacional, Centro de Investigación e Innovación Tecnológica, Cerrada de Cecati S/N, Santa Catarina, Azcapotzalco, Mexico City 02250, Mexico; angarciam@ipn.mx (A.G.-M.); fcarrillo@ipn.mx (F.C.-R.)

* Correspondence: jortizla@ipn.mx; Tel.: +52-5557296000 (ext. 54267)

Abstract: Recently, ceramic–carbonate membrane reactors have been proposed to selectively separate CO_2 at elevated temperatures and to valorize this pollutant gas by coupling a catalyzed reaction. This work explores using a membrane reactor to perform the oxidative reforming of propane by taking advantage of the CO_2- and O_2-permeating properties of a $LiAlO_2$/Ag–carbonate membrane. The fabricated membrane showed excellent permeation properties, such as CO_2/N_2 and O_2/N_2 selectivity, when operating in the 725–850 °C temperature range. The membrane exhibited remarkable stability during the long-term permeation test under operating conditions, exhibiting minor microstructural and permeation changes. Then, by packing a Ni/CeO_2 catalyst, the membrane reactor arrangement showed efficient syngas production, especially at temperatures above 800 °C. A hydrogen-rich syngas mixture was obtained by the contributions of the oxidative reforming and cracking reactions. Specific issues observed regarding the membrane reactor's performance are attributed to the catalyst that was used, which experienced significant poisoning by carbon deposition during the reaction, affecting syngas production during the long-term test. Thermodynamic calculations were performed to support the experimental results.

Keywords: membrane; CO_2 separation; O_2 separation; oxidative reforming of propane; syngas

1. Introduction

Over the last 20 years, liquid fuel production, as a primary energy source, has increased significantly to cover global industrial demands [1]. Fischer–Tropsch synthesis is a well-known process to synthesize fuels and other chemical products; this technology can produce ammonia, dimethyl ether, and low-carbon olefins such as ethylene and propylene from non-petroleum resources; for instance, syngas, a mixture of H_2 and CO [2–4]. The key point is that syngas can be derived by converting light hydrocarbons with CO_2, reducing greenhouse gas emissions from conventional industrial processes. Recently, reforming processes for hydrocarbons with CO_2 have been reinvestigated because of the potential to obtain syngas mixtures exhibiting different stoichiometric ratios of H_2/CO. For example, dry reforming of CH_4 with CO_2 has been used a case study for obtaining a low syngas ratio of about ~1 [5]. Like CH_4 dry reforming, propane (C_3H_8) reforming has recently attracted attention as a new route for decarbonization. Propane occurs during the petroleum refining process, and it is produced in a liquid state at room temperature under pressure; moreover, its storage and distribution are practical. Based on the above, hydrogen production through propane reforming has been raised [6–8]; specifically, the reaction of C_3H_8 with CO_2 is currently being investigated. Propane reforming can guarantee a syngas ratio of about 1.5, as shown in the following Reaction (1) [9].

$$C_3H_8 + 3CO_2 \rightarrow 6CO + 4H_2 \Delta H^{\circ} = 621.3 \text{ kJ·mol}^{-1} \quad (1)$$

The main drawback of a dry reforming reaction is that it is highly endothermic and, therefore, energy-consuming. The utilization of air/oxygen mixtures has been investigated through the so-called oxidative reforming reaction by combining a partial oxidation process, as represented by Reaction (2), to decrease the energy costs and produce syngas during C_3H_8 reforming with CO_2 (Reaction (3)) [8,10,11].

$$C_3H_8 + 1.5O_2 \rightarrow 3CO + 4H_2 \Delta H^{\circ} = -227.7 \text{ kJ·mol}^{-1} \quad (2)$$

$$C_3H_8 + 1.5CO_2 + 0.75O_2 \rightarrow 4.5CO + 4H_2 \Delta H^{\circ} = 196.7 \text{ kJ·mol}^{-1} \quad (3)$$

In the state of the art, the reforming of catalysts based on precious metal and metal oxides has gained attention in propane reforming reactions because these materials can show a selective transformation [12]. However, the catalysts' lifetimes, costs, and performances are currently still subjects of study [13]. In recent studies, catalytic membrane reactors have been proposed to produce syngas through reforming and other related reactions; these technologies rely on the design and fabrication of an appropriate membrane that offers good separation performance, high thermal and chemical stabilities, and high selectivity toward obtaining the gaseous species of interest, for example, using CO_2 over N_2 for its subsequent conversion into syngas [14,15]. Therefore, catalytic membrane reactors have emerged as a new technology to carry out the gas separation process and chemical reactions simultaneously. Different approaches to developing catalytic membrane reactors have been considered in studying a new class of membranes based on supported molten salts, e.g., metal–carbonates and ceramic–carbonate membranes [16–18]. For example, silver–molten-carbonate membranes have been investigated to selectively separate CO_2 at high temperatures. Silver exhibits the highest electrical conductivity among metals and negligible chemical reactivity against alkaline molten- salt compositions [19]. Using these properties, supported molten-salt membranes based on silver can simultaneously separate CO_2 and O_2 with high permeation fluxes of 0.82 and 0.43 mL·min^{-1}·cm^{-2} at intermediate temperatures (650–700 °C), respectively [19]. Briefly, the species transport in this kind of membrane is based on the surface formation and subsequent ionic conduction of ${CO_3}^{2-}$ through the bulk of the carbonate phase. The proposed mechanism has been extensively described previously [16]. Briefly, the mechanism involves the electronic conductivity of the solid phase in the membrane according to Reaction (4).

$$CO_{2\,(g)} + 1/2O_2 + 2\,e^- \leftrightarrow CO_3^{2-} \quad (4)$$

Recently, various attempts to improve the thermal stability of silver–molten-carbonate membranes have included the surface modification of the silver matrix, for example, with ceramics using conventional impregnation methods and sophisticated techniques, such as chemical dealloying, chemical vapor deposition, and atomic layer deposition [20–23]. The refractory properties of the deposited metal oxides and the effectiveness of the surface protection at the silver matrix's nanometric scale have improved the membrane's thermal stability and stabilized the permeation flux at high temperatures. More recently, the fabrication of cermets by integrating silver with ceramic oxides, such as SDC and $LiAlO_2$, has been subject to research on developing supported molten-salt membranes [24,25]. Using SDC and γ-$LiAlO_2$ as ceramic fillers improves the wettability properties of the silver matrix against the molten-carbonate phase and hinders its thermal densification at high temperatures. These features guarantee the highly selective separation of CO_2 and O_2 at high temperatures, reaching CO_2-permeation flux values of as high as 0.49 and 0.78 mL·min^{-1}·cm^{-2} for SDC/Ag and γ-$LiAlO_2$/Ag molten-salt membranes at 850 °C, respectively. Moreover, cermet–molten-salt membranes extend the operational lifetime, taking the conventional silver–molten-carbonate membranes as a baseline [19–23]. The improvement in the wettability properties against molten carbonates and the protraction of the operating time of the silver-based membranes has opened the possibility to explore their use in hydrogen production by coupling reforming reactions [25,26]. For example, cermet–molten-salt membranes, composed of γ-$LiAlO_2$/Ag and a molten-carbonate mixture of Li_2CO_3, Na_2CO_3, and K_2CO_3, have been studied to obtain syngas through the oxidative methane reforming reaction. The membrane is shown to be highly stable under reaction conditions for up to 120 h, achieving a total syngas production (H_2+CO) of 4.94 mL·min^{-1}·cm^{-2} [25]. Taking advantage of the high performance of cermet–molten-carbonate membranes, in the present work, a γ-$LiAlO_2$/Ag-based membrane was exploited to elucidate the feasibility of obtaining syngas through the oxidative reforming of propane at elevated temperatures (725–850 °C).

2. Materials and Methods

2.1. Membrane Preparation and Characterization

To obtain the cermet–molten-carbonate membrane, first, cermet powders of γ-$LiAlO_2$/Ag were prepared from Ag atomized commercial powders of 400 mesh (99.9%, Stannum, Mexico City, Mexico) and γ-$LiAlO_2$ ceramic powders; the latter were chemically synthesized using the EDTA/citrate method, as previously described in detail [25,27]. Cermet powders with a composition of 50:50 vol. % were prepared by ball milling using a Spex-Mixer/Mill apparatus (Chemplex Instruments, Urbana, IL, USA). The grinding medium was alumina balls, using a ratio of balls to powders of 10 to 1 in weight. The powders were dry milled for 15 min, recovered from the milling jar, and mixed with polyvinyl alcohol as a polymeric binder to subsequently be uniaxially pressed into disk-shaped membrane supports. An incipient sintering process was conducted at 900 °C for 20 h using a heating and cooling rate of 3 °C/min to obtain porous supports with a final porosity of around 33 vol. %. Finally, a dense membrane was obtained by direct infiltration of the supports with a eutectic ternary mixture of $Li_2CO_3/Na_2CO_3/K_2CO_3$ using a molar ratio of 43.5/31.5/25, respectively, at 550 °C. After infiltration, the cermet–molten-carbonate membrane was slowly cooled, and the carbonate excess was removed using sandpaper until a final membrane thickness of 1.2 mm.

The membrane was characterized by the X-ray diffraction (XRD) technique to identify the crystalline phase structure; for this purpose, the diffractometer model D8 Focus (Bruker, Billerica, MA, USA), equipped with a Cu-Ka radiation source, was used. The XRD diffraction patterns were recorded in a 2-theta range of 10–90° using a scanning rate of 2° min^{-1} and a step size of 0.02°. The morphology and microstructural features of the samples were characterized using a JSM-6400 scanning electron microscope (SEM) from JEOL, Tokyo, Japan. The analyses were conducted using both secondary electron and backscattered-electron techniques. Archimedes measurements, based on the international standard ASTM C373-88 [28], were performed to determine the effective pore volume of the supports. In addition, in the unsteady state, He-permeation measurements were also used to test the open and interconnected porosity in the supports and to corroborate the membrane's infiltration by molten carbonates.

2.2. Permeation Measurements and Oxidative Reforming of Propane Test

The CO_2- and O_2-permeation measurements were conducted using a commercial Probostat high-temperature permeation system (NORECs, Norway) equipped with gas mixers and flow controllers, as described in previous works [14,25]. Briefly, the membrane was tightly attached to an inner alumina tube with a silver seal (99% purity) using special spring accessories. The system was completely enclosed with an outer highly dense alumina tube and heated at 950 °C using a ramping rate of 2° min^{-1} to soften the silver seal and, therefore, the membrane sealing. The system was fed with a gas mixture composition of $CO_2/O_2/N_2$ in a 15/17/68 vol. % and a constant flow rate of 50 mL·min^{-1}. N_2 was used as a diluent to monitor the presence of leaks through the membrane defects. The permeation measurements were conducted in a temperature range from 725 to 850 °C with increases of 25 °C. Each reading was recorded when the system reached steady state conditions after 60 min of reaching the target temperature. On the permeate side, a constant argon flow rate (50 $mL{\cdot}min^{-1}$) was used as a sweep gas to sweep the permeated species.

To perform the oxidative reforming reaction in the system, a catalyst bed of 10 wt.% Ni supported on CeO_2 was placed 1 cm underneath the membrane on the sweep side. In the membrane's permeate side, the sweep gas was switched from pure Ar to a mixture of 5 vol. % of C_3H_8 in Ar with a total flow of 50 $mL{\cdot}min^{-1}$. This propane concentration means an excess of the stoichiometric amount of this gas required to perform the oxidative reforming (Reactions (3) and (4)) based on the CO_2- and O_2-permeation measurements. Before the reaction test, the catalyst was held under a gas mixture of 50/50 vol. % H_2 and Ar at 850 °C for 60 min to reduce the Ni active phase. The permeate species and the reaction products were continuously analyzed in the whole range of temperatures studied by gas chromatography, using a chromatograph model GC-2014 (SHIMADZU, Kyoto, Japan) equipped with a TCD detector and a Carboxen-1010 PLOT capillary column. Moreover, in situ analysis by an online FT-IR was coupled with the sweep gas effluent to monitor the presence of water concentration during the reaction test. For this purpose, an FT-IR equipment model ALPHA II (Bruker, USA) was used.

The permeation flux of CO_2 and O_2 were calculated considering the total flux and the N_2 leak contribution by the Knudsen diffusion (if the case); however, if N_2 was not detected, a minimum lake value of 0.0013 $mL{\cdot}min^{-1}{\cdot}cm^{-2}$ was considered for the correction of the fluxes for CO_2 and O_2 based on the chromatograph detection limit. In the same sense, the conversion rates for C_3H_8, CO_2, and O_2 were estimated based on the chromatography analysis (Supplementary Materials). Thermogravimetric analysis of the fresh and spent catalysts was performed in a dry air atmosphere using a thermobalance model Regulus STA 2500 (Netzsch, Waldkraiburg, Germany). Finally, to elucidate the syngas production through the oxidative reforming of propane, thermodynamic calculations were conducted using the FactSage 7.2 thermochemical software [29].

3. Results and Discussion

3.1. Synthesis and Characterization of γ-$LiAlO_2$/Ag Powders and Fabrication of Cermet–Molten-Carbonate Membranes

Figure 1 shows the XRD analysis of the γ-$LiAlO_2$/Ag cermet prepared by the ball milling process. The γ-$LiAlO_2$/Ag diffraction peaks indicate that both cermet components are highly stable during milling because neither structural changes nor secondary phases were identified because of cermet formation starting from the single components. It is notable that the reflection peaks of the ceramic phase are less intense in relation to the metallic phase; this fact can be attributed to the difference in the atomic weights of the elements present in each phase of the cermet and, therefore, the X-ray mass absorption coefficient of the pristine materials [25,30]. The characteristic reflection peaks of the materials were identified using the ICSD database. For the γ-$LiAlO_2$ phase, a PDF 98-003-0249 card was used, and it matched a *P41212* space group, whereas, for silver, the PDF 98-006-4994 card with an *Fm3m* space group was used.

The morphology, particle size, and microstructural features of the γ-$LiAlO_2$/Ag cermet differ significantly from the pristine materials (Figure 2a–c). On the one hand, the silver powders were mainly composed of spheroidal particles exhibiting particle sizes in a range of 4–35 μm; on the other hand, γ-$LiAlO_2$ was composed of soft, irregular agglomerates on the submicrometric scale. In the case of cermet, the powders showed irregular morphologies and particle sizes of about 5–30 μm. Finally, the SEM image obtained by the backscattering technique (Figure 2d) shows that milling processing promotes the integration of both single materials, ceramic, and metal to form the expected composite. This technique agrees with the results of the XRD, since it successfully indicated the formation of the cermet. Similar behavior was observed during the formation of the $Ce_{0.9}Sm_{0.1}O_{2-\delta}$/Ag and $LiAlO_2$/Ag cermets, which have previously been studied [24,25]; therefore, these results evidence the feasibility and reproducibility of the ball milling procedure for cermets' formation.

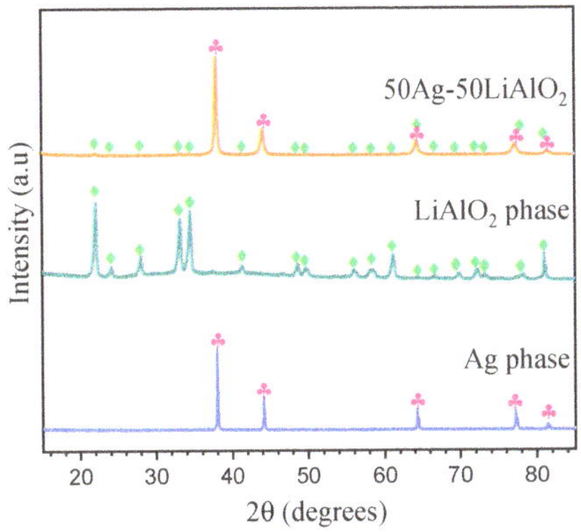

Figure 1. XRD analysis of the pristine materials (γ-$LiAlO_2$, Ag) and γ-$LiAlO_2$/Ag cermet (50:50 vol. %) synthetized by the ball milling process.

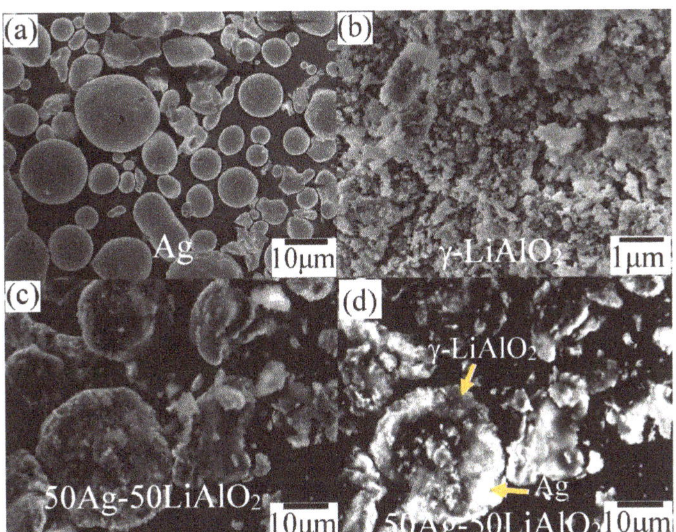

Figure 2. Microstructural analyses by SEM of (**a**) silver, (**b**) γ-$LiAlO_2$, and (**c**,**d**) γ-$LiAlO_2$/Ag cermet powders (50:50 vol. %) synthetized by the milling process.

Figure 3 shows a SEM image of the cross-section of the porous support obtained by pressing and subsequent sintering at 900 °C, as well as the cermet–molten-carbonate membrane obtained after infiltration. As shown in Figure 3a, the cermet support has a highly porous microstructure, consisting of different irregular pore sizes. This porosity represents an open porosity of around 33 vol. %, estimated by Archimedes' method. Moreover, the steady-state helium permeation test indicated that the porosity-to-tortuosity ratio $\left(\frac{\varepsilon}{\tau}\right)$ of the cermet support was 0.27, with a helium permeance value of 2.24×10^{-5} mol·Pa^{-1}·m^{-2}·s^{-1}. These results suggest that the porosity of the support is also highly interconnected. Figure 3b shows details of the cross-section of the infiltrated membrane. In this case, the open porosity of the support was wholly filled with the molten-carbonate phase, leading to a dense microstructure typical for this type of membrane. In addition, the helium-permeation measurement corroborated a successful densification process and indicates that the membrane is free of defects, because the permeance value decreased to 0.94×10^{-10} mol·Pa^{-1}·m^{-2}·s^{-1}, which means five orders of magnitude lower than the porous support. Figure 3c shows the EDS spectrum of the infiltrated sample. The observed signals evidence the presence of all of the chemical constituents after infiltration.

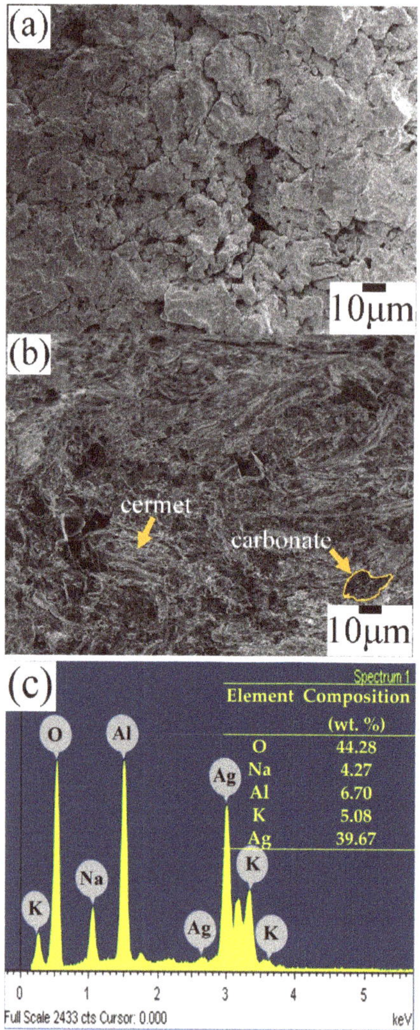

Figure 3. Microstructural analyses by SEM: (**a**) porous support sintered at 900 °C for 20 h; (**b**) cermet–molten-carbonate membrane obtained by infiltration of the $Li_2CO_3/Na_2CO_3/K_2CO_3$ mixture; (**c**) EDS spectrum shown as evidence of the carbonates' incorporation into the membrane.

3.2. Permeation Measurements and Oxy-CO_2 Reforming of Propane at High Temperatures in the Membrane Reactor Arrangement

Figure 4 shows the permeance properties of the dense membrane in a temperature range of 725–850 °C. Before carrying out a series of measurements as a function of temperature, a stabilization period was maintained in the membrane for the first 24 h at

825 °C (Figure 4b). Afterward, the permeance data were recorded once it was corroborated that the CO_2 and O_2 fluxes were stable. As shown in Figure 4a, the CO_2 and O_2 permeations through the membrane increased because of the ionic conduction of the surface-formed CO_3^{2-} species (Reaction (4)), which relies on the ionic conduction properties of the molten-carbonate phase, that is, a thermally activated process [14]. Overall, the CO_2 and O_2 transport was remarkable—between 725 and 825 °C—and the permeance increased from 0.56×10^{-7} to 1.65×10^{-7} mol·Pa^{-1}·m^{-2}·s^{-1} for CO_2 and from 0.26×10^{-7} to 1.01×10^{-7} mol·Pa^{-1}·m^{-2}·s^{-1} for O_2, respectively. In the whole temperature range studied, no N_2 concentration was detected, i.e., a negligible leak was observed. Hence, considering the detection limit of the GC, if a minimum concentration of N_2 is present, this would be equivalent to a permeation value of 0.4×10^{-10} mol·Pa^{-1}·m^{-2}·s^{-1}, which would not change significantly the observed membrane selectivity.

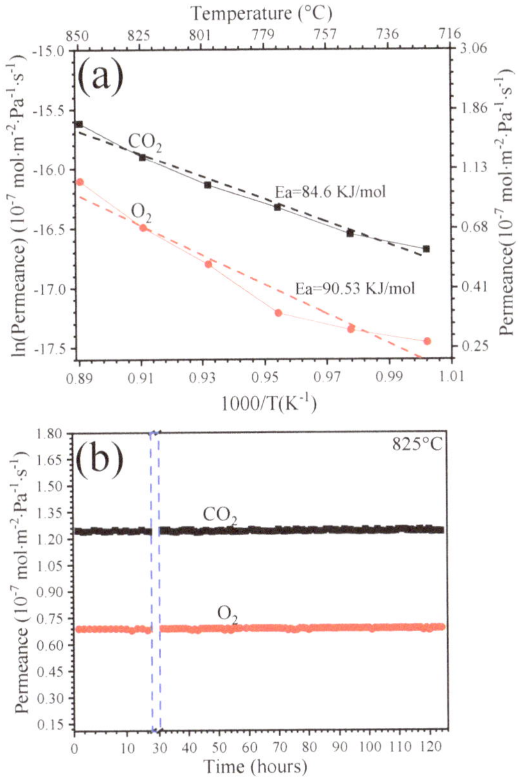

Figure 4. CO_2- and O_2-permeation measurements at high temperature of the cermet–molten-carbonate membrane: (**a**) CO_2- and O_2-permeance properties as a function of temperature after the first 24 h of stabilization; (**b**) long-term stability of the membrane under permeance conditions at 825 °C. Dash lines indicate the period wherein the measurements against temperature where performed.

As expected, the apparent activation energy values were calculated by constructing Arrhenius plots (Figure 4a). The CO_2 and O_2 permeances had similar E_a values of 84.6 kJ·mol^{-1} for CO_2 and 90.5 kJ·mol^{-1} for O_2. This fact results from the transport of both species being a concurrent phenomenon related to Reaction (4) (backward). The long-term stability of the membrane under permeation conditions was examined at 825 °C.

Figure 4b shows the permeation experiments as a function of time. When the temperature reached 825 °C and was maintained for the next 95 h, the CO_2- and O_2-permeance fluxes were highly stable, at this point completing a total of 125 h under operating conditions. Remarkably, the membrane achieved concomitant separation with a total permeance value of $J_{CO_2} + J_{O_2}$ equal to 2.66×10^{-7} mol·Pa^{-1}·m^{-2}·s^{-1} at 825 °C.

As mentioned, at least on the laboratory scale, the cermet–molten-carbonate membrane showed high thermal and chemical stability over the long term. Owing to this feature, the oxidative reforming of propane was conducted using the same membrane. Figure 5 shows the syngas production obtained when C_3H_8 diluted in Ar was fed as the sweep gas in the experimental arrangement to perform the reaction. As shown in Figure 5a, the H_2 production rate improved as the temperature increased, achieving a flux rate from 0.12 to 4.0 mL·min^{-1}·cm^{-2} when the membrane reached 850 °C. Similarly, the CO increased as a function of the temperature, reaching a total production rate of 1.56 mL·min^{-1}·cm^{-2} at 850 °C. Considering that the excess of propane present in the sweep was consumed, it can be concluded that the amount of hydrogen produced was higher than predicted by the stoichiometric oxidative reforming reactions described in Equations (3) and (4), and, therefore, other side reactions must be involved. As a result, the H_2/CO ratio was about 3.9 at lower temperatures (725 °C), whereas it was 2.5 at 850 °C. Actually, based on the GC analysis, the presence of C_3H_8, CO_2, and O_2 was not detected in the sweep out for the whole temperature range studied, which means the total conversion of the reactants was achieved.

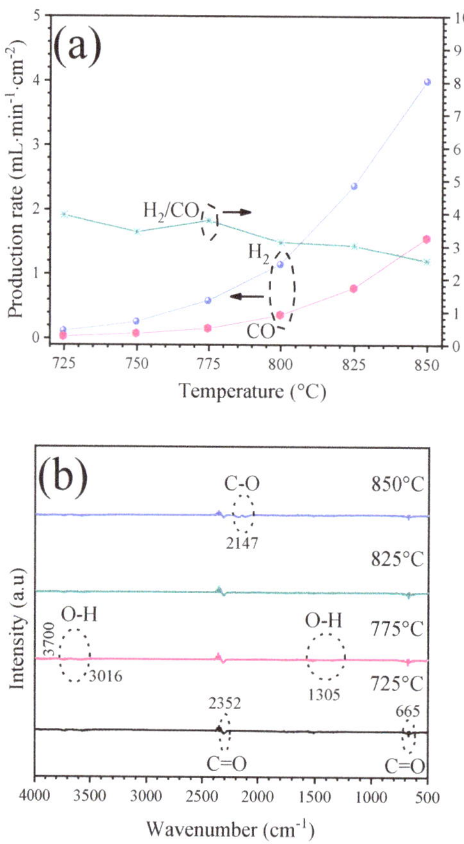

Figure 5. Syngas production when using a sweep of 5 vol. % of C_3H_8 in argon: (**a**) H_2 and CO production rates as a function of temperature; (**b**) FT-IR in situ analysis of the syngas at different temperatures. Arrows indicate the corresponding Y axis.

A complementary FT-IR analysis in situ was also used to analyze the sweep gas composition. In Figure 5b, the FT-IR spectrum at different temperatures shows very weak bands of trace species. The O-H from water molecules was between 3016 and 3700 cm^{-1} and symmetric and asymmetric stretching of CO_2 occurred at 665 and 2352 cm^{-1}. On the other hand, the FT-IR spectrum shows bands at 2147 cm^{-1} assigned to the stretching vibrational modes of the CO. Therefore, the results suggest that propane was consumed to favor syngas production through both dry reforming and partial oxidation reactions at high temperatures, but a reverse water–gas shift reaction may be involved in syngas production, as has also been reported for oxygen-transporting membranes studied under oxidative reforming of propane [31,32].

Afterward, thermochemical simulations of syngas were conducted using the FactSage 7.2 software [29]. As shown in Figure 6a, oxidative reforming of propane is highly viable, following both the partial oxidation and dry reforming reactions in the temperature range at which the membrane was studied. Furthermore, thermodynamic calculations support the possible influence of other reactions occurring during syngas production. Different reaction pathways can also involve thermal dehydrogenation, thermal cracking, complete combustion of propane, and even methanation and partial oxidation of methane, as reported in [11,31–34].

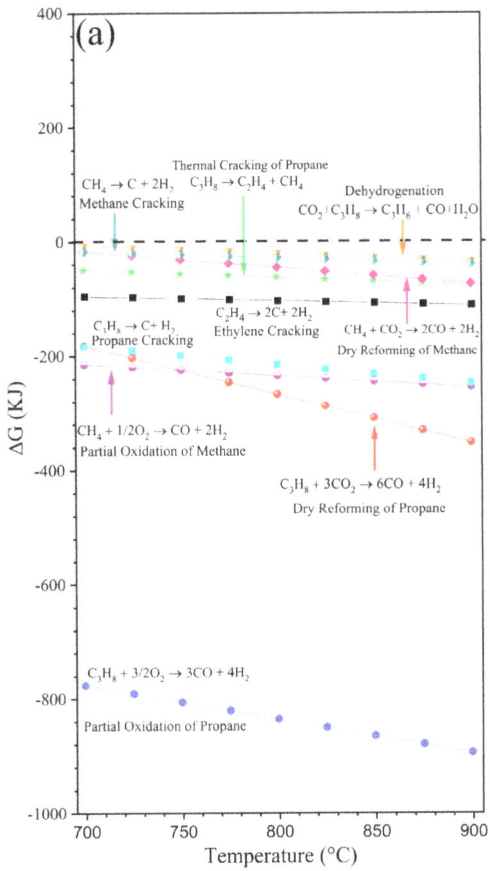

Figure 6. *Cont.*

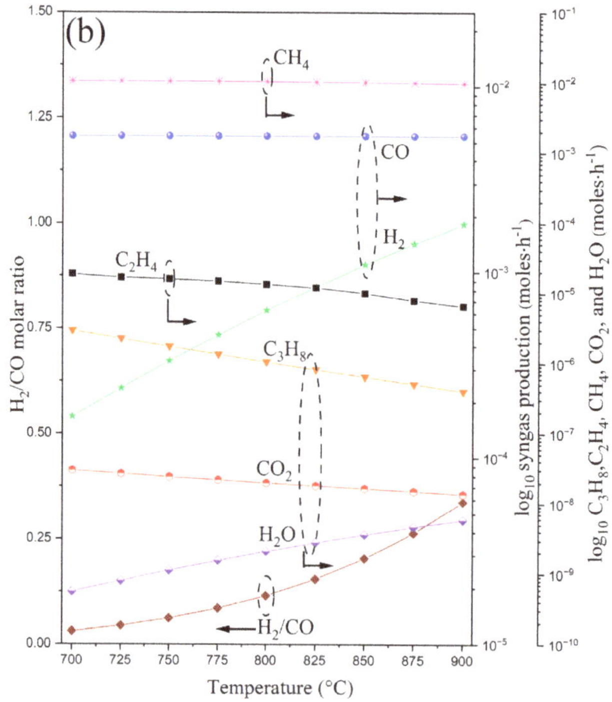

Figure 6. Thermochemical simulations of syngas using FactSage 7.2 software: (**a**) ΔG° versus temperature of the main reactions involved in the syngas production; (**b**) thermodynamic equilibrium calculus of the produced species at high temperatures (600–900 °C). Arrows indicate the corresponding Y axis.

On the other hand, the equilibrium composition of the syngas was calculated as a function of the temperature (Figure 6b). The thermodynamic equilibrium calculus indicates that, certainly, when the temperature increases from 700 to 900 °C, CO_2 and C_3H_8 are consumed to favor syngas production. However, based on thermodynamics, CO production was higher than hydrogen for the whole temperature range. This fact does not match the experimental data, and it could be attributed to different factors, including catalyst performance in terms of catalyst selectivity and significant contributions by the cracking reactions that can result in an increase in the H_2 yield and, therefore, the H_2/CO ratio observed experimentally. Moreover, thermodynamic estimations also suggest the formation of small quantities of water, corroborating the experimental data from this work.

3.3. Stability Test Under Oxidative Reforming of Propane

The membrane was kept under reaction conditions for 10 h longer at 825 °C, for a total of 151 h of operation, to determine its stability under the oxidative reforming of propane. Figure 7 shows that the H_2 and CO production rates were stable for about the first 6 h. The H_2 production and the CO production during the stability test were 2.37 mL·min^{-1}·cm^{-2} and 0.77 mL·min^{-1}·cm^{-2}, respectively. Moreover, the H_2/CO molar ratio was about 3.4 during the reaction time. After, a rapid decline in syngas production was observed. This fact is believed to be due to catalyst poisoning by carbon deposition. TGA analysis was performed to elucidate this fact, and the results show significant differences between the fresh and spent catalysts (Figure 8). Assuming that the observed weight loss in the TGA curve was due to the carbon combustion, the carbon content was, therefore, negligible for the fresh catalyst. In contrast, the spent catalyst showed a high total carbon content of 28 wt. %. These results corroborate that the observed hydrogen production involves oxidative reforming reactions and a significant contribution from the catalytic thermal cracking. Therefore, the latter reactions consumed the excess propane supply to the system, wherein the thermal decomposition of the possible intermediate products, as in Reactions (5)–(7), leads to high H_2/CO molar ratios. In this sense, the propane concentration in the sweep and the stability of Ni-CeO_2 catalysts need further investigation to enhance the obtainment of syngas through the proposed oxidative reforming of propane approach.

$$C_3H_8 \rightarrow 3C + 4H_2 \Delta H^{\circ} = 119.5 \text{ kJ·mol}^{-1} \quad (5)$$

$$C_2H_4 \rightarrow 2C + 2H_2 \Delta H^{\circ} = -52.5 \text{ kJ·mol}^{-1} \quad (6)$$

$$CH_4 \rightarrow C + 2H_2 \Delta H^{\circ} = 74.85 \text{ kJ·mol}^{-1} \quad (7)$$

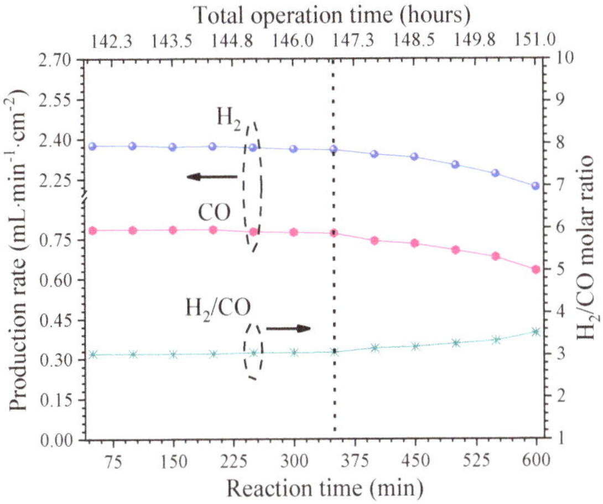

Figure 7. Stability of syngas production through the cermet–molten-carbonate membrane under oxidative reforming of propane. Stability of membrane with a total continuous operation time of 151 h. Arrows indicate the corresponding Y axis.

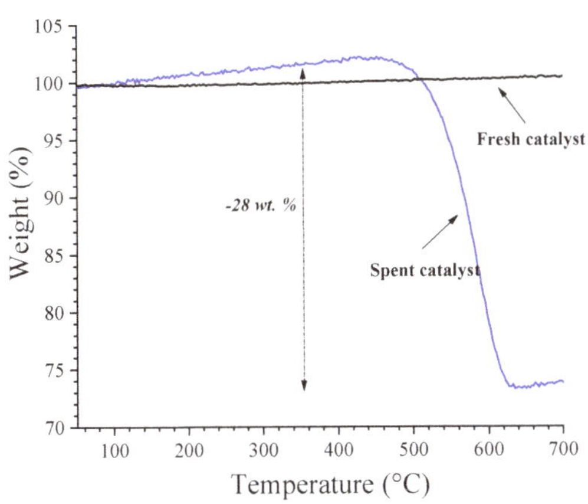

Figure 8. TGA analysis of the fresh and spent catalysts. The weight change indicates the removal by combustion of the carbon formed due to the thermal cracking reactions.

After studying the oxidative reforming, the membrane surface was analyzed by SEM. Figure 9 shows the results of the postmortem analysis for both sides of the membrane. The surface that was in contact with C_3H_8 exhibited significant changes in its microstructure, primarily related to a growth in the size of the grains and porosity, while the surface that was in contact with the feed mixture of $CO_2/O_2/N_2$ (15/17/68 vol. %) only showed minor microstructural changes. Although propane promotes certain changes on the membrane's surface, its stability is remarkable because no N_2 leaks were attributed to the membrane's failure, nor were there significant fluctuations in the permeation performance. Finally, elemental analysis by EDS of the sweep side of the used membrane (Figure 9c) corroborates the presence of all of the original elements constituting the fresh membrane (Figure 3c). Overall, the results suggest the high potential of the proposed cermet–molten-carbonate membrane to couple the high-temperature CO_2 and O_2 separation with the oxy-CO_2 reforming of propane as a potential conversion and valorization alternative for CO_2.

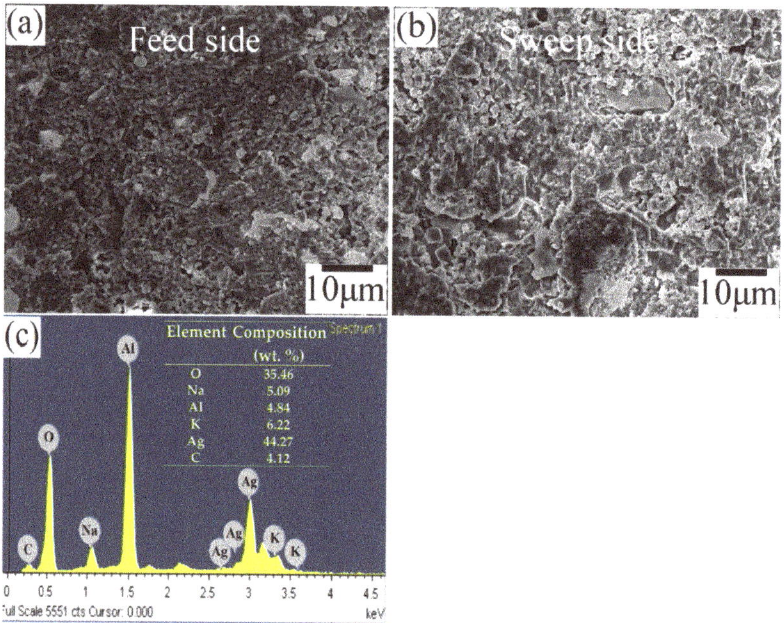

Figure 9. SEM postmortem of the cermet–molten-carbonate membrane: (**a**) feed-side surface in contact with the feed mixture of $CO_2/O_2/N_2$ (15/17/68 vol. %); (**b**) surface in contact with a 5 vol. % of C_3H_8 in argon, at a total flow of 50 mL·min^{-1}; (**c**) elemental analysis by EDS of the sweep side of the membrane.

4. Conclusions

A ceramic–carbonate membrane reactor was successfully probed to simultaneously perform the selective separation of CO_2 and O_2 at elevated temperatures with the subsequent production of syngas by coupling the oxidative reforming of propane. The results are conclusive and evidence the high capability of the studied ceramic–carbonate membrane to combine its perm-selectivity properties, exhibited between 725 and 850 °C, to perform the proposed catalyzed reactions. The studied operating temperatures promote the contribution of propane cracking to the desirable dry reforming and partial oxidation of the propane reactions. Indeed, as the catalyst's deactivation resulted in the syngas production's detriment, it is clear that further improvements must be made to the catalyst's selection, looking to the future achievement of a higher carbon deposition resilience, especially over long-term operations.

Supplementary Materials: The following supporting information can be downloaded at: https://www.mdpi.com/article/10.3390/membranes14110238/s1, S1: Conversion calculations.

Author Contributions: Conceptualization, J.O.-L.; methodology, J.A.F.-A.; formal analysis, J.O.-L.; investigation, J.A.F.-A. and L.G.C.-H.; data curation, J.A.F.-A., I.C.R.-I. and J.O.-L.; writing—original draft preparation, J.A.F.-A. and J.O.-L.; writing—review and editing, J.A.R.-S., A.G.-M., F.C.-R. and I.C.R.-I.; funding acquisition, J.O.-L. All authors have read and agreed to the published version of the manuscript.

Funding: This research was supported by SIP-IPN, Project No. 20240215, and CONAHCYT, "Ciencia Básica y de Frontera 2023–2024", Project No. CBF2023–2024-1992.

Institutional Review Board Statement: Not applicable.

Data Availability Statement: The original contributions presented in the study are included in the article/Supplementary Materials, further inquiries can be directed to the corresponding author.

Acknowledgments: The authors express their appreciation to the SNII-CONAHCYT, SIBE-IPN, EDI-IPN, and BEIFI-IPN programs for financial support.

Conflicts of Interest: The authors declare no conflicts of interest.

References

1. International Energy Agency (IEA). *World Energy Outlook Report 2022*; International Energy Agency (IEA): Paris, France, 2022; pp. 1–759.
2. Inderwildi, O.R.; King, D.A.; Jenkins, S.J. Fischer-Tropsch synthesis of liquid fuels: Learning lessons from homogeneous catalysis. *Phys. Chem. Chem. Phys.* **2009**, *11*, 11110–11112. [CrossRef] [PubMed]
3. Lin, T.; An, Y.; Yu, F.; Gong, K.; Yu, H.; Wang, C.; Sun, Y.; Zhong, L. Advances in Selectivity Control for Fischer–Tropsch Synthesis to Fuels and Chemicals with High Carbon Efficiency. *ACS Catal.* **2022**, *12*, 12092–12112. [CrossRef]
4. Xu, H.; Cao, M.; Li, Z.; Li, W.; Meng, S.; Song, H. Production of low carbon number olefins from natural gas: Methane-involved catalytic non-oxidative propane dehydrogenation. *Chem. Eng. J.* **2023**, *454*, 140508. [CrossRef]
5. Alhassan, M.; Jalil, A.A.; Bahari, M.B.; Owgi, A.H.K.; Nabgan, W.; Hassan, N.S.; Tran, T.V.; Abdulrasheed, A.A.; Hamid, M.Y.S.; Ikram, M.; et al. Profitable Fischer Tropsch realization via CO_2-CH_4 reforming; an overview of nickel-promoter-support interactions. *RSC Adv.* **2023**, *13*, 1711–1726. [CrossRef]
6. Kokka, A.; Katsoni, A.; Yentekakis, I.V.; Panagiotopoulou, P. Hydrogen production via steam reforming of propane over supported metal catalysts. *Int. J. Hydrogen Energy* **2020**, *45*, 14849–14866. [CrossRef]
7. Barzegari, F.; Kazemeini, M.; Rezaei, M.; Farhadi, F.; Keshavarz, A.R. Syngas production through CO_2 reforming of propane over highly active and stable mesoporous NiO-MgO-SiO_2 catalysts: Effect of calcination temperature. *Fuel* **2022**, *322*, 124211. [CrossRef]
8. Liao, M.; Chen, Y.; Cheng, Z.; Wang, C.; Luo, X.; Bu, E.; Jiang, Z.; Liang, B.; Shu, R.; Song, Q. Hydrogen production from partial oxidation of propane: Effect of SiC addition on Ni/Al_2O_3 catalyst. *Appl. Energy* **2019**, *252*, 113435. [CrossRef]
9. Wang, X.; Wang, N.; Zhao, J.; Wang, L. Thermodynamic analysis of propane dry and steam reforming for synthesis gas or hydrogen production. *Int. J. Hydrogen Energy* **2010**, *35*, 12800–12807. [CrossRef]
10. Lintz, H.-G.; Müller, S.P. The partial oxidation of propane on mixed metal oxides—A short overview. *Appl. Catal. A Gen.* **2009**, *357*, 178–183. [CrossRef]
11. Althenayan, F.M.; Adesina, A.A. Oxy-dry reforming of propane over Ce-promoted Co-Ni/Al_2O_3 catalyst. *Appl. Petrochem. Res.* **2018**, *8*, 239–251. [CrossRef]
12. Al-Shafei, E.; Aljishi, M.; Albahar, M.; Alahmed, A.; Sanhoob, M. Effect of CO_2/propane ratio and trimetallic oxide catalysts on maximizing dry reforming of propane. *Mol. Catal.* **2023**, *537*, 112945. [CrossRef]
13. Arora, S.; Prasad, R. An overview on dry reforming of methane: Strategies to reduce carbonaceous deactivation of catalysts. *RSC Adv.* **2016**, *6*, 108668–108688. [CrossRef]
14. Fabián-Anguiano, J.; Mendoza-Serrato, C.; Gómez-Yáñez, C.; Zeifert, B.; Ma, X.; Ortiz-Landeros, J. Simultaneous CO_2 and O_2 separation coupled to oxy-dry reforming of CH_4 by means of a ceramic-carbonate membrane reactor for in situ syngas production. *Chem. Eng. Sci.* **2019**, *210*, 115250. [CrossRef]
15. Abdallah, H. A Review on Catalytic Membranes Production and Applications. *Bull. Chem. React. Eng. Catal.* **2017**, *12*, 136–156. [CrossRef]
16. Chung, S.J.; Park, J.H.; Li, D.; Ida, J.-I.; Kumakiri, I.; Lin, J.Y.S. Dual-Phase Metal-Carbonate Membrane for High-Temperature Carbon Dioxide Separation. *Ind. Eng. Chem. Res.* **2005**, *44*, 7999–8006. [CrossRef]
17. Ovalle-Encinia, O.; Lin, Y.S. High-pressure CO_2 permeation properties and stability of ceramic-carbonate dual-phase membranes. *J. Memb. Sci.* **2022**, *646*, 120249. [CrossRef]
18. Sun, S.; Wen, Y.; Huang, K. A New Ceramic–Carbonate Dual-Phase Membrane for High-Flux CO_2 Capture. *ACS Sustain. Chem. Eng.* **2021**, *9*, 5454–5460. [CrossRef]

19. Xu, N.; Li, X.; Franks, M.A.; Zhao, H.; Huang, K. Silver-molten carbonate composite as a new high-flux membrane for electrochemical separation of CO_2 from flue gas. *J. Membr. Sci.* **2012**, *401–402*, 190–194. [CrossRef]
20. Zhang, L.; Gong, Y.; Yaggie, J.; Wang, S.; Romito, K.; Huang, K. Surface modified silver-carbonate mixed conducting membranes for high flux CO_2 separation with enhanced stability. *J. Membr. Sci.* **2014**, *453*, 36–41. [CrossRef]
21. Fang, J.; Tong, J.; Huang, K. A superior mixed electron and carbonate-ion conducting metal-carbonate composite membrane for advanced flue-gas carbon capture. *J. Membr. Sci.* **2016**, *505*, 225–230. [CrossRef]
22. Tong, J.; Si, F.; Zhang, L.; Fang, J.; Han, M.; Huang, K. Stabilizing electrochemical carbon capture membrane with Al_2O_3 thin-film overcoating synthesized by chemical vapor deposition. *Chem. Commun.* **2015**, *51*, 2936–2938. [CrossRef]
23. Zhang, P.; Tong, J.; Jee, Y.; Huang, K. Stabilizing a high-temperature electrochemical silver-carbonate CO_2 capture membrane by atomic layer deposition of a ZrO_2 overcoat. *Chem. Commun.* **2016**, *52*, 9817–9820. [CrossRef]
24. Mendoza-Serrato, C.G.; López-Juárez, R.; Reyes-Montero, A.; Serrano, J.A.; Fabián-Anguiano, C.G.-Y.J.A.; Ortiz-Landeros, J. Performance of membranes based on novel $Ce_{0.8}Sm_{0.2}O_{2-\delta}$/Ag cermet and molten carbonates for CO_2 and O_2 separation. *Chem. Eng. Sci.* **2022**, *255*, 117673. [CrossRef]
25. Fabián-Anguiano, J.; Ramírez-Moreno, M.; Balmori-Ramírez, H.; Romero-Serrano, J.; Romero-Ibarra, I.; Ma, X.; Ortiz-Landeros, J. Syngas production with CO_2 utilization through the oxidative reforming of methane in a new cermet-carbonate packed-bed membrane reactor. *J. Membr. Sci.* **2021**, *637*, 119607. [CrossRef]
26. Zhang, P.; Tong, J.; Huang, K. Dry-Oxy Methane Reforming with Mixed e-/CO_3^{2-} Conducting Membranes. *ACS Sustain. Chem. Eng.* **2017**, *5*, 5432–5439. [CrossRef]
27. Ovalle-Encinia, O.; Pfeiffer, H.; Fabián-Anguiano, J.A.; Ortiz-Landeros, J. Nanosized lithium aluminate (γ-$LiAlO_2$) synthesized by EDTA-citrate complexing method, using different thermal conditions. *J. Mex. Chem. Soc.* **2019**, *63*, 229–245. [CrossRef]
28. *ASTM C373-88;* Standard Test Method for Water Absorption, Bulk Density, Apparent Porosity, and Apparent Specific Gravity of Fired Whiteware Products. ASTM International: West Conshohocken, PA, USA, 2006. [CrossRef]
29. Bale, C.; Chartrand, P.; Degterov, S.; Eriksson, G.; Hack, K.; Ben Mahfoud, R.; Melançon, J.; Pelton, A.; Petersen, S. FactSage thermochemical software and databases. *Calphad* **2002**, *26*, 189–228. [CrossRef]
30. Scardi, P.; Leoni, M. Diffraction line profiles from polydisperse crystalline systems. *Acta Crystallogr. Sect. A Found. Crystallogr.* **2001**, *57*, 604–613. [CrossRef]
31. Yeyongchaiwat, J.; Matsumoto, H.; Ishihara, T. Oxidative reforming of propane with oxygen permeating membrane reactor using $Pr_2Ni_{0.75}Cu_{0.25}Ga_{0.05}O_4$ perovskite related mixed conductor. *Solid State Ionics* **2017**, *301*, 23–27. [CrossRef]
32. Gomez, E.; Kattel, S.; Yan, B.; Yao, S.; Liu, P.; Chen, J.G. Combining CO_2 reduction with propane oxidative dehydrogenation over bimetallic catalysts. *Nat. Commun.* **2018**, *9*, 1398. [CrossRef]
33. Markova, E.B.; Cherednichenko, A.G.; Smirnova, S.S.; Sheshko, T.F.; Kryuchkova, T.A. Features of the Catalytic Cracking of Propane with a Step-Wise Change $Pr_xYb_{2-x}Zr_2O_7$. *Catalysts* **2023**, *13*, 396. [CrossRef]
34. Azara, A.; Benyoussef, E.-H.; Mohellebi, F.; Chamoumi, M.; Gitzhofer, F.; Abatzoglou, N. Catalytic Dry Reforming and Cracking of Ethylene for Carbon Nanofilaments and Hydrogen Production Using a Catalyst Derived from a Mining Residue. *Catalysts* **2019**, *9*, 1069. [CrossRef]

membranes

MDPI

Article

Mixed-Matrix Organo-Silica–Hydrotalcite Membrane for CO_2 Separation Part 1: Synthesis and Analytical Description

Lucas Bünger *, Krassimir Garbev *, Angela Ullrich, Peter Stemmermann and Dieter Stapf

Institute for Technical Chemistry, Karlsruhe Institute of Technology, 76344 Karlsruhe, Germany; angela.ullrich@kit.edu (A.U.); peter.stemmermann@kit.edu (P.S.); dieter.stapf@kit.edu (D.S.)
* Correspondence: lucas.buenger@kit.edu (L.B.); krassimir.garbev@kit.edu (K.G.); Tel.: +49-721-6082-4405 (L.B.); +49-721-6082-6878 (K.G.)

Abstract: Hydrotalcite exhibits the capability to adsorb CO_2 at elevated temperatures. High surface area and favorable coating properties are essential to harness its potential for practical applications. Stable alcohol-based dispersions are needed for thin film applications of mixed membranes containing hydrotalcite. Currently, producing such dispersions without the need for delamination and dispersing agents is a challenging task. This work introduces, for the first time, a manufacturing approach to overcoming the drawbacks mentioned above. It includes a synthesis of hydrotalcite nanoparticles, followed by agent-free delamination of their layers and final dispersion into alcohol without dispersing agents. Further, the hydrotalcite-derived sorption agent is dispersed in a matrix based on organo-silica gels derived from 1,2-bis(triethoxysilyl)ethane (BTESE). The analytical results indicate that the interconnection between hydrotalcite and BTESE-derived gel occurs via forming a strong hydrogen bonding system between the interlayer species (OH groups, $CO_3{}^{2-}$) of hydrotalcite and oxygen and silanol active gel centers. These findings lay the foundation for applications involving incorporating hydrotalcite-like compounds into silica matrices, ultimately enabling the development of materials with exceptional mass transfer properties. In part 2 of this study, the gas separation performance of the organo-silica and the hydrotalcite-like materials and their combined form will be investigated.

Keywords: hydrotalcite; BTESE gel; delamination; CO_2 separation; mixed-matrix membrane

1. Introduction

Gas separation processes play an important part in reaching carbon-neutral production. Especially high-temperature applications that depend on suitable materials that can efficiently capture CO_2 above 200 °C receive more attention [1]. Fortunately, sorbent research has made tremendous efforts in the past to investigate the adsorption properties of inorganic materials such as Hydrotalcite-like compounds (HTlcs).

The mineral hydrotalcite, $[Mg_6Al_2(OH)_{16}](CO_3)(H_2O)_4$, was first described in 1842 by Hochstetter in serpentine–magnesite rocks in Snarum, Buskerud, Norway [2]. It belongs to the hydrotalcite supergroup of minerals, also known as layered double hydroxides (LDHs). More than 40 structurally and chemically related varieties are known so far [3]. All LDH crystal structures share the similarity of being built up of positively charged brucite-type layers. Their octahedral sites are mainly occupied by M^{2+} and M^{3+} cations, where M^{2+} are Mg, Fe, Mn, Ni, Cu, Ca, and Zn, whereas M^{3+} are Al, Fe, Mn, Co, and Cr. The octahedral layers alternate with negatively charged interlayers occupied by $(CO_3)^{2-}$, H_2O, Cl^-, etc. The crystal structures of hydrotalcite-like compounds and their polytypes have been the object of numerous studies: Allmann and Jepsen (1969) [4] and Bellotto et al. (1996) [5] for the variety quintinite ($M^{2+}/M^{3+} = 2$) and Zhitova et al. (2019) for the "true" hydrotalcite ($M^{2+}/M^{3+} = 3$) [6].

Synthetic materials on an LDH basis are in great demand due to their useful physicochemical properties [7]. Their ability to function as a high-temperature adsorption agent for

CO_2 has been studied extensively [8–11]. However, the bulk of these investigations focuses on HTlc materials in their layered and powdery forms rather than technical applications. To bridge this gap and realize their potential in real-world applications, a transformation of HTlc is essential. The preferred approach involves converting them into a delaminated state, which offers an enlarged surface area for more effective interactions [12]. Delaminated HTlcs have a broad range of applications, including their use as catalysts [13], enhanced drug delivery [14], biosensors, and adsorbents [15]. Their use as pure material membranes has not shown convincing results due to the resulting porosity that facilitates only Knudsen diffusion [16,17]. However, their utilization as adsorptive mass transfer agents in polymeric membranes has already demonstrated efficacy at lower temperatures [18].

For high-temperature membrane applications, a mixed-matrix approach seems to be promising. Attempts to incorporate hydrotalcite-derived substances in water-based γ-alumina matrices yielded mesoporous microstructures with unselective Knudsen diffusion [19]. Amorphous organo-silica has proven to be a suitable material as it forms a microporous microstructure that enables selective mass transport and withstands harsh process conditions [20–23]. These silica matrices are prepared from alcohol-based dispersions, which are highly sensitive to water. In the preparation of the mixed-matrix membrane, the hydrotalcite-derived adsorption agent must, therefore, be in a stable alcohol-based form, completely devoid of water content, while possessing similar drying characteristics to the matrix. Gels derived from 1,2-bis(triethoxysilyl)ethane (BTESE) are considered suitable matrices as their linking units could be used to control the membrane pore sizes. However, these membranes experience a loss of selectivity at higher temperatures due to a decrease in adsorptive affinity for CO_2 [24,25] and an increase in N_2 permeance caused by activated permeation [26–28]. An excellent review of the state of the scientific knowledge and technology regarding the application of organo-silica membranes for gas separation purposes is given by I. Agirre et al. [29]. The high-temperature adsorption agent HTlc combined with a microporous organo-silica xerogel matrix is envisioned to offset their respective weaknesses.

The use of delaminating and dispersing agents presents challenges as they tend to persist on the substance post-drying. While these agents can be eliminated during thermal treatment, they create porosity, thereby compromising the selectivity of the resulting membrane for gas separation applications.

Successful delamination processes using formamide as a solvent have shown promising results; however, there is an effort to find alternatives due to its toxicity. Recent research suggests positive outcomes with a combination of dodecyl sulfate and n-butanol as an alcoholic dispersant [30]. Additionally, experiments with sodium acetate have demonstrated stable HTlc dispersions in ethanol [31]. However, all these processes combine the disadvantage of either utilizing large amounts of dispersing agents, a solvent with different drying properties than ethanol, or the HTlc not being in its active CO_2 adsorptive form. The synthesis of the precursor for the CO_2 adsorbing compound, hydrotalcite, offers various routes described in the literature [32]. However, for the specific application of creating stable ethanol-based dispersions, the focus lies particularly on preparing particles of a nanometer size.

Therefore, this study shows an approach to creating an HTlc-derived adsorption agent in ethanol without the use of a dispersing agent. Firstly, the preparation method of the hydrotalcite is presented, then its delamination in water, and finally, the transformation of the calcined active form into the ready-for-use ethanol dispersion. Furthermore, the combination of the produced dispersion is investigated after mixing it with the aforementioned silica matrix based on BTESE gel derivates. All described steps are analyzed using various techniques to determine their respective composition and structural properties and how they vary during processing.

2. Materials and Methods

2.1. Preparation

2.1.1. Hydrotalcite

The synthesis of hydrotalcite is adapted from the work of Gardner et al. [33,34]. To control the particle size, water is changed to methanol as a solvent. In 100 mL methanol, 6.1 g of magnesium chloride hexahydrate and 2.41 g of aluminum chloride hexahydrate, with a molar ratio of Mg/Al = 3, are dissolved. This mixture is heated to 65 °C and stirred under reflux. To start the LDH-forming reaction, 3.8 g of sodium hydroxide is dissolved in 100 mL of methanol and added dropwise to the metal salt solution. For a small particle size distribution, the synthesis time from Gardner's work is reduced from 3 days to 1 h. The synthesized product is rigorously washed with methanol to dissolve the by-product NaCl and subsequently dried at 50 °C. The resulting sample is further referred to as HTlc-pure (1) throughout the manuscript.

After allowing the reaction mixture to reach room temperature, it is subjected to centrifugation, and the resulting pellet is dispersed in water. This procedure is repeated until the pH level reaches neutrality. The dispersed material is then left in water overnight to undergo delamination, resulting in a transparent dispersion. This specific form of the delaminated substance is subsequently dried and referred to as HTlc-delaminated (2a). The membrane coating process yields structures with oriented platelets, as seen in the SEM images in part two [1]. To investigate the influence of this orientation, oriented samples are produced by dropping single droplets of the transparent dispersion onto a sample carrier. This process is repeated several times, resulting in platelets that are oriented parallel to the carrier surface. Those samples are further referred to as HTlc-delaminated-oriented (2b).

The delaminated dispersion is subjected to thermal treatment at 400 °C for 3 h in air to activate the agent. Following this treatment, the resulting powder is placed in an evacuated desiccator until the subsequent analysis. This calcined form is labeled as HTlc-calcined (3).

2.1.2. Organo-Silica Matrix

For the synthesis of the matrix material, which enables CO_2 separation, the approach is based on the work of Van Gestel [22]. He based the synthesis on pioneering work conducted at the University of Twente in the group of Ten Elshof [20]. The synthesis method used in this work is a simplified version. It involves mixing 16.655 mL of the membrane-forming precursor (BTESE) with 28.14 mL of ethanol (both thermo scientific chemicals), 0.63 mL of nitric acid (65 wt%), and 4.57 mL of water. Respective amounts result in a water-to-hydrolyzable ethoxy group ratio of one. No ice bath is used due to the rapid mixing of the mixture. The resulting mixture is heated to 60 °C and refluxed for 90 min, followed by cooling to room temperature. To stabilize the sol for storage, an additional 50 mL of ethanol is added, and the mixture is stored in the refrigerator (stock sol). To produce the gels, the stock sol was dried at room temperature to produce the lyogel (BTESE-derived lyogel). Additional thermal treatment was carried out at 300 °C for 3 h under an N_2 atmosphere to produce the xerogel. This xerogel is further referred to as BTESE-derived xerogel and is also analyzed for comparison.

2.1.3. Mixed-Matrix

To mix the HTlc with the organo-silica, the HTlc must first be converted into an ethanol-based dispersion. Two grams of the calcined and ground Htlc (3) were mixed with 50 mL of ethanol and redispersed using an ultrasonic device for 20 min. Afterward, the dispersion was centrifuged for 20 min at 4000 rpm to remove any undispersed particles. This process was repeated until no undispersed particles were found and a stable dispersion was formed. Then, 5 mL of this sol was mixed with 5 mL of the organo-silica stock-sol and filled up with 40 mL of ethanol. Subsequently, this sol was dried into a powder for analysis. This form is denoted as HTlc-modified organo-silica (4) and the final form for the high-temperature membrane application.

The membrane-forming process and resulting SEM images of the prepared membranes and schematic drawings of the microstructure are given in part 2 of this publication [1].

2.2. Analytical Methods

Structural information is obtained by a combination of infrared (IR) and Raman spectroscopy and X-ray diffraction (XRD). Additionally, its thermal behavior is investigated by thermogravimetric (TGA) experiments.

2.2.1. X-ray Diffraction (XRD)

X-ray diffraction measurements were conducted on an Empyrean diffractometer (Malvern-PANalytical, Almelo, The Netherlands) with a Cu-Kα tube operated at 40 mA and 40 kV equipped with a multistrip PIXcel3D detector (255 channels, simultaneously covering 3.347° 2θ) and a Bragg–Brentano HD device. The powder patterns were measured in the range of 5 to 90° 2θ with step size 0.013° 2θ, time per step 1s, slits of 0.125°, soller slits of 2.3°, and sample rotation. The powder patterns were evaluated with HighScorePlus v.5.1 software. The phases were identified using the ICDD (2004) and COD (2021) databases.

Rietveld refinements were performed with TOPAS v.7 for the estimation of the unit cell parameters and the size of coherent scattering domains of hydrotalcite. The structure data of Zhitova et al. [5], ICSD Nr.-133744 and ICSD Nr.-133742 have been used as a starting model.

2.2.2. IR Analysis

IR measurements in Attenuated Total Reflection (ATR) arrangement were performed on powder samples in a Golden Gate ATR cell with a diamond crystal (Specac LTD, Orpington, UK) positioned in a Tensor II spectrometer (Bruker Optics, Ettlingen, Germany) equipped with a deuterated triglycine sulfate (DTGS) detector. Spectra with 64 scans and a spectral resolution of 2 cm^{-1} in the range of 400–4000 cm^{-1} were acquired.

2.2.3. Raman Spectroscopy (RS)

Raman spectroscopic analyses were performed on the powdered samples dusted on glass slides. A WITec alpha300 R equipped with a UHTS300 spectrometer (300 mm focal length) and a Zeiss confocal microscope were employed for single-spot measurements. As an excitation source, a 533 nm laser operated at 50 mW (on sample) was used. The measurements were performed with a 100× objective with a numerical aperture of 0.9 using 600 and 1800 groves/mm holographic gratings with spectral resolution better than 3 cm^{-1} and 1 cm^{-1}, respectively. A high-performance back-illuminated CCD camera with 96% quantum efficiency was used for detection. Typical acquisition times were between 5 and 20 s with 5–10 scans. The phase identification was aided by the RRUFF mineral database and an own database integrated into the WITec True Match program. The data processing and preparation of the Raman images were performed with the Project 5.3+ software from WITec (Ulm, Germany). The single spectra were corrected by the cosmic ray removal algorithm, and the consequent subtraction of the background was calculated using the shape algorithm.

2.2.4. Thermal Analysis (TA)

The thermal analyses (TA) were performed on an STA 403 Jupiter F3 (Netzsch, Selb, Germany) equipped with a Pt oven, coupled to a Tensor 27 Fourier transform infrared (FTIR) spectrometer (Bruker-Optics, Ettlingen, Germany). The TA device was purged with N_2 (2 × 60 mL/min) and was connected via a transfer gas line to an FTIR gas cell (both kept at 200 °C to prevent condensation), where the detection of the gases released upon heating was performed with a liquid nitrogen cooled MCT (mercury cadmium telluride) detector. The samples were heated at a rate of 10 K/min, and IR spectra of released gases were acquired with 32 scans and spectral resolution of 4 cm^{-1} in the range of 600–4000 cm^{-1}. The calibration of the FTIR spectrometer was carried out with certified test gases in the validity range of Beer–Lambert's law. The quantification of H_2O and CO_2 was performed using the method developed by Merz et al. [35].

3. Results and Discussion

3.1. XRD Results

Figure 1 shows the X-ray powder patterns of samples 1, 2a, 2b, 3, BTESE, and 4. Table 1 shows refined unit cell parameters considering the 3R and 2H polytypes and calculated sizes of the coherent scattering domains (results from the double Voigt approach) for hydrotalcite in the corresponding samples.

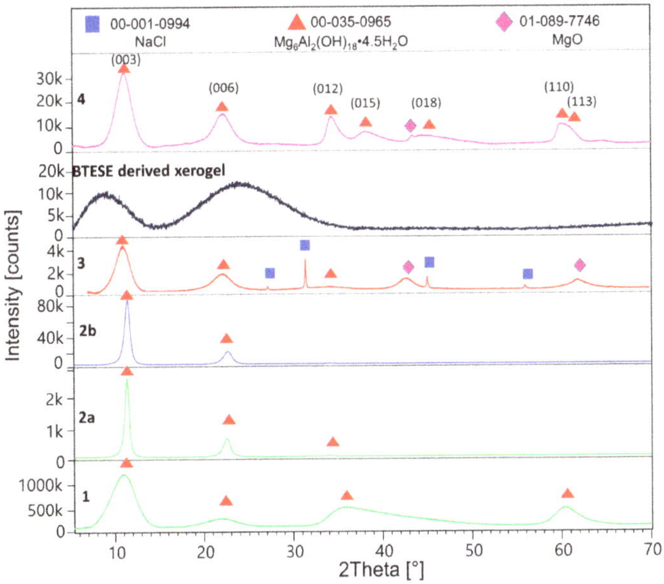

Figure 1. Powder XRD patterns of the methanol-washed HTlc-pure sample (1), HTlc-delaminated hydrotalcite (2a), HTlc-oriented sample (2b), HTlc-calcined sample (3), BTESE-derived xerogel, and HTlc-modified organo-silica (4).

Table 1. Unit cell parameters and size of the coherent scattering domains of hydrotalcite in samples 1–4 refined for 3R (space group R-3m H) and 2H ($P6_3$/mmc) polytypes.

XRD Results	Samples				
u.c.p.	1	2a	2b	3	4
a (Å)	3.07(1)	3.066(1)	3.14(1)	3.10(5)	3.064(1)
c (Å) 3R	25.25(1)	23.683(2)	23.557(2)	24.22(1)	24.25(1)
c (Å) 2H	16.833(7)	15.788(1)	15.705(2)	16.19(1)	16.181(7)
crystal size (nm)	1.6(1)	10.6(2)	8.42(7)	3.2(1)	3.4(1)

After being washed with methanol, the synthesized hydrotalcite HTlc-pure shows very broad reflections pointing to a very small size of the coherent scattering domains of about 1.6 nm (further referred to as crystal size). This confirms the findings of Gardner et al. [32] pointing to a tiny crystal size of hydrotalcite precipitated from methanol solutions of Mg and Al salts. A crystal size of 1.6 nm can also suggest a thickness of one unit cell parameter in the z direction of the 2H polytype consisting of two brucite layers (Table 1). The delaminated and oriented samples (2a and 2b), on the other hand, show very sharp reflections, pointing to crystal sizes of 10.6 and 8.4 nm, respectively. The significantly higher values for samples 2a and 2b are expected, given the complete hydrolysis of the methoxide ions and the stacking of the hydrotalcite platelets along the [001] direction after drying. The term "delamination" refers to the fact that upon drying, although almost perfect stacking along the z direction of hydrotalcite takes place, a strong disorder in the xy plane is observed. Therefore, only 00l reflections are seen in the patterns of samples 2a and 2b. A tendency for decreasing the size of the *c*-unit cell parameter from 25.2 to 23.68 to 23.56 Å is also observed for samples 1, 2a, and 2b, respectively. This is consistent with the expected exchange of methoxide with H_2O. The largest c parameter of sample 1 proves the intercalation of the alkoxide anions between the brucite layers [32]. Samples 2a and 2b show the typical unit cell parameters for H_2O exchanged hydrotalcites (basal spacing of 003 = 7.9Å). The calcined sample (3) once again shows broad reflections of hydrotalcite due to the expected disorder accompanied by additional broad peaks at around 43 and 62 °2Theta, which could be assigned to MgO (ICDD: 01-089-7746). The *c*-unit cell parameter shows a small increase to 24.22 Å. The crystal size decreases to 3.2 nm. Therefore, the calcination at 400 °C leads to partly dehydrated, disordered "hydrotalcite" with similar crystal features as the starting material with disrupted long-range order. In addition, the sample shows some tiny impurities of NaCl, which is a by-product of the synthesis of hydrotalcite. BTESE organo-silica gel shows two broad "humps" at about 8 and 24 °2Theta. The HTlc-modified organo-silica membrane sample (4) shows similar features as sample 3 in its powder pattern but is accompanied by additional 0 kl and 11 l reflections, thus witnessing a random orientation of the "hydrotalcite" platelets which seem to be homogenously distributed in the gel matrix. Considering the crystal size of samples 3 and 4, they consist of four brucite layers.

3.2. IR Spectroscopy

The IR spectra are shown in Figure 2, and relevant bands are summarized in Table 2. Spectrum 1 and 2a are hydrotalcite synthesized and washed with methanol and the delaminated sample, respectively. The assignment is made based on the results of Kloprogge et al. [36] and Frost et al. [37], as well as [38–40]. Sample 1 shows broader bands than 2a and 2b, which is consistent with the XRD results. The main band situated at 1365 cm^{-1} is due to the asymmetrical ν_3 (CO_3) mode. The broad high-frequency slope is due to the absorption of ν_2 CH_3 bending modes, which are expected in the range of 1440–1480 cm^{-1} [41], and the possible interaction of the CO_3 groups of methanol [42]. Additionally, it may witness a large $\Delta\nu_3$ CO_3 splitting typical for adsorbed CO_2 as monodentate or bidentate carbonate [38]. The corresponding ν_2 (CO_3) and ν_1 (CO_3) modes are seen at about 870 cm^{-1} and 1075 cm^{-1}, respectively. The relatively high intensity of the latter band in the spectrum of sample 1 indicates that this carbonate is due to monodetate and bidentate adsorption of CO_2 [38]. It possibly shows contributions of C-O stretching from $CO2CH_3OH$ (in the range of 1020–1060 cm^{-1}), which is known to be sensitive to interactions with hydrogen bonding [39].

The broad band at 1645 cm^{-1} belongs to the bending vibration of H_2O. The broad absorption is possibly due to Mg(Al)-OH interactions. The high-frequency region, 2800–3800 cm^{-1}, is marked by a broad band due to ν_1 OH (centered at 3425 cm^{-1}) and a very broad low-frequency slope due to ν_1 CH_3 and its interaction with OH. Thus, the presence of methoxide anions is once again confirmed.

Samples 2a and 2b show quite similar spectra with bands at 775 cm^{-1} (translation Al-OH) and 860 cm^{-1} (ν_2 CO_3). The strongest band is once again the ν_3 CO_3 (1365 cm^{-1}), lacking the contribution of CH_3 bending modes. It may also indicate smaller $\Delta\nu_3$ CO_3 splitting due to the presence of polydentate CO_3 tightly bonded in the hydrotalcite structure. The band at 1075 cm^{-1} (ν_1 CO_3) also lacks the contribution of CO stretching of methanol. The delamination with H_2O also leads to well-pronounced H_2O bands at 1635 cm^{-1} (bending) and 3410 cm^{-1} (ν_1 OH in H_2O) with a shoulder at 3610 cm^{-1} (ν_1 OH). The presence of the OH bands at 775 cm^{-1} highlights the difference between sample 1 on one side and samples 2a and 2b on the other side.

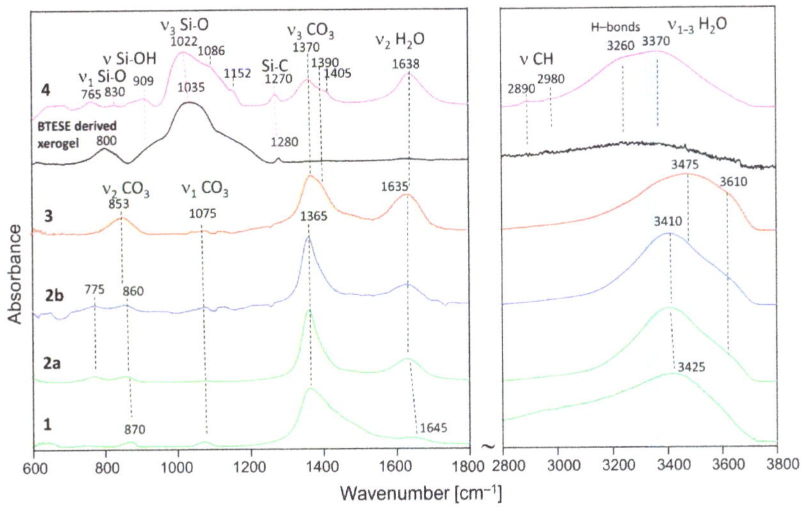

Figure 2. ATR-IR spectra of methanol-washed hydrotalcite (HTlc-pure: 1), HTlc-delaminated (2a), HTlc-delaminated-oriented (2b), HTlc-calcined hydrotalcite (3), BTESE-derived xerogel, and HTlc-modified organo-silica (4).

Table 2. Band assignment of the IR spectra in Figure 2.

Hydrotalcite				BTESE, HTlc-Modified Organo-Silica Membrane		
Sample	**1**	**2a,b**	**3**	**Sample**	**BTESE**	**4**
Band Assignment	**Band Positions (cm^{-1})**			**Band Assignment**	**Band Positions (cm^{-1})**	
ν Mg(Al)-O	630			ν Mg(Al)-O		650–700
OH libration		775		OH libration		765
ν_2 (CO$_3$)	870	860	853	ν_1 Si-O	800	830
ν_1 (CO$_3$)	1075	1075	1075	Si–OH	950sh	909
ν_3 (CO$_3$)	1365	1365	1370	ν_3 Si-O	1035s	1022
M,Bν_3 (CO$_3$), ν_2 CH$_3$	1400–1500		1390	ν_1 (CO$_3$)		1086
δ H$_2$O	1645	1635	1635	ν_3 Si-O	1150sh	1152
H-bonding	3000–3400	3100–3400	3200–3400	ν Si-C	1280	1270
					ν_3 (CO$_3$)	
					1370	
ν_1 OH(H$_2$O)	3425	3420	3475		M,Bν_3 (CO$_3$)	
					1405	
				δ H$_2$O		1638
				ν-CH$_2$		2890
				ν-CH$_3$		2980
				H-bonding		3260
				ν_1 OH(H$_2$O		3370

M,Bν_3 (CO$_3$)—Monodentate, Bidentate CO$_3$, sh—shoulder, s—strong band.

H_2O interacting with interlayer carbonate in the hydrotalcite structure gives rise to increased absorption between 3000 and 3400 cm^{-1}.

The HTlc-calcined sample (3) shows, once again, very broad bands due to increased disorder. The ν_2 bending band of CO$_3$ shows a shift to a lower frequency (853 cm^{-1}), whereas the asymmetrical stretching band ν_3 (CO$_3$) shifts to a higher frequency [1370 cm^{-1}]. The broader ν_3 CO$_3$ band (with a shoulder at 1390 cm^{-1}) may also indicate the presence of different environments of carbonate having on one side polydentate character but also monodentate and bidentate surface carbonate on the other side. Similar behavior is observed for the OH stretching bands (shift to 3475 cm^{-1}), pointing to the presence of more slightly bonded hydroxyls as a result of rehydration.

The IR spectrum of the organo-silica xerogel BTESE is characterized by strong and broad multiple bands in the range of 950–1200 cm^{-1} with a maximum intensity located at about 1035 cm^{-1}. These bands could be attributed to asymmetric Si-O stretching modes (ν_3 Si-O) [26] in ring-like siloxane moieties formed via the condensation reaction of silanol groups. Compared to pure SiO_2 gels (TEOS, Aerosil), which show frequencies centered at about 1070–1100 cm^{-1} [43], the lower frequency in BTESE could be explained by lower silicate polymerization induced by the presence of organic moieties in the silica network [44,45]. Additionally, it may be due to the lower rate of hydrolysis, as the synthesis was performed with low H_2O content to keep the particle size as low as possible. The band centered around 800 cm^{-1} is characteristic of symmetric Si-O stretching (ν_1 Si-O). A broad absorption seen as a shoulder at about 910 cm^{-1} is ascribed to the stretching vibration of silanol bonds (Si–OH). Its intensity is rather low, which is expected in fired xerogels where enhanced condensation has taken place. The band at 1280 cm^{-1} is characteristic of Si-C stretching vibrations [46]. The range of 2800–3800 cm^{-1} is dominated by broad bands due to ν_1 and ν_3 O-H symmetric and asymmetric stretching of H_2O and OH groups. Due to the firing at 300 °C, a very low intensity of H_2O and OH-related bands is observed. In addition, the low-intensity bands at 2890 and 2980 cm^{-1}, typical for C-H stretching vibrations, prove the presence of CH_2 and CH_3 in the Si–$(CH_2)_2$–Si and (–OCH_2CH_3) groups, respectively. The low intensity of the C-H bands marks a high degree of hydrolysis and consequent condensation of silanol groups to siloxanes.

The spectrum of the HTlc-modified organo-silica membrane sample (4) retains similar features characteristic for both BTESE and HTlc-calcined hydrotalcite (3) but with some frequency shifts of the original bands. For example, the organo-silica rings seem to be strongly influenced by the presence of hydrotalcite fragments showing shifts of ν_1 Si-O from 1035 to 1022 cm^{-1} and of Si-C from 1280 to 1270 cm^{-1}. The band at 800 cm^{-1} in BTESE is also shifted, pointing to a change in the Si-O-Si bridging angle. The spectrum of sample 4 shows more pronounced silanol groups (909 cm^{-1}). The formation of silanol groups correlates with indications for a strong hydrogen-bonded system, which is manifested by a low-frequency shift of the stretching vibrations of H_2O and OH in the spectrum of sample 4 compared to the rest of the samples. The silanol band frequency is higher than that reported by Moriyama et al. for pure BTESE gels (890 cm^{-1}) [26]. These observations imply that the interaction between hydrotalcite and BTESE gel most probably takes place via enhanced hydrogen bonding between OH and H_2O of hydrotalcite and Si-O-Si, SiOH, and Si-CH_2(3) moieties of BTESE gel. The CO_3 modes clearly show the presence of two bands at 1370 and 1405 cm^{-1} with $\Delta\nu_3$ splittings once again higher (35 cm^{-1}) than that observed for the calcined sample (20 cm^{-1}). It possibly shows an interaction between the surface CO_3 groups of hydrotalcite and the xerogel.

3.3. Raman Spectroscopy

The Raman spectra of the 1, 2a, and 2b samples are presented in Figure 3. All samples show bands at 549 cm^{-1} and 457 cm^{-1}, characteristic for hydrotalcite, which could be assigned to stretching and bending vibrations (*M*–O, *M*–O–*M*, and O–*M*–O) within the octahedral layers [36]. Most probably, the band at a lower frequency is due to Al-O vibrations, whereas the band at 549 cm^{-1} could be assigned to Mg-O due to the difference in the atomic weight and corresponding quantities. These bands are sharper in samples 2a and 2b. Very pronounced is the band at 1078 cm^{-1} (sample 1) assigned to the symmetrical stretching vibration of CO_3 (ν_1 CO_3). The corresponding band in samples 1b and 2 is much sharper and shows a frequency shift to 1062 cm^{-1}. Additionally, the spectrum of sample 1 shows the presence of bands at 1467 cm^{-1} and 2820 cm^{-1}, 2840 cm^{-1}, and 2950 cm^{-1} belonging to bending and ν_1 and ν_3 stretching of CH_3 groups in methoxide anions [47,48], respectively. Therefore, it is evident that methoxide anions (OR) are grafted to the brucite layers, partially exchanging OH groups after the formula $[M^{2+}_{(1-x)}\, M^{3+}_{x}\, (OH)_{2-y}]\, (OR)_y]$ $A^{n-}_{(x/n)}\, (OR)_y\, .mH_2O$ [32], where A^{n-} represents additional anions like CO_3^{2-}, which are inevitably present due to the contact with CO_2 from the air.

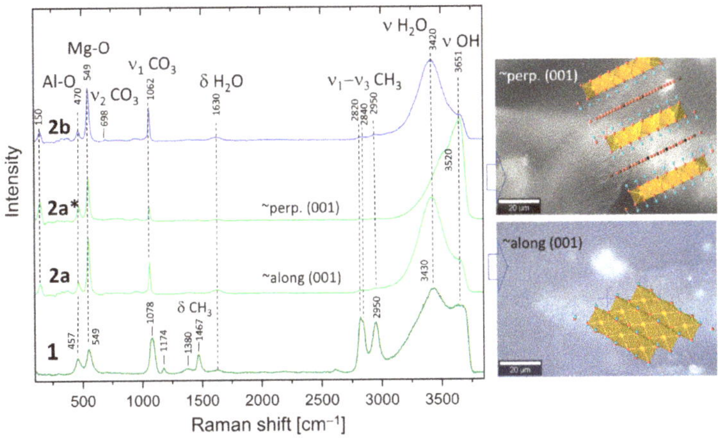

Figure 3. Raman spectra of synthesized and washed HTlc-pure sample (1), HTlc-delaminated (2a, 2a*), and HTlc-oriented hydrotalcite (2b). Spectrum 2a of the delaminated sample is taken almost along the [001] direction, thus enhancing the H_2O bending (1630 cm^{-1}) and stretching bands (3420 cm^{-1}), whereas the spectrum 2a* is measured along the octahedral layers. As a result, the intensity of the OH bands (3651 cm^{-1}) is enhanced. The optical images overlaid with the corresponding approximate orientations of the hydrotalcite structure are shown on the right-hand side. Crosses mark the exact positions of the Raman measurements. Completely oriented sample (2b) shows a similar spectrum to 2a (Red dots for oxygen atoms and blue dots for hydrogen atoms).

Low-intensity bands at 1174 and 1380 cm^{-1} are also due to methanol. Broad bands belonging to bending and stretching modes of H_2O and OH groups are present at 1630 cm^{-1} (δ H_2O), 3430 cm^{-1} (ν_1/ν_3 H_2O), and 3650 cm^{-1} (ν_3 OH). Spectra marked as 2a and 2a* are taken in two orientations of the delaminated sample. The former is taken almost along the [001] direction, leading to an enhanced H_2O stretching band (3420 cm^{-1}) compared to that taken along the octahedral layers (perpendicular to [001]). Spectrum 2a* shows enhanced intensity of the OH bands (3651 cm^{-1}) due to the larger vector components of the OH bonds seen in this direction. The right-hand side of Figure 3 shows optical images overlaid with the corresponding approximate orientations of the hydrotalcite structure. The band at 150 cm^{-1} also shows a higher intensity in the spectrum of sample 2a*, which implies that it may also be due to an OH mode. The opposite trends are observed for both Mg-O modes (549 cm^{-1}), CO_3 bands (ν_1-CO_3), and δ-H_2O, which is consistent with the proposed orientation of the crystals. The spectrum of the oriented sample (2b) is strikingly similar to that of 2a, which confirms these considerations once again. There are remarkable differences in the frequencies of Al-O and CO_3 bands in the methanol-washed sample (1) and samples 2a and 2b. Whereas the former shifts to higher frequency (470 cm^{-1}), the latter shifts to 1062 cm^{-1}. Therefore, methanol strongly influences the carbonate groups in the interlayer and possibly selectively the Al from the octahedral layers.

Figure 4 shows a comparison of the Raman spectra of HTlc-calcined hydrotalcite (3), BTESE, and mixed HTlc-modified organo-silica membrane (4). Sample 3 shows broader bands than the delaminated and oriented samples. The frequencies of the octahedral Mg-O vibrations change to a minimally higher frequency (551 cm^{-1}), whereas those typical for Al-O shift to the lower (450 cm^{-1}). The most intense band of MgO is expected also in this region [49]. Additionally, a broad band occurs centered at about 670 cm^{-1}, possibly due to the partial formation of disordered spinel, which needs to be avoided as it has no relevant CO_2 adsorption capacity [50].

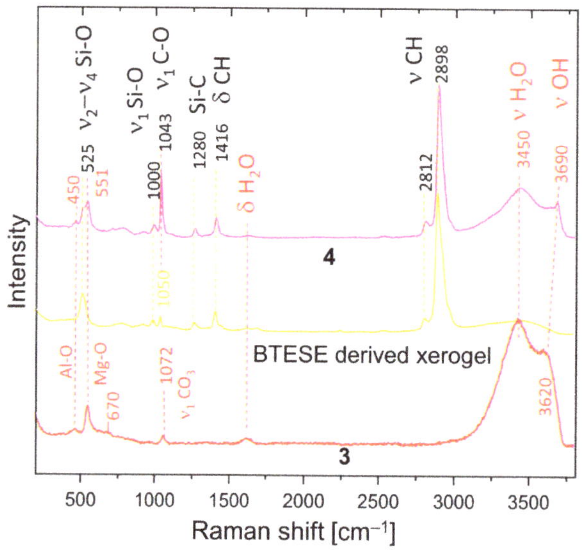

Figure 4. Raman spectra of HTlc-calcined sample (3), BTESE xerogel, and HTlc-modified organo-silica (4).

Both are expected products from the breakdown of hydrotalcite [51]. The symmetrical stretching of CO_3 is observed at 1072 cm^{-1} instead of 1078 cm^{-1}, pointing to a change in the environment of the carbonate anion. In addition, the intensity of this band is considerably lower than in the spectra of not calcined samples (Figure 3). The difference in the vibrational frequencies of the stretching bands of H_2O and OH in the spectrum of the calcinated sample (3450 cm^{-1}, 3620 cm^{-1}) compared to the spectra of samples 1b and 2 (3420 cm^{-1}, 3650 cm^{-1}) also witness a change in the interlayer environment. A partial rehydration of the sample from air moisture seems possible.

BTESE shows typical CH stretching bands in the range of 2800–3000 cm^{-1}. They consist of a very intense band at 2898 cm^{-1} with high-frequency shoulders at 2930 and 2980 cm^{-1} and a small band at a lower frequency (2812 cm^{-1}). A band at 1416 cm^{-1} manifests the corresponding bending vibrations. A question arises about the difference in the intensity of the C-H stretching bands in the IR and Raman spectra. A possible explanation is that the ethoxide groups in the xerogel are oriented to the centers of the siloxane rings. Therefore, a relatively low number of such groups is observed at the surface, thus becoming "invisible" for ATR measurements. Si-C stretching is observed at 1280 cm^{-1}. In addition, bands typical for breathing vibrations of siloxane rings appear at 525 cm^{-1}. They are at a higher frequency than the four-membered siloxane rings of the gels derived from TEOS (490 cm^{-1}) [52]. Broad bands at about 790 cm^{-1} and 925 cm^{-1} are attributed to Si-O-Si and Si-OH vibrations, respectively. At 1000 cm^{-1}, (ν_1 Si-O) are present. The assignment of the band at 1050 cm^{-1} is somewhat ambiguous. It is also present in non-fired BTESE gels but with lower intensity (not present). Therefore, it could be due to the C-O stretching mode of ethoxy groups H_3C-O-R on one side or to Si-O stretching modes in environments different than siloxanes (Si-O-Si). These might be the (C)Si-O environments typical for the BTESE rings.

The Raman spectrum of the mixture of BTESE and calcined hydrotalcite (4) shows complementary features characteristic for both constituents and, in addition, strong indications of interactions between the two. The band at 1043 cm^{-1} could be due to C-O modes of methanol used as a dispersion medium or ethanol remnants from the synthesis of the

BTESE, but an influence of (C)Si-O, as discussed above, is also possible. The intensity of this band is much higher in spectrum 4 than in spectra 3 and BTESE, respectively.

Therefore, the more probable explanation is that this band is due to adsorbed CO_2 as a carbonate, which may be shared between BTESE and calcined hydrotalcite. The small shoulder on the high-frequency side of this band shows the presence of additional environments of CO_3 in the sample, also witnessed by the $\Delta\nu_3$ splitting in the IR spectra. A new feature is the pronounced OH band shifted to 3690 cm^{-1}. The higher frequency indicates a lower dependence on the environment, which is characteristic of unassociated OH groups on the surface of the mixture. The unchanged frequencies of the Mg-O and Al-O bands give a hint that the interaction between hydrotalcite and BTESE gel takes place over OH and CO_3 species characteristic for the interlayer and not over bridging oxygens from the main octahedral layers. Whatever assignment of the band at 1043 cm^{-1} is correct (ethoxy groups or CO_3 groups), the observed frequency shift and intensity might be an indication for cross-linking between hydrotalcite and BTESE gel.

3.4. *Thermal Analysis*

The TGA curves are presented in Figure 5A. Figure 5B,C shows the traces of CO_2 and H_2O releases, respectively, as a function of temperature. A comparison between A, B, and C allows the assignment of the observed weight loss. Sample 1 shows an early inflection point (at 70 °C) of the weight loss between 30 and 200 °C probably due to two factors: (1) a large quantity of adsorbed H_2O because of the very high specific area and (2) decomposition of the methoxide anions. Compared to the H_2O exchanged and delaminated sample (2a), the weight loss in the range of 30–250 °C of sample 1 takes place not only earlier but is also higher: 14.7% compared to 12.2%. The latter correlates with the ideal stoichiometry of hydrotalcite, which considers it a loss of absorbed and interlayer H_2O. Between 300 and 500 °C, a strong mass loss takes place attributed to almost simultaneous dehydroxylation of the brucite layers and decarbonation of hydrotalcite [53,54]. A comparison of Figure 5B,C for sample 2a allows the interpretation that the H_2O could be separated into two events, one of which precedes the CO_2 release marginally.

The observed weight loss between 300 and 500 °C for samples 1 and 2a (27.8% and 26.5%) is way too low compared to the expected theoretical value for full decarbonation and dehydroxylation of hydrotalcite with formula ($Mg_6Al_2(OH)_{16}(CO_3).4H_2O$): CO_2 + $8H_2O$ = 7.3% + 23.9% = 31.2%, based on an M = 604 g/mol. On the other hand, sample 1 is composed of methoxide instead of pure H_2O in the interlayer, and the detected carbonate should only be due to a surface reaction with the CO_2 from the air, as shown by the large $\Delta\nu_3$ shift in the IR spectrum. In addition, this carbonate persists at higher temperatures, as observed in [38]. Therefore, the full decomposition of hydrotalcite is reached only at higher temperatures. The small step observed above 900 °C evidences an additional weight loss, especially considering the TG curves of samples 2a and 3. The total mass loss of samples 1 and 2 is 47.1% and 45.3%, respectively, the latter being very close to the theoretical value of 43.1%.

The higher loss of sample 1 is expected, given the large presence of methoxide groups. Calculation of the molar mass considering full substitution of OH by H_3CO^{1-}, the presence of CO_3^{2-} and $4H_2O$ (M = 693 g/mol) would result in a total mass loss of 50.4 %, which seems to be exaggerated but explains the higher loss of sample 1. Other authors also reported similar observations [49]. A small contribution to the weight loss above 900 °C could be the melting of residual NaCl, but its content is too low to account for the whole effect. It cannot be ruled out that some Cl is bonded in the hydrotalcite interlayer [55] and references therein].

Sample 3 (HTcl-calcined) shows a total mass loss of 17.2%. It is subdivided into three events taking place between room temperature and 200 °C (6%), 200–400 °C (4.9%), 400–700 °C (2.6%), and 700–1000 °C (3.7%). This shows once again that, on the one hand, the complete decomposition of hydrotalcite takes place at higher temperatures, and on the other hand, particular rehydration and recarbonation take place after calcination.

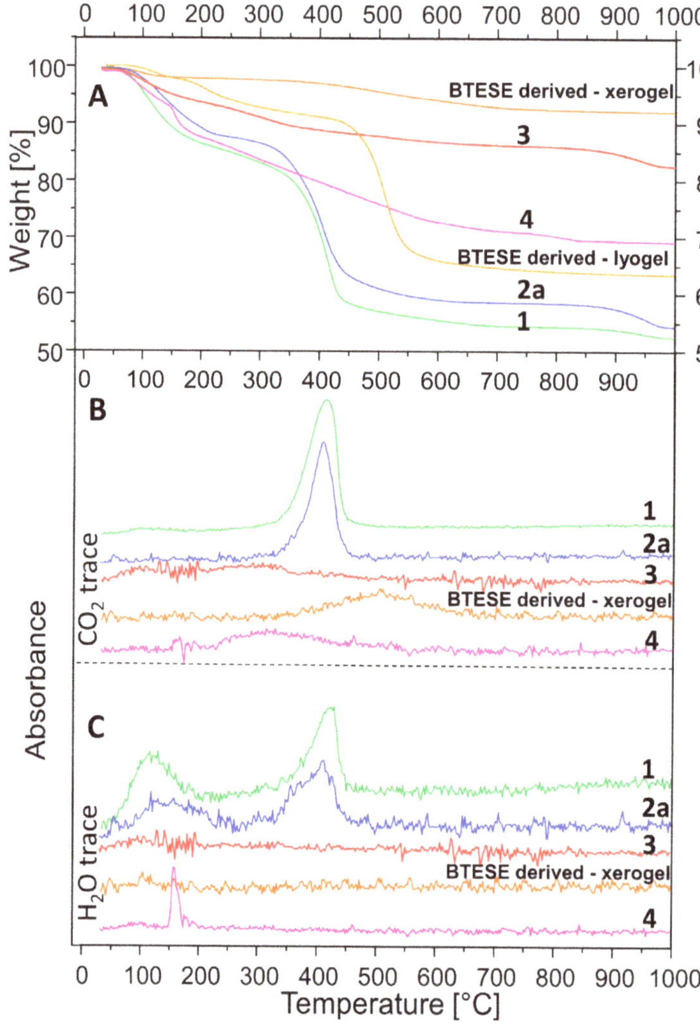

Figure 5. TGA coupled with simultaneous FTIR analysis. Thermogravimetric analysis (**A**); CO_2 release (**B**); H_2O release (**C**). Samples: HTlc-pure (1); HTlc-delaminated (2a); HTlc-calcined (3), BTESE-derived xerogel; BTESE-derived lyogel, HTlc-modified organo-silica (4).

The thermal behavior of the BTESE-derived xero- and lyogel show a minor weight loss due to adsorbed water below 150 °C. The lyogel, which has not been exposed to temperatures above 50 °C, shows condensation reactions above 200 °C, which results in further water loss. The fired xerogel shows no further condensation reactions. At temperatures above 450 °C, the lyogel undergoes degradation and combustion of ethoxide groups and ethylen bridges with C_2H_4, C_2H_6, and C_2H_5OH as products [56]. The total weight loss for the lyogel accumulates to 37%.

The continuous weight loss for the BTESE-derived xerogel between 300 and 700 °C could be attributed to non-reacted degradation and combustion of ethoxide groups ($-OCH_2CH_3$) and ethylene bridges ($-CH_2-CH_2-$). In this range, a release of CO_2 was detected by the IR spectroscopy (Figure 5B). The total weight loss adds up to 7.6%.

Sample 4, as a combination of (3) and BTESE-derived lyogel, shows a total weight loss of 31%. Up to 150 °C, a 5.3% loss is due to the release of adsorbed H_2O and possibly ethanol. Between 150 and 200 °C, a sharp weight loss of 6.9% is observed, which corresponds to the water loss from the condensation of the Si–OH groups upon the formation of Si–O–Si bonds. In the range of 200–650 °C, a gradual weight loss takes place, which is mostly due to the release of CO_2, as seen in Figure 5B. It seems to be a combination of CO_2 release from hydrotalcite remnants and combustion of ethoxides, as described above. The sharp loss between 150 and 200 °C is not observable in the TG curves of the "pure" constituents BTESE gel and calcined hydrotalcite. This shows unequivocally that both are not only chemically connected in the mixture but also that this happens via hydrogen bonding between silanol groups and active oxygen centers of the brucite layers on one side and OH groups of the brucite layers and oxygens from the silicate network on the other side following the reaction MgO-H + Si-OH → MgOSi + H_2O. Given the basicity of hydrotalcite on the one hand and the variability of the degree of hydrolysis of the BTESE gel in dependence on pH [26] on the other hand, a variable degree of cross-linking both within BTESE siloxane rings and between siloxane rings and brucite layers could be achieved for the design of membranes with different properties.

4. Conclusions

This study presents a novel approach for the preparation of mixed-matrix membranes for gas separation consisting of hydrotalcite dispersed in a matrix of an organo-silica gel derived from 1,2-bis(triethoxysilyl)ethane (BTESE). The procedure includes synthesis and subsequent agent-free delamination of hydrotalcite nanoparticles. This is followed by the dispersion of the delaminated and calcined hydrotalcite in the organo-silica gel matrix. The materials were systematically examined after each preparation step by XRD, Raman, IR spectroscopy, and TG analysis. The results show substantial evidence for the successful preparation, delamination, dispersion, and calcination processes. XRD provides valuable information about the layer thickness, unit cell parameters, and size of coherent scattering domains of hydrotalcite after synthesis and every treatment step. The synthesis in methanol produces nano-sized hydrotalcite with an average of two brucite layer thicknesses. The consequent hydrolysis and delamination are proven by the lowering of the c-unit cell parameter and the formation of stacking faults along the z direction. The results from IR and Raman spectroscopy reveal the nature of the bonding mechanisms between hydrotalcite and BTESE-derived gel. These include forming a strong hydrogen bonding system between the interlayer species (OH groups, CO_3^{2-}) of hydrotalcite and siloxane oxygens on one hand and silanol active gel centers and bridging oxygens from brucite layers on the other. The interaction between ethoxide groups and hydrotalcite can also not be excluded, but further investigations are necessary to elucidate their role. These findings may give an impetus for further development of mixed-membranes containing inorganic sorption or catalytic agents combined with organo-silica networks. Part 2 of this work will present a study of all the investigated materials for their permeation and separation behavior.

Author Contributions: Conceptualization, L.B., K.G., and P.S.; methodology, L.B., K.G., P.S., and D.S.; formal analysis, K.G., A.U., and L.B.; investigation, L.B., K.G., A.U., and P.S.; writing—original draft preparation, K.G. and L.B.; writing—review and editing, K.G., P.S., A.U., and D.S.; supervision, D.S. All authors have read and agreed to the published version of the manuscript.

Funding: Program-oriented funding for this research was provided through the German Helmholtz Association in the frame of the "Materials and Technologies for the Energy Transition" program.

Institutional Review Board Statement: Not applicable.

Data Availability Statement: The data that support the findings of this study are available from the corresponding author upon request.

Conflicts of Interest: The authors declare no conflicts of interest.

References

1. Bünger, L.; Kurtz, T.; Garbev, K.; Stemmermann, P.; Stapf, D. Mixed-Matrix Organo–Silica–Hydrotalcite Membrane for CO_2 Separation Part 2: Permeation and Selectivity Study. *Membranes* **2024**, *14*, 156. [CrossRef] [PubMed]
2. Hochstetter, C. Untersuchung über die Zusammensetzung einiger Mineralien. *J. Prakt. Chem.* **1842**, *27*, 375–378. [CrossRef]
3. Mills, S.J.; Christy, A.G.; Schmitt, R.T. The creation of neotypes for hydrotalcite. *Mineral. Mag.* **2016**, *80*, 1023–1029. [CrossRef]
4. Allmann, R.; Jepsen, H.P. Die Struktur des Hydrotalkits. *Neues Jb. Miner. Monat.* **1969**, *1969*, 544–551.
5. Bellotto, M.; Rebours, B.; Clause, O.; Lynch, J.; Bazin, D.; Elkaim, E. A reexamination of hydrotalcite crystal chemistry. *J. Phys. Chem.* **1996**, *100*, 8527–8534. [CrossRef]
6. Zhitova, E.; Krivovichev, S.; Pekov, I.; Greenwell, H. Crystal chemistry of natural layered double hydroxides. 5. Single-crystal structure refinement of hydrotalcite, $[Mg_6Al_2(OH)_{16}](CO_3)(H_2O)_4$. *Mineral. Mag.* **2019**, *83*, 269–280. [CrossRef]
7. Evans, D.G.; Slade, R.C.T. Structural Aspects of Layered Double Hydroxides. In *Layered Double Hydroxides. Structure and Bonding*; Duan, X., Evans, D.G., Eds.; Springer: Berlin/Heidelberg, Germany, 2005; Volume 119. [CrossRef]
8. Ram Reddy, M.K.; Xu, Z.P.; Lu, G.Q.; Diniz da Costa, J.C. Layered Double Hydroxides for CO_2 Capture: Structure Evolution and Regeneration. *Ind. Eng. Chem. Res.* **2006**, *45*, 7504–7509. [CrossRef]
9. León, M.; Díaz, E.; Bennici, S.; Vega, A.; Ordóñez, S.; Auroux, A. Adsorption of CO_2 on Hydrotalcite-Derived Mixed Oxides: Sorption Mechanisms and Consequences for Adsorption Irreversibility. *Ind. Eng. Chem. Res.* **2010**, *49*, 3663–3671. [CrossRef]
10. Wang, Q.; Tay, H.H.; Ng, D.J.W.; Chen, L.; Liu, Y.; Chang, J.; Zhong, Z.; Luo, J.; Borgna, A. The effect of trivalent cations on the performance of Mg-M-CO_3 layered double hydroxides for high-temperature CO_2 capture. *Chem. Sus. Chem.* **2010**, *3*, 965–973. [CrossRef] [PubMed]
11. Bublinski, M. CO2-Abtrennung aus Synthesegasen mit Hydrotalciten unter Hochtemperatur-Hochdruckbedingungen. Ph.D. Thesis, University of Stuttgart, Stuttgart, Germany, 2017. [CrossRef]
12. Adachi-Pagano, M.; Forano, C.; Besse, J.-P. Delamination of layered double hydroxides by use of surfactants. *Chem. Commun.* **2000**, *1*, 91–92. [CrossRef]
13. Xie, J.; Khalid, Z.; Oh, J.-M. Recent advances in the synthesis of layered double hydroxides nanosheets. *B. Kor. Chem. Soc.* **2022**, *44*, 100–111. [CrossRef]
14. Shirin, V.A.; Sankar, R.; Johnson, A.P.; Gangadharappa, H.; Pramod, K. Advanced drug delivery applications of layered double hydroxide. *J. Control. Release* **2021**, *330*, 398–426. [CrossRef]
15. Mao, N.; Zhou, C.H.; Tong, D.S.; Yu, W.H.; Lin, C.C. Cynthia Lin, Exfoliation of layered double hydroxide solids into functional nanosheets. *Appl. Clay Sci.* **2017**, *144*, 60–78. [CrossRef]
16. Kim, T.W.; Sahimi, M.; Tsotsis, T.T. The Preparation and Characterization of Hydrotalcite Thin Films. *Ind. Eng. Chem. Res.* **2009**, *48*, 5794–5801. [CrossRef]
17. Fajrina, N.; Yusof, N.; Ismail, A.; Jaafar, J.; Aziz, F.; Salleh, W.; Nordin, N. MgAl-CO_3 layered double hydroxide as potential filler in substrate layer of composite membrane for enhanced carbon dioxide separation. *J. Environ. Chem. Eng.* **2021**, *9*, 106164. [CrossRef]
18. Huang, N.; Wang, C.; Chen, C. Ethylene vinyl acetate copolymer/Mg-Al-layered double hydroxide nanocomposite membranes applied in CO_2/N_2 gas separation. *Polym. Compos.* **2021**, *42*, 4065–4072. [CrossRef]
19. Wiheeb, A.D.; Shakir, S.W.; Othman, M.R. Synthesis and Characterization of Mesoporous Hydrotalcite-Alumina Membrane for Carbon Dioxide Enrichment. *IOP Conf. Ser. Mater. Sci. Eng.* **2018**, *454*, 12107. [CrossRef]
20. de Vos, R.M.; Verweij, H. High-selectivity, high-flux silica membranes for gas separation. *Science* **1998**, *279*, 1710–1711. [CrossRef]
21. Elshof, J.E.T. Hybrid Materials for Molecular Sieves. In *Handbook of Sol-Gel Science and Technology, Living Reference Work*; Klein, L.C., Aparicio, M., Jitianu, A., Eds.; Springer International Publishing: Cham, Switzerland, 2016; pp. 1–27. [CrossRef]
22. Kanezashi, M.; Matsugasako, R.; Tawarayama, H.; Nagasawa, H.; Tsuru, T. Pore size tuning of sol-gel-derived triethoxysilane (TRIES) membranes for gas separation. *J. Membr. Sci.* **2017**, *524*, 64–72. [CrossRef]
23. Van Gestel, T.; Velterop, F.; Meulenberg, W.A. Meulenberg, Zirconia-supported hybrid organosilica microporous membranes for CO_2 separation and pervaporation. *Sep. Purif. Technol.* **2021**, *259*, 118114. [CrossRef]
24. Castricum, H.L.; Kreiter, R.; van Veen, H.M.; Blank, D.H.; Vente, J.F.; Elshof, J.E.T. High-performance hybrid pervaporation membranes with superior hydrothermal and acid stability. *J. Membr. Sci.* **2008**, *324*, 111–118. [CrossRef]
25. Yu, L.; Kanezashi, M.; Nagasawa, H.; Guo, M.; Moriyama, N.; Ito, K.; Tsuru, T. Tailoring Ultramicroporosity To Maximize CO_2 Transport within Pyrimidine-Bridged Organosilica Membranes. *ACS Appl. Mater. Interfaces* **2019**, *11*, 7164–7173. [CrossRef]
26. Castricum, H.L.; Paradis, G.G.; Mittelmeijer-Hazeleger, M.C.; Kreiter, R.; Vente, J.F.; Elshof, J.E.T. Tailoring the separation behavior of hybrid organosilica membranes by adjusting the structure of the organic bridging group. *Adv. Funct. Mater.* **2011**, *21*, 2319–2329. [CrossRef]
27. Kanezashi, M.; Yada, K.; Yoshioka, T.; Tsuru, T. Design of silica networks for development of highly permeable hydrogen separation membranes with hydrothermal stability. *J. Am. Chem. Soc.* **2009**, *131*, 414–415. [CrossRef]
28. Moriyama, N.; Nagasawa, H.; Kanezashi, M.; Ito, K.; Tsuru, T. Bis(triethoxysilyl)ethane (BTESE)-derived silica membranes: Pore formation mechanism and gas permeation properties. *J. Sol-Gel Sci. Technol.* **2018**, *86*, 63–72. [CrossRef]

29. Agirre, I.; Arias, P.L.; Castricum, H.L.; Creatore, M.; Elshof, J.E.T.; Paradis, G.G.; Ngamou, P.H.; van Veen, H.M.; Vente, J.F. Hybrid organosilica membranes and processes: Status and outlook. *Sep. Purif. Technol.* **2014**, *121*, 2–12. [CrossRef]
30. Leroux, F., Adachi-Pagano, M.; Intissar, M.; Chauvière, S.; Forano, C.; Besse, J.-P. Delamination and restacking of layered double hydroxides. *J. Mater. Chem.* **2001**, *11*, 105–112. [CrossRef]
31. Piccinni, M.; Bellani, S.; Bianca, G.; Bonaccorso, F. Nickel-Iron Layered Double Hydroxide Dispersions in Ethanol Stabilized by Acetate Anions. *Inorg. Chem.* **2022**, *61*, 4598–4608. [CrossRef]
32. Othman, M.R.; Helwani, Z.; Martunus; Fernando, W.J.N. Synthetic hydrotalcites from different routes and their application as catalysts and gas adsorbents: A review. *Appl. Organometal. Chem.* **2009**, *23*, 335–346. [CrossRef]
33. Gursky, J.A.; Blough, S.D.; Luna, C.; Gomez, C.; Luevano, A.N.; Gardner, E.A. Gardner, Particle-particle interactions between layered double hydroxide nanoparticles. *J. Am. Chem. Soc.* **2006**, *128*, 8376–8377. [CrossRef]
34. Gardner, E.; Huntoon, K.M.; Pinnavaia, T.J. Direct Synthesis of Alkoxide-Intercalated Derivatives of Hydrocalcite-like Layered Double Hydroxides: Precursors for the Formation of Colloidal Layered Double Hydroxide Suspensions and Transparent Thin Films. *Adv. Mater.* **2001**, *13*, 1263. [CrossRef]
35. Merz, D.; Dregert, O.; Garbev, K.; Stemmermann, P. A reliable quantitative TA-FTIR method for cementitious material characterization. In Proceedings of the GEFTA-STK-Joint Meeting on Thermal Analysis and Calorimetry, Saarbrücken, Germany, 10–12 October 2012; Volume 53.
36. Kloprogge, J.T.; Wharton, D.; Hickey, L.; Frost, R.L. Infrared and Raman study of interlayer anions CO_3^{2-}, NO_3^-, SO_4^{2-} and ClO_4^- in Mg/Al-hydrotalcite. *Am. Mineral.* **2002**, *87*, 623–629. [CrossRef]
37. Frost, R.L.; Spratt, H.J.; Palmer, S.J. Infrared and near-infrared spectroscopic study of synthetic hydrotalcites with variable divalent/trivalent cationic ratios. *Spectrochim. Acta Part A Mol. Biomol. Spectrosc.* **2009**, *72*, 984–988. [CrossRef]
38. Hernandez-Moreno, M.J.; Ulibarri, M.A.; Rendon, J.L.; Serna, C.J. IR characteristics of hydrotalcite-like compounds. *Phys. Chem. Miner.* **1985**, *12*, 34–38. [CrossRef]
39. Mališová, M.; Horňáček, M.; Mikulec, J.; Hudec, P.; Jorík, V. FTIR study of hydrotalcite. *Acta Chim. Slov.* **2018**, *11*, 147–166. [CrossRef]
40. Coenen, K.; Gallucci, F.; Mezari, B.; Hensen, E.; van Sint Annaland, M. An in-situ IR study on the adsorption of CO_2 and H_2O on hydrotalcites. *J. CO2 Util.* **2018**, *24*, 228–239. [CrossRef]
41. Dawes, A.; Mason, N.J.; Fraser, H. Using the C-O stretch to unravel the nature of hydrogen bonding in low-temperature solid methanol-water condensates. *Phys. Chem. Chem. Phys. PCCP* **2016**, *18*, 1245–1257. [CrossRef] [PubMed]
42. Plyler, E.K. Infrared spectra of methanol, ethanol, and n-propanol. *J. Res. Nat. Bur.Stand.* **1952**, *48*, 281–286. [CrossRef]
43. Rosales-Reina, B.; Cruz-Quesada, G.; Padilla-Postigo, N.; Irigoyen-Razquin, M.; Alonso-Martínez, E.; López-Ramón, M.V.; Espinal-Viguri, M.; Garrido, J.J. Tunability of Hybrid Silica Xerogels: Surface Chemistry and Porous Texture Based on the Aromatic Precursor. *Gels* **2023**, *9*, 382. [CrossRef]
44. Kim, Y.-H.; Hwang, M.S.; Kim, H.J.; Kim, J.Y.; Lee, Y. Infrared spectroscopy study of low-dielectric-constant fluorine-incorporated and carbon-incorporated silicon oxide films. *J. Appl. Phys.* **2001**, *90*, 3367–3370. [CrossRef]
45. Meng, L.; Kanezashi, M.; Wang, J.; Tsuru, T. Permeation properties of BTESE–TEOS organosilica membranes and application to O_2/SO_2 gas separation. *J. Membr. Sci.* **2015**, *496*, 211–218. [CrossRef]
46. Wahab, M.A.; Kim, I.; Ha, C.-S. Hybrid periodic mesoporous organosilica materials prepared from 1,2-bis (triethoxysilyl) ethane and (3-cyanopropyl) triethoxysilane. *Micropor. Mesopor. Mat.* **2004**, *69*, 19–27. [CrossRef]
47. Keefe, C.D.; Gillis, E.A.L.; MacDonald, L. Improper Hydrogen-Bonding CH Center Dot Y Interactions in Binary Methanol Systems As Studied by FTIR and Raman Spectroscopy. *J. Phys. Chem. A* **2009**, *113*, 2544–2550. [CrossRef] [PubMed]
48. Yu, Y.; Wang, Y.; Lin, K.; Hu, N.; Zhou, X.; Liu, S. Complete Raman spectral assignment of methanol in the C-H stretching region. *J. Phys. Chem. A* **2013**, *117*, 4377–4384. [CrossRef]
49. Ishikawa, K.; Fujima, N.; Komura, H. First—order Raman scattering in MgO microcrystals. *J. Appl. Phys.* **1985**, *57*, 973–975. [CrossRef]
50. Slotznick, S.P.; Shim, S.-H. In situ Raman spectroscopy measurements of $MgAl_2O_4$ spinel up to 1400 °C. *Am. Mineral.* **2008**, *93*, 470–476. [CrossRef]
51. Kloprogge, J.T.; Frost, R.L. Infrared emission spectroscopic study of the thermal transformation of Mg-, Ni- and Co-hydrotalcite catalysts. *Appl. Catal. A* **1999**, *184*, 61–71. [CrossRef]
52. Matsui, K.; Satoh, H.; Kyoto, M. Raman Spectra of Silica Gel Prepared from Triethoxysilane and Tetraethoxysilane by the Sol-Gel Method. *J. Ceram. Soc. Jpn* **1998**, *106*, 528–530. [CrossRef]
53. Kanezaki, E. Effect of atomic ratio Mg/Al in layers of Mg and Al Layered double hydroxide on thermal stability of hydrotalcite-like layered structure by means of in situ high temperature powder X-ray diffraction. *Mat. Res. Bull.* **1998**, *33*, 773–778. [CrossRef]
54. Frost, R.L.; Martens, W.; Ding, Z.; Kloprogge, J.T. DSC and high-resolution TG of synthesized hydrotalcites of Mg and Zn. *J. Therm. Anal. Calorim.* **2003**, *71*, 429–438. [CrossRef]

55. Ke, X.; Bernal, S.A.; Provis, J.L. Uptake of chloride and carbonate by Mg-Al and Ca-Al layered double hydroxides in simulated pore solutions of alkali-activated slag cement. *Cem. Concr. Res.* **2017**, *100*, 1–13. [CrossRef]
56. Kanezashi, M.; Yada, K.; Yoshioka, T.; Tsuru, T. Organic–inorganic hybrid silica membranes with controlled silica network size: Preparation and gas permeation characteristics. *J. Membr. Sci.* **2010**, *348*, 310–318. [CrossRef]

membranes

Article

Mixed-Matrix Organo–Silica–Hydrotalcite Membrane for CO_2 Separation Part 2: Permeation and Selectivity Study

Lucas Bünger *, Tim Kurtz, Krassimir Garbev, Peter Stemmermann and Dieter Stapf

Institute for Technical Chemistry, Karlsruhe Institute of Technology, Kaiserstrasse 12, 76131 Karlsruhe, Germany; tim.kurtz@kit.edu (T.K.); krassimir.garbev@kit.edu (K.G.); peter.stemmermann@kit.edu (P.S.); dieter.stapf@kit.edu (D.S.)

* Correspondence: lucas.buenger@kit.edu; Tel.: +49-721-608-24405

Abstract: This study introduces an innovative approach to designing membranes capable of separating CO_2 from industrial gas streams at higher temperatures. The novel membrane design seeks to leverage a well-researched, high-temperature CO_2 adsorbent, hydrotalcite, by transforming it into a membrane. This was achieved by combining it with an amorphous organo-silica-based matrix, extending the polymer-based mixed-matrix membrane concept to inorganic compounds. Following the membrane material preparation and investigation of the individual membrane in Part 1 of this study, we examine its permeation and selectivity here. The pure 200 nm thick hydrotalcite membrane exhibits Knudsen behavior due to large intercrystalline pores. In contrast, the organo-silica membrane demonstrates an ideal selectivity of 13.5 and permeance for CO_2 of 1.3×10^{-7} mol m^{-2} s^{-1} Pa^{-1} at 25 °C, and at 150 °C, the selectivity is reduced to 4.3. Combining both components results in a hybrid microstructure, featuring selective surface diffusion in the microporous regions and unselective Knudsen diffusion in the mesoporous regions. Further attempts to bridge both components to form a purely microporous microstructure are outlined.

Keywords: inorganic membrane; silica membrane; hydrotalcite membrane; CO_2/N_2 separation; microporous mixed-matrix membrane

1. Introduction

Providing captured CO_2 from industrial exhaust streams is a basic premise for circular processing e.g., in methanization or reverse water–gas shift. A continuous and concentrated CO_2 stream at high temperatures is necessary for efficient process operation [1]. However, no separation process currently operates continuously without experiencing a temperature drop [2].

State-of-the-art separation processes for carbon capture, such as methanol scrubbing, operate at low temperatures, while pressure-swing adsorption (PSA) yields CO_2 pseudo-continuously [3]. Current developments facilitate adsorption at higher temperatures by employing appropriate high-temperature adsorbents and implementing intelligent process interconnections. However, this approach requires relatively high operating costs and capital investment. Membrane processes might offer a solution by utilizing separation layers that incorporate adsorbents, allowing for the separation of CO_2 at elevated temperatures [4]. These processes can provide continuous volume flows by spatially separating adsorption and desorption through diffusion across the membrane (see Figure 1). Therefore, membrane processes can be viewed as spatially separated PSA applications.

Gas-separation membranes, both pure polymeric and mixed-matrix membranes with inorganic fillers, show great CO_2/N_2 separation quality; however, they fail to maintain that behavior at high temperatures [5,6]. This work tries to offer a new route in developing high-temperature gas-separation membranes by combining high-temperature adsorbents within well-known inorganic xerogel matrices. Therefore, the focus is on the design process and the general feasibility of the preliminary membrane.

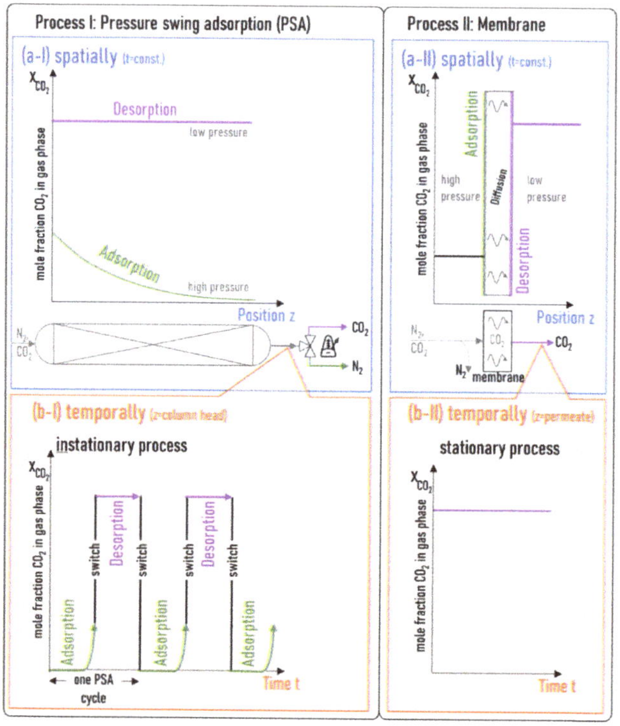

Figure 1. Conceptual comparison of PSA and membrane processes for CO_2/N_2 separation. Panels (**a**) show the CO_2 mole fraction X_{CO2} in the gas phase as a function of the position z: (**a-I**) in the adsorption column and (**a-II**) in the membrane at a point in time t. Panels (**b**) show the CO_2 mole fraction in the gas phase as a function of the time: (**b-I**) at the column head for the PSA and (**b-II**) at the permeate side for the membrane. In the process I (PSA), adsorption and desorption take place at the same location (**a-I**) but at different times (**b-I**). In process II (membrane), adsorption and desorption are spatially separated (**a-II**) but occur simultaneously (**b-II**).

To design selective membranes, the mass transport mechanisms of the respective gas components in the porous membrane comprised the critical design criteria. The mass transport characteristics in porous materials are primarily influenced by the size of the pore diameter, d_p. In macro pores (dp > 50 nm), the primary mechanism is viscous flow. Smaller pores within the mesopore range (2 nm < dp < 50 nm) experience an increased prominence of Knudsen diffusion. Mass transport kinetics in these pores are influenced by the molecular weight. In membranes containing micropores (dp < 2 nm), the impact of the internal surface becomes relevant, facilitating the adsorption/desorption behavior of gas components as an additional mass transport effect. In such cases, the membrane material becomes the dominant factor influencing mass transport kinetics, given the significance of mass transport over the surface. Within the ultramicroporous range (dp << 1 nm), the selective exclusion of gaseous molecules begins, resulting in a decline in permeation or molecules larger than the pore opening, eventually approximating zero. Membranes featuring pores within the molecular sieving range can demonstrate significant selectivities [7,8]. Nevertheless, such membranes often display very low fluxes for the permeating component relative to membranes with larger pores [9]. The same low-flux phenomena are observed for molten hydroxides and carbonate membranes operating at high temperatures [10], only showing good permeances above 650 °C–850 °C [11]. For efficient membrane design, only membranes with pore sizes below 2 nm, made of a CO_2 adsorption agent, are the most efficient solution for CO_2 gas separation [7,8].

Several researched membrane materials exhibit microporosity. Various zeolites [12], metal–organic frameworks (MOFs) [13], carbon-based membranes [14], and sol–gel-derived silica gels [15,16] are suitable materials for this separation challenge. In this study, silica membranes were chosen due to their relatively simple sol–gel coating production process and their ability to achieve CO_2 selectivity while maintaining adequate permeability.

Figure 2a portrays the inorganic xerogel organo–silica structure, wherein mass transport occurs along the interior of the silica network through pores of different sizes (pores with different sizes marked by (1), (2), and (3)). The selectivity of silica membranes is primarily based on their adsorptive affinity for CO_2. However, the utilization of silica membranes is limited to lower temperature ranges (<200 °C) due to a strong decrease in their adsorption capacity upon increasing temperatures [17,18]. For a membrane to effectively separate CO_2 at high temperatures, the active membrane material needs to have adsorption sites with an exothermic heat of adsorption at the operating temperature [19]. Strong interaction of adsorption sites can hinder the mobility of the adsorbed gaseous component, leading to reduced diffusion [20]. Therefore, it is crucial to find the right adsorption agent.

The field of high-temperature adsorbents has received a lot of attention, leading to the identification of layered double hydroxides (LDH) as a suitable precursor for CO_2 adsorption applications at elevated temperatures [21–24]. The used LDH can differ in the composition of cations (e.g., Mg, Al, Fe, Co, Mo, etc.), anions (e.g., sulfate, chlorides, and nitrates), structural order (crystalline, partially disordered, or amorphous), as well as in content and bonding of hydrogen (hydrous/anhydrous, partially dehydroxylated up to hydrogen-free, or "mixed oxides") [25]. While there is extensive research on the use of calcined hydrotalcite-like components (HTlcs) as an adsorbent, there have been relatively few investigations regarding its application as an active separation agent, particularly as the sole membrane material for gas separation. Membranes solely made of LDHs yielded low selectivity for CO_2, as the fabricated membranes showed intercrystalline pore structures with average diameters of permeation that favored Knudsen diffusion [26,27]. Figure 2b shows the delaminated calcined HTlc membrane microstructure schematically as a result of thermal treatment and delamination. The permeation of the gases is through the volume that is opened by the effective average diameter between the HTlc sheets.

The approach to overcome the selectivity problem is demonstrated by mixing a crystalline inorganic compound with high selective surface diffusion properties into an amorphous matrix, as shown in Figure 2c. This approach has already been demonstrated for polymeric matrices in a wide range of combinations with different fillers [28]. There are several approaches for the use of LDH as a filler in polymeric matrices forming classic mixed-matrix membranes (MMM). They all show good separation behavior but have limited temperature stability and low permeance [29]. However, MMMs with matrices made of inorganic xerogels are not well investigated yet. Embedding a sorption-enhancing agent into a silica membrane was studied by the group of Othman and Wiheeb with tetraethylorthosilicate as a precursor for silica, which is very sensitive to moisture for applications in flue gases with water content [30–35].

This work presents a possible way to combine the adsorptive effect of calcined HTlc with an amorphous silica matrix to enhance the separation performance at high temperatures. First, a description of the individual preparation methods for silica and hydrotalcite membranes as well as for the combination of the two materials is given. Second, the experimental setup for conducting mass transport experiments with the prepared membrane is illustrated. Finally, the preparation techniques and the corresponding mass transport performance are evaluated.

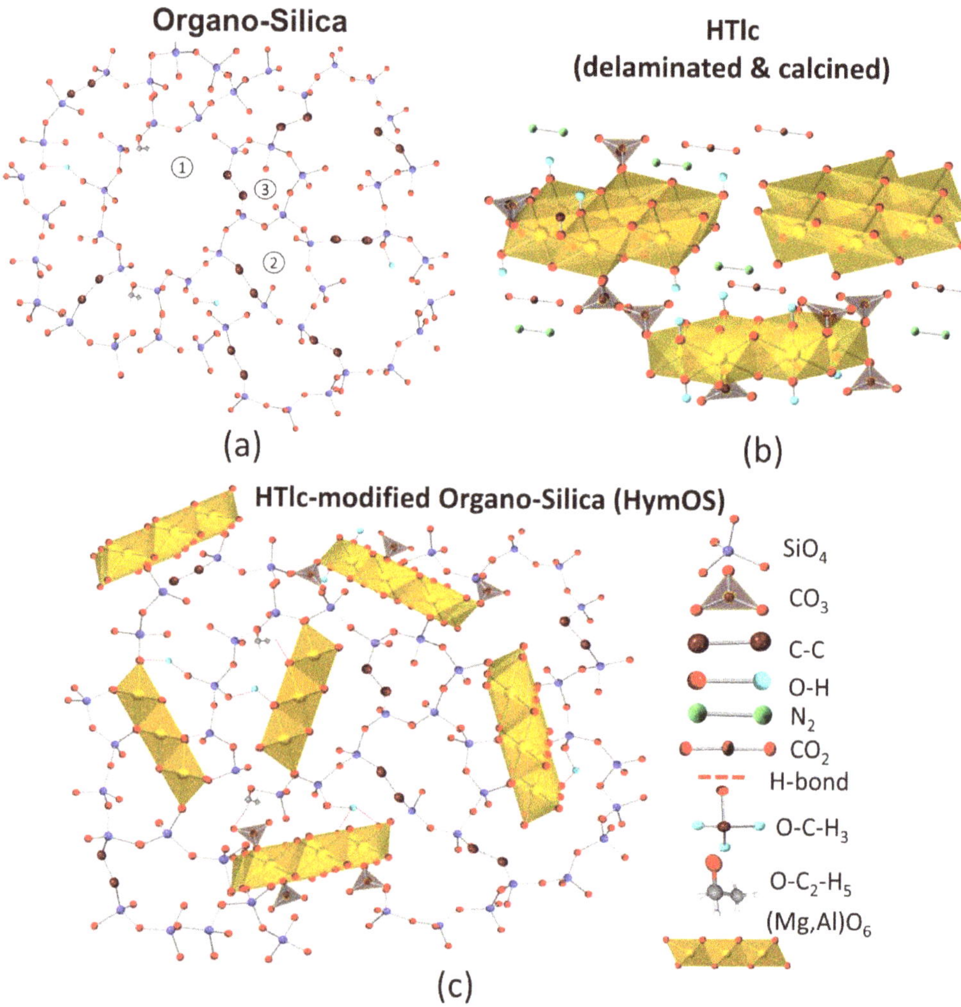

Figure 2. Schematic visualization of the microstructures of (**a**) organo–silica (pores with different sizes and resulting mass transport mechanism are marked by (1), (2) and (3), (**b**) the pure calcined hydrotalcite, (**c**) and the hydrotalcite-modified organo–silica (HymOS) as the combination of both (**c**).

2. Materials and Methods

2.1. Membrane Fabrication

As a substrate for membrane deposition, a macroporous α-alumina disc with a thickness of 2.2 mm and a diameter of 39 mm and an approximate pore size of 80 nm of its top layer (Pervatech, The Netherlands) was used. On the top of the α-Al_2O_3 substrate, a supportive intermediate mesoporous γ-Al_2O_3 layer was deposited, produced by sol–gel synthesis, using a thermally treated boehmite (γ-AlO(OH)) sol after the method of Yoldas [36]. To obtain reproducible layer thickness, 3.5 g of polyvinylalcohol (PVA) was dissolved in 100 mL water and mixed with the boehmite sol in a ratio of 2:3.

A detailed description of the mass transport-enhancing compound synthesis, hydrotalcite, and respective analysis of all steps are given in Part 1 of this work.

The HTlc used in this work was synthesized by dissolving 6.1 g of magnesium chloride hexahydrate and 2.41 g of aluminum chloride hexahydrate, with a molar ratio of Mg/Al = 3, in 100 mL methanol. This mixture was heated to 65 °C and stirred under reflux. The pH was set by a solution of 3.8 g of sodium hydroxide in 100 mL of methanol that was added dropwise to the metal–salt solution [37,38]. For a small particle size distribution, the synthesis time was reduced from 3 days to 1 h.

After cooling the reaction mixture to room temperature, it was centrifuged, and the remaining pellet was dispersed in water. This process was repeated until the pH became neutral. The remaining pellet was then dispersed in water and left overnight to delaminate. As a result, a clear and stable dispersion of delaminated hydrotalcite-like nanosheets was produced. This sol was used to fabricate the pure HTlc membrane. After coating and drying, thermal treatment to activate the agent was carried out at 400 °C for 3 h in air [21].

The synthesis of the microporous organo–silica membrane (Org-Sil) for CO_2 separation was performed after the approach of van Gestel et al. and Castricum et al. [18,39]. It involved mixing 16.655 mL of the membrane-forming precursor, 1,2-bis(triethoxysilyl)-ethane (BTESE), with 28.14 mL of ethanol (both Thermo Fisher (Kandel) GmbH, Kandel, Germany), 0.63 mL of nitric acid (65 wt%), and 4.57 mL of water. These amounts result in a water-to-hydrolysable ethoxy group ratio of one. The resulting mixture was heated to 60 °C and refluxed for 90 min, followed by cooling to room temperature. The sol was stabilized for storage by the addition of 50 mL ethanol. The resulting mixture was stored in the refrigerator (stock sol). For the coating process in terms of the formation of defect-free films, the mixture was diluted in a ratio of 1 to 20 with ethanol. After coating and drying at room temperature for 1 h, thermal treatment of the membrane was conducted under a N_2 atmosphere at 300 °C for 3 h [39,40].

A detailed description of the process to transfer the water-based HTlc sol into a mixable solution with silica is given in Part 1 of this study. The stable delaminated HTlc sol was first dried and then calcined at 400 °C. Two grams of the resulting calcined and ground hydrotalcite were redispersed in ethanol with the assistance of an ultrasonic device.

To ensure the removal of undispersed particles, the dispersion was subjected to centrifugation for 20 min at 4000 rpm, resulting in a stable dispersion of calcined delaminated hydrotalcite-like nanosheets.

Various mixing ratios of both sols were explored, but only the lower and higher ends of the mixing range are presented here.

At the lower end of the mixing range, 100 mL of the same silica sol used for the pure organo–silica (Org-Sil) membrane was combined with 5 mL of ethanol-based hydrotalcite sol to form the HymOS1-sol. At the higher end of the mixing range, 100 mL of the hydrotalcite dispersion was mixed with 5 mL of the silica stock sol, resulting in the HymOS2-sol. After 30 min of mixing, the mixtures were ready for coating onto the substrate to form the mixed-matrix membranes. The resulting as-deposited film was then allowed to air dry for approximately 1 h. Subsequently, thermal treatment was carried out at 300 °C for 3 h under a N_2 atmosphere.

The coating process for all the aforementioned sols was performed uniformly. A 39 mm disc, used as the substrate, was attached to a vacuum suction cup located at the end of a rotating pendulum. A petri dish containing 50 mL of the respective sol was positioned at the lowest point of the pendulum. The coating process involved immersing the polished surface of the substrate into the sol with a controlled and defined velocity.

Once the pendulum reached a perpendicular position, its movement was halted for 15 s. After this pause, the substrate was rotated out of the sol with the same velocity as before. To prevent cracking caused by thermally induced tension, both the heating and cooling rates during thermal treatment were controlled to be 1 K/min. To ensure the production of crack-free membranes, all coating and thermal treatment processes were conducted twice. Afterward, the membranes were ready for structural and permeation analysis.

2.2. Structural Characterization

A high-resolution scanning electron microscope (SEM) was used for qualitative assessment of the prepared membrane layers. Images were captured using the Zeiss Supra 55 VP microscope with an acceleration voltage of 18 kV.

To prepare the membrane for SEM analysis, a crucial step involved freezing the silica layer by placing the membrane disc in liquid N_2. This step was essential for obtaining high-quality images, as attempts to analyze the membrane without this freezing step resulted in images with varying quality. After spending a few minutes in liquid N_2, the membrane was cut into small 1 mm^2 pieces using a side cutter. In a further step, the pieces were then sputtered with gold to ensure electrical conductivity.

Additional analyses by X-ray diffraction, IR, and Raman spectroscopy and thermal analysis of the products after every step of preparation are extensively described in Part 1 of this study.

2.3. Gas Permeation and Separation Experiments

A custom-made cell designed to hold 39 mm membrane discs was used for single-gas experiments. The top of the membrane was sealed via a 32 mm Viton O-ring to the cell, resulting in an effective permeation area (A_m) of 804 mm^2. The thermal stability of the Viton limited the use of this setup to temperatures up to 200 °C.

The gases used in this study were delivered from gas bottles with reducers, allowing measurements at pressures up to 6 bar. They were fed into the cell via an inlet on the top side. The setup also included a retentate outlet (top side) as well as a permeate outlet at the bottom side of the cell. A Bronkhorst El Press pressure controller valve was installed in the retentate line. Pressure measurements were performed in the feed (p_f) and the permeate (p_p) line.

The mass flow of the permeating stream was quantified by three Bronkhorst EL-Flow Prestige mass flow meters with different measurement ranges (0–1 g/h, 0–10 g/h, and 200 g/h) to reduce the measurement error (error: ±0.5% of the measured value; ±0.1% of endpoint value). A membrane pump (KNF LABORPORT N938.50) installed in the permeate line allowed vacuums as low as 70 mbar. To measure the membrane performance at different temperatures, the module was placed in an oven. The temperature was controlled with a thermocouple installed in the proximity of the membrane. Each measurement was taken in steady-state for above an hour. For each membrane type, multiple membranes were tested. However, only a representative single membrane was used for the displayed results to reduce the impact of cracks and voids.

The permeance (Q_i) for a gaseous component i is calculated from the measured membrane area, transmembrane pressure difference, and mole flow J_i, according to Equation (1):

$$Q_i = \frac{J_i}{A_m \cdot \left(p_{i,f} - p_{i,p}\right)} = \frac{P_i}{d_{layer}} \tag{1}$$

The ideal selectivity S_{CO_2/N_2} as the ratio of the single-gas permeances is calculated by Equation (2):

$$S_{CO_2/N_2} = \frac{Q_{CO_2}}{Q_{N_2}} \tag{2}$$

A resistance-in-series model was utilized to determine the successive mass transport resistances in the composite membrane [26]. The resistance of each layer is inversely proportional to the intrinsic mass transport kinetics of the respective layer, expressed by its permeability $P_{i,layer}$. Sequential measurements of each membrane layer provide the permeance (Equation (1)), and using the corresponding membrane thickness (d_{layer}) obtained from SEM images, the permeability is calculated according to Equation (3).

$$Q_{i,composite} = \left(\frac{d_{\alpha-Al_2O_3}}{P_{i,\alpha-Al_2O_3}} + \frac{d_{\gamma-Al_2O_3}}{P_{i,\gamma-Al_2O_3}} + \frac{d_{org-sil}}{P_{i,org-sil}} \right)^{-1} \quad (3)$$

3. Results

3.1. HTlc Membrane

Figure 3 presents SEM images of a pure calcined delaminated hydrotalcite-like membrane, showing the surface (a) and the cross-section (b). Panel a shows that all crystals are preferentially oriented parallel to the substrate, which is a result of the membrane-coating process. Figure 3b shows a defect-free calcined HTlc layer with a thickness of 200 nm on top of a 2.2 µm thick γ-Al_2O_3 layer.

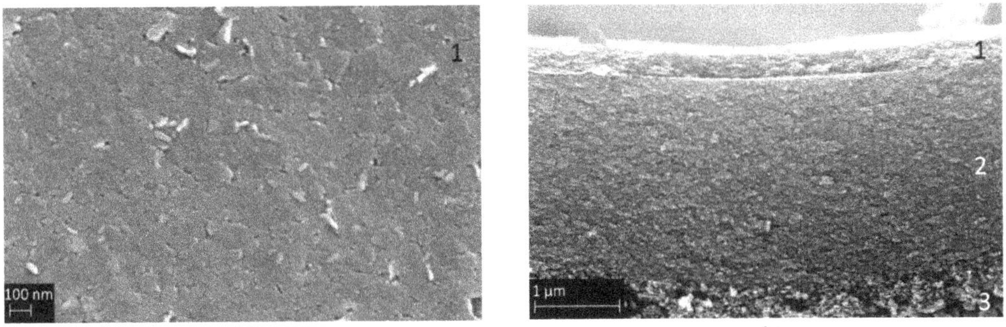

(**a**) (**b**)

Figure 3. SEM secondary electron images of (**a**) the surface of calcined HTlc membrane (1) and (**b**) a cross-section of the composite membrane with 200 nm calcined HTlc on top of the mesoporous γ-Al_2O_3 mesolayer (2) and the pervatech macroporous support (3).

For a better comparison and understanding of the influence of the support layers of the membrane, a consecutive investigation of the membrane layers was conducted. Figure 4 shows the permeance of the substrate and the substrate + γ-Al_2O_3 layer, respectively. The supporting layers show Knudsen behavior; the pure support shows slightly higher permeance than the combination of the support with the γ-Al_2O_3 layer. This is as expected since the γ-Al_2O_3 layer introduces an additional resistance.

The permeance in the HTlc membrane is lower for both components (2.5×10^{-7} mol m^{-2} s^{-1} Pa^{-1} for CO_2 and 3.1×10^{-7} mol m^{-2} s^{-1} Pa^{-1} for N_2) than in the supporting layers, demonstrating a further resistance for the permeating gas molecules. Additionally, N_2 is the primary permeating component, displaying a similar behavior to γ-Al_2O_3 and the substrate, albeit at a lower level. This suggests that the pure calcined HTlc layer possesses a structure with fewer or smaller pores than the other layers. The selectivity for CO_2 at 20 °C is 0.81, indicating that the diffusion through this layer is primarily governed by Knudsen diffusion. Based on the selectivity, the size of the pores is in the same range as in the support. However, the lower flux is a consequence of the orientation of the crystals. They are arranged parallel to the substrate surface, thereby blocking a portion of the flow, as there is no direct path for diffusion through the crystal.

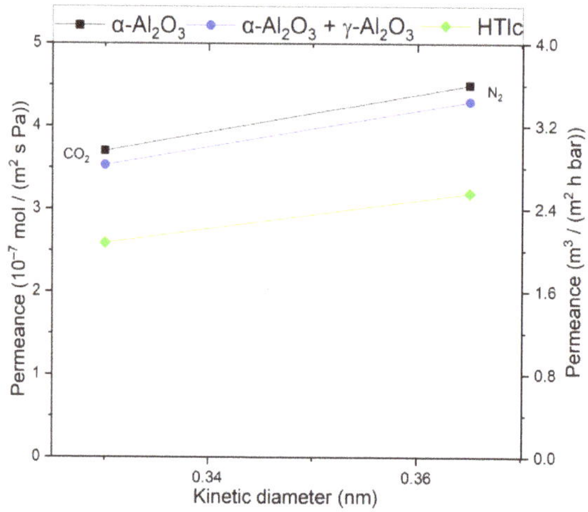

Figure 4. Single-gas permeation through the α-Al_2O_3 support (squares), the support + γ-Al_2O_3 (circle), and the pure hydrotalcite membrane on top of both (diamond) at 25 °C and 3.2 bar. Permeances for both CO_2 and N_2 are in scientific and technical units.

To our knowledge, no pure hydrotalcite membrane with surface diffusion as the predominant mass transport mechanism has been reported yet. However, its theoretical feasibility is convincing. For a material to have surface diffusion that can impact the overall mass transport its heat of adsorption is crucial. Calcined hydrotalcite exhibits a heat of adsorption for CO_2 in the range of 40–50 kJ mol^{-1} at temperatures between 573 K and 623 K [41,42]. This enables surface diffusion according to the correlation presented by Yang [20]. The resulting surface diffusion coefficient ranges in the range of 10^{-4} to 10^{-5} cm^{2}s^{-1}. This is in the same range as the diffusion of CO_2 on silica at 300 K [43]. This shows that calcined HTlcs are a suitable material for enhancing mass transport at elevated temperatures.

However, the way the crystals are arranged always leads to inter-crystalline pores in the range where Knudsen diffusion is dominating (see Figure 2b). Therefore, it is not possible to take advantage of the CO_2 affine surface. Several studies have observed the similar diffusion through pure hydrotalcite layers with permeabilities in the same range (see Table 1) but always with selectivities in the Knudsen range [26,27]. To shift the mass transport from diffusion through the pore volume to surface diffusion, the pore volume must be converted into a microporous structure. Approaches to achieve this include stacking the hydrotalcite layers up to 300 μm thick packages that show selectivity for CO_2 but lead to very low permeances due to the enhanced layer thickness [26]. More promising is the effort to fill the pore volume with a permeable material to reduce the effective average diameter of the permeation path that still allows for economically feasible flows. This leads to mixed-matrix membranes that utilize polymers as the permeable matrix. Tsotsis et al. tried silicone as a matrix material with success, creating membranes with high selectivities that were, however, still on a low permeance level and with no application at higher temperatures [27].

The novel approach in this work focuses on exchanging the polymer with a temperature-stable inorganic material (see Figure 2c). The challenge lies in creating a microporous microstructure where an inorganic organo–silica matrix fills up the free space between the HTlc phase. This tries to transfer the idea of organic matrices with an inorganic filler to a membrane where the matrix is made of an inorganic material, too.

Table 1. Calculated single-gas, stand-alone, single-layer permeances; permeabilities; and respective ideal selectivities at 25 °C. Measured layer thickness derived from SEM images are 2.2 mm for α-Al_2O_3, 4 µm for γ-Al_2O_3, and 100 nm for Org-Sil, HymOS1, and HymOS2.

Single Layer	Permeance (10^{-7} mol (m^2 s Pa)$^{-1}$)		Permeability /10^{-14} mol (m s Pa)$^{-1}$		Perm-Selectivity CO_2/N_2
	N_2	CO_2	N_2	CO_2	
α-Al_2O_3	4.5	3.6	99,000	79,200	0.8
γ-Al_2O_3	36.0	61.2	1440	2448	1.7
HTlc	14.7	12.3	29.5	24.6	0.8
Organo–silica	0.1	2.1	0.1	2.1	20.5
HymOS1	0.7	3.8	0.7	3.8	5.2
HymOS2	3.3	19.7	3.3	19.7	6.0

3.2. Pure Organo–Silica Membrane (Org-Sil)

To understand the inorganic matrix material, its permeating properties as a single layer were investigated. Figure 5 displays SEM images of an organo–silica membrane in a cross-section with the surface visible in the upper part. Three distinct layers are visible. In the insert, the organo–silica top layer 4 is shown, with a thickness of about 100 nm and a defect-free, smooth surface.

Figure 5. SEM image of the microporous organo–silica top layer (4) on top of the γ-Al_2O_3 layer (2) and the α-Al_2O_3 support (3). The insert shows the same membrane at a higher magnification for better visualization of the organo–silica layer thickness.

Single-gas permeation experiments were conducted with CO_2 and N_2 to determine the permeation kinetics (see Figure 6). The organo–silica layer deposited on the top of the γ-Al_2O_3 layer shows different permeation behavior compared to the other membranes. While both gases traverse micropores, surface diffusion emerges as the predominant mass transport mechanism. N_2 can solely diffuse through the free pore volume because its adsorptive interactions are neglectable. The permeance decreases to approximately ~0.1×10^{-7} mol m^{-2} s^{-1} Pa^{-1} at room temperature. CO_2, having an additional diffusion path along the inner pore surface, undergoes a smaller decrease, resulting in a permeance of around ~1.3×10^{-7} mol m^{-2} s^{-1} Pa^{-1}. Consequently, the ideal selectivity reaches 13.

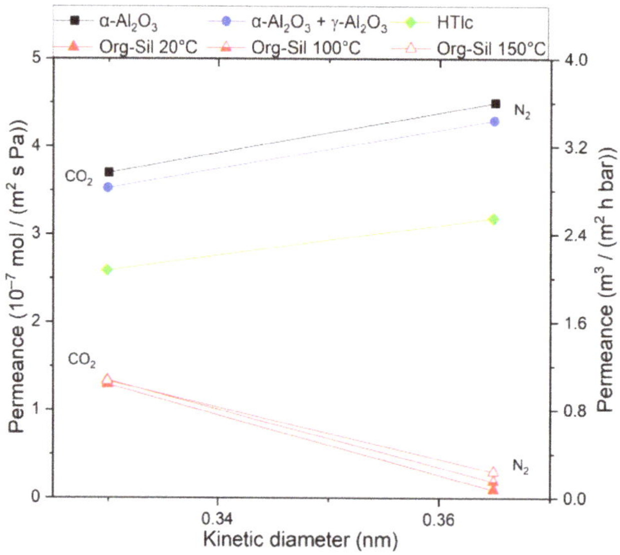

Figure 6. Single-gas permeation of CO_2 and N_2 through an Org-Sil membrane (red) at different temperatures, in reference to the permeances shown in Figure 2 at 25 °C. The transmembrane pressure was kept constant at 3.2 bar. Permeances for both CO_2 and N_2 are in scientific and technical units.

As temperature rises during organo–silica measurements, two distinct behaviors become apparent. Firstly, the permeance for CO_2 appears to remain unaffected. As temperature increases, activated permeation rises, while surface diffusion, which relies on active adsorption sites, decreases [18]. These two mass transport mechanisms seem to counterbalance each other for this pore size in the case of CO_2, resulting in a relatively consistent permeance as temperature rises. Secondly, the permeance for N_2 increases. With increasing temperature, more of the required activation energy is provided for the activated permeation in ultramicropores. Surface diffusion does not participate in the overall N_2 transport; thus, no observable decrease, as in the case of CO_2, counters the increase. A comprehensive description of activated permeation can be found in the study conducted by Lange et al. [19]. Overall, because of the increasing N_2 permeance, the selectivity diminishes to 4.3 at 150 °C.

While an efficient separation of N_2 and CO_2 is achievable at low temperatures, a decrease in separation efficiency at elevated temperatures is observed. This could be explained by the fact that whereas the CO_2 permeance remains relatively constant at higher temperatures, the N_2 permeance increases. The combination of organo–silica and HTlcs is expected to overcome the respective limitations in a novel MMM.

3.3. *HTlc Modified Organo–Silica Membrane (HymOS)*

Figure 7 displays two SEM images of the modified membrane with different HTlc concentrations: lower concentration (a) and higher concentration (b). The preparation of mixed membranes proved to be challenging in terms of HTlc/sol ratio. Too little hydrotalcite in the silica sol would not notably alter the mass transport behavior, while an excess would enlarge the pores, creating voids and thus enhancing Knudsen diffusion, as seen in Figure 2c [32,44].

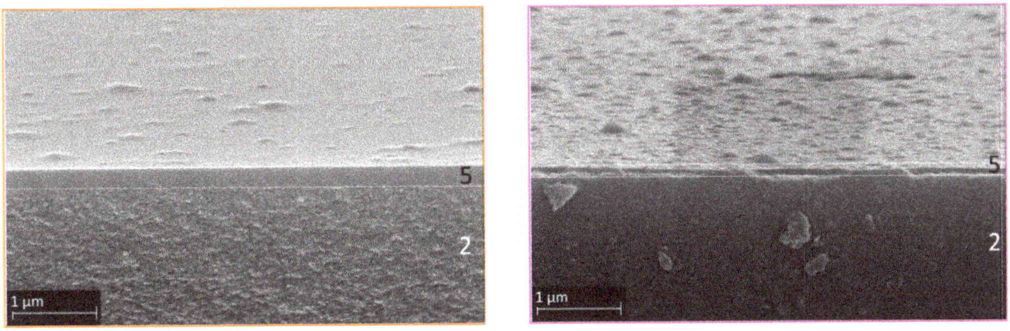

(**a**) (**b**)

Figure 7. SEM secondary electron images of the HymOS1 membrane (5) on top of the γ-Al_2O_3 layer (2) (**a**) with a low hydrotalcite concentration and (**b**) the HymOS2 membrane with a higher concentration.

Both SEM images exhibit continuous and defect-free membranes with clearly seen differences in the hydrotalcite content. To assess the functionality of the prepared membranes, single-gas experiments were conducted. Figure 8 illustrates the permeance of HymOS1 and HymOS2. For a better comparison with other membranes. Table 1 lists the calculated (Equation (3)) single-layer, stand-alone permeabilities for all manufactured membranes in this work.

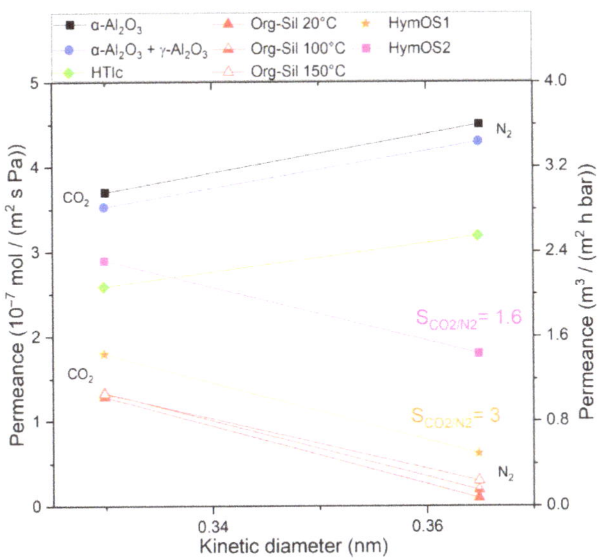

Figure 8. Single-gas permeation at 20 °C (other temperatures are indicated) through two modified membranes (HymOS1, orange star; HymOS2, pink square) in relation to the permeances of the single-component membranes described before. The transmembrane pressure was kept constant at 3.2 bar. Permeances for both CO_2 and N_2 are in scientific and technical units.

Both membranes exhibit higher permeance values for both components compared to the pure organo–silica membrane. This outcome aligns with expectations, as the presence of HTlcs disrupts the silica microstructure, increasing porosity and pore size. For HyMOS1, with lower hydrotalcite content, the permeance increased from around

1.4×10^{-7} mol m^{-2} s^{-1} Pa^{-1} to 1.9×10^{-7} mol m^{-2} s^{-1} Pa^{-1} for CO_2 and from 0.1×10^{-7} mol m^{-2} s^{-1} Pa^{-1} to 0.6×10^{-7} mol m^{-2} s^{-1} Pa^{-1} for N_2. Consequently, the selectivity dropped from 13 to 3. The second membrane (HymOS2), with higher hydrotalcite content, demonstrates similarly elevated permeance for both components, albeit at a higher level. With more hydrotalcite crystals further disrupting the microstructure, the ratio favoring Knudsen diffusion is intensified, leading to a further reduced selectivity of 1.6. The results indicate that the expected improvement in mass transfer along the HTlc surface was not observed. The increased Knudsen diffusion, caused by the disrupted microstructure, overrides the effect of additional surface diffusion. For membranes with a Knudsen diffusion microstructure, temperature-dependent mass transfer experiments were not conducted because Knudsen diffusion does not exhibit temperature-dependent selectivity.

4. Discussion

Research efforts have yielded various materials that potentially facilitate the selective separation of N_2 from CO_2. This study focused on utilizing a hydrotalcite-derived component in a way that has not been investigated yet. Numerous studies prepared membranes by use of some LDH-derived components [45]. Most are designed to facilitate selectivity by molecular sieving through the interlayer gallery. Some have prepared LDHs with CO_2 transport channels that have some sort of carrier mechanic that is highly selective for CO_2 [46]. Former membranes lack the applicability for flue gas cleaning, as the permeances are low and expected to decrease further if mixed-gas permeances were investigated. The transport channel membranes, however, show enormous permeances while having high selectivities. To our knowledge, the transport mechanism is highly pressure-dependent and reduces at technical operation points.

Therefore, this study attempted to introduce a membrane with a third option of facilitating high flux and high selectivity at elevated temperatures by utilizing calcined hydrotalcite, which has proven its application in high-temperature CO_2-adsorbent processes. Creating selective CO_2 diffusion paths by providing high-temperature adsorption sites within a matrix is a membrane design concept with little research yet.

For the insertion of an active separation agent into a permeable matrix to form a functioning membrane, an imagined resistance interconnection needs to be considered [47]. As proposed by Zimmermann et al., MMMs with polymeric matrices can be viewed as parallel resistance of the respective permeating components [48]. This approach is transferred to inorganic–inorganic hybrid microstructures. Equation (4) shows the effective permeance Q for CO_2 and Equation (5) for N_2 through an imagined MMM with the volume fraction Φ of the calcined HTlc. The permeance through the silica matrix is described by $(Q_{i,org-sil})$, exhibiting a decreasing portion of the overall permeance with increasing volume fraction of the HTlc phase.

$$Q_{CO_2,\,MMM} = (1 - \phi_{HTlc})Q_{CO_2,\,org-sil} + \phi_{HTlc}\left(Q_{CO_2,HTlc,S} + Q_{CO_2,HTlc,Kn}\right) \quad (4)$$

$$Q_{N_2,\,MMM} = (1 - \phi_{HTlc})Q_{N_2,\,org-sil} + \phi_{HTlc}\left(Q_{N_2,HTlc,S} + Q_{N_2,HTlc,Kn}\right) \quad (5)$$

For the mixture to be a selective membrane, the mass transport of CO_2 along the surface of the HTlc ($Q_{CO2,HTlc,S}$) needs to reach higher values than for N_2 ($Q_{N2,HTlc,S}$) at operating temperatures. Increasing the volume fraction of HTlcs improves the overall surface for CO_2 diffusion. However, at the same time, the filler material causes the creation of voids that favor Knudsen diffusion ($Q_{i,HTlc,Kn}$), thus rendering the membrane less selective. Therefore, the membrane design needs to focus on finding the right amount of surface diffusion agent incorporated in the matrix to increase the selectivity. Figure 9 displays Equations (4) and (5) schematically to show the proposed design criteria. The Knudsen diffusion dominates the overall mass transport for high filler contents; therefore, Figure 9 only depicts low volume fractions and should be interpreted as an excerpt of the range from 0 to 100%.

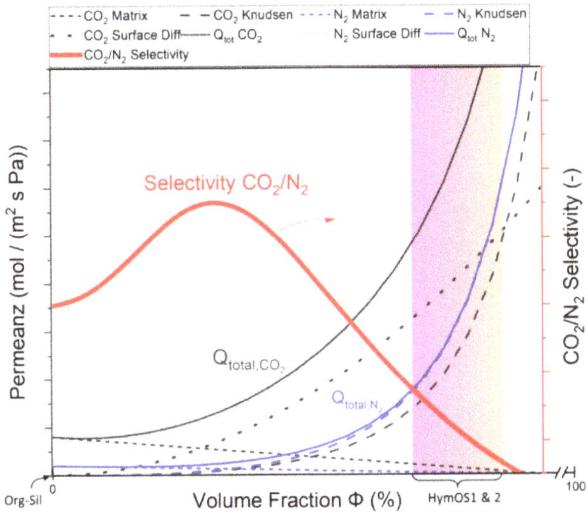

Figure 9. Visualization of the total MMM permeance described in Equation (2) for CO_2 (black) and Equation (3) for N_2 (blue) as a function of the added HTlc content (only the part for low volume fractions is shown here). The total permeance and the permeances for different mass transport mechanisms that comprise the total permeance are given in dark for CO_2 and blue for N_2. The resulting selectivity is shown in red. The prepared membranes HymOS1 and 2 lie within in the marked area.

For low HTlc content, CO_2 surface diffusion increases more quickly than N_2 Knudsen diffusion due to the additional free volume. In small pores, the added surface favors surface diffusion. This pattern reverses as the pores become larger because the volume of the pores increases more rapidly than the surface, leading to a preference for Knudsen diffusion with higher permeances. This results in an overall decrease in selectivity.

The prepared membranes in this work are located on the right side of Figure 9, indicated by the marked area. Both membranes show lower selectivities than the pure matrix ($\Phi = 0$) and have higher permeances. Further investigations with lower adsorption agent content or smaller particles that create less voluminous voids can shift the selectivity to its optimum. Furthermore, for the design of a functioning membrane, the bridging between matrix and HTlcs needs to be ensured. Part 1 of this study showed evidence of those bridges forming; however, this process seems to be a challenge deserving optimization. This could be achieved with further manipulation of the boundary surfaces for faster polymerization via bridging oxygens between octahedral layers of the HTlc and silica. Combining this research with the recent efforts in organo–silica membranes can improve the performance of the matrix 10-fold, as the newest organo–silica membranes show better performance than the one used here [49]. In addition, the allocation of third-party bridging agents to help close the voids seems a possible approach [44].

5. Conclusions

This work offers a new design approach to fabricating membranes for high-temperature gas separation. LDH of hydrotalcite type was synthesized, delaminated, and made into a 200 nm thick membrane, which provides highly CO_2-affinitive adsorption sites after calcination. As expected, the resulting pore size distribution favors unselective Knudsen diffusion. The microporous membrane, made of amorphous organo–silica, shows selective permeation behavior up to 150 °C. Combining both materials into a novel mixed-matrix membrane comprised of two inorganic materials proves to be possible. Although the functionality of the mixed organo–silica–HTlc membrane seems still to be temperature-limited for gas separation, these materials provide a large field for optimization. A combination of unselective Knudsen diffusion and selective surface diffusion explained the observed permeation behavior of the MMM. Incorporating the crystalline adsorption agent into the microporous structure is assumed to create voids at the interfaces that favor Knudsen diffusion. As a main objective for further research, the optimization of both pore size distribution of the gel matrix as well as the improvement of the connectivity between the gel bridging sites and octahedral layers will be identified.

This work establishes the foundation for membranes of this kind and fosters research for finding ways to overcome the bridging problems between both components.

Author Contributions: Conceptualization, L.B.; methodology, L.B.; validation, L.B. and T.K.; formal analysis, L.B.; investigation, L.B.; data curation, L.B.; writing—original draft preparation, L.B.; writing—review and editing T.K., K.G., P.S. and D.S.; visualization, T.K. and L.B.; supervision, K.G., P.S. and D.S.; funding acquisition, D.S. All authors have read and agreed to the published version of the manuscript.

Funding: Program-oriented funding for this research was provided through the German Helmholtz association in the frame of the "Materials and Technologies for the Energy Transition" program.

Institutional Review Board Statement: Not applicable.

Data Availability Statement: The data that support the findings of this study are available from the corresponding author upon request.

Conflicts of Interest: The authors declare no conflicts of interest.

References

1. Zhu, M.; Ge, Q.; Zhu, X. Catalytic Reduction of CO_2 to CO via Reverse Water Gas Shift Reaction: Recent Advances in the Design of Active and Selective Supported Metal Catalysts. *Trans. Tianjin Univ.* **2020**, *26*, 172–187. [CrossRef]
2. Hong, W.Y. A techno-economic review on carbon capture, utilisation and storage systems for achieving a net-zero CO_2 emissions future. *Carbon Capture Sci. Technol.* **2022**, *3*, 100044. [CrossRef]
3. Sircar, S. Pressure Swing Adsorption Technology. In *Adsorption: Science and Technology*; Rodrigues, A.E., LeVan, M.D., Tondeur, D., Eds.; Springer: Dordrecht, The Netherlands, 1989; pp. 285–321.
4. Kenarsari, S.D.; Yang, D.; Jiang, G.; Zhang, S.; Wang, J.; Russell, A.G.; Wei, Q.; Fan, M. Review of recent advances in carbon dioxide separation and capture. *RSC Adv.* **2013**, *3*, 22739. [CrossRef]
5. Amooghin, A.E.; Sanaeepur, H.; Pedram, M.Z.; Omidkhah, M.; Kargari, A. New advances in polymeric membranes for CO_2 separation. In *Polymer Science: Research Advances, Practical Applications and Educational Aspects*; Formatex Research Center: Norristown, PA, USA, 2016; pp. 354–368.
6. Barooah, M.; Kundu, S.; Kumar, S.; Katare, A.; Borgohain, R.; Uppaluri, R.V.S.; Kundu, L.M.; Mandal, B. New generation mixed matrix membrane for CO_2 separation: Transition from binary to quaternary mixed matrix membrane. *Chemosphere* **2024**, *354*, 141653. [CrossRef]
7. Baker, R.W. Membrane Transport Theory. In *Membrane Technology and Applications*; Baker, R.W., Ed.; John Wiley & Sons, Ltd.: Chichester, UK, 2004; pp. 15–87.
8. Burggraaf, A.J. Transport and separation properties of membranes with gases and vapours. In *Fundamentals of Inorganic Membrane Science and Technology*; Burggraaf, A.J., Cot, L., Eds.; Elsevier: Amsterdam, The Netherlands, 1996; pp. 331–434.
9. Robeson, L.M. Correlation of separation factor versus permeability for polymeric membranes. *J. Membr. Sci.* **1991**, *62*, 165–185. [CrossRef]
10. Cerón, M.R.; Lai, L.S.; Amiri, A.; Monte, M.; Katta, S.; Kelly, J.C.; Worsley, M.A.; Merrill, M.D.; Kim, S.; Campbell, P.G. Surpassing the conventional limitations of CO_2 separation membranes with hydroxide/ceramic dual-phase membranes. *J. Membr. Sci.* **2018**, *567*, 191–198. [CrossRef]
11. Lu, B.; Lin, Y.S. Synthesis and characterization of thin ceramic-carbonate dual-phase membranes for carbon dioxide separation. *J. Membr. Sci.* **2013**, *444*, 402–411. [CrossRef]
12. Bernal, M.P.; Coronas, J.; Menéndez, M.; Santamaría, J. Separation of CO_2 /N_2 mixtures using MFI-type zeolite membranes. *AIChE J.* **2004**, *50*, 127–135. [CrossRef]
13. Liu, Y.; Wang, N.; Diestel, L.; Steinbach, F.; Caro, J. MOF membrane synthesis in the confined space of a vertically aligned LDH network. *Chem. Commun.* **2014**, *50*, 4225–4227. [CrossRef] [PubMed]
14. Richter, H.; Voss, H.; Kaltenborn, N.; Kämnitz, S.; Wollbrink, A.; Feldhoff, A.; Caro, J.; Roitsch, S.; Voigt, I. High-Flux Carbon Molecular Sieve Membranes for Gas Separation. *Angew. Chem. Int. Ed.* **2017**, *56*, 7760–7763. [CrossRef]
15. de Vos, R.M.; Verweij, H. High-selectivity, high-flux silica membranes for gas separation. *Science* **1998**, *279*, 1710–1711. [CrossRef]

16. ten Elshof, J.E. Hybrid Materials for Molecular Sieves. In *Handbook of Sol-Gel Science and Technology*; Klein, L., Aparicio, M., Jitianu, A., Eds.; Springer International Publishing: Cham, Switzerland, 2018; pp. 1–27.
17. Kanezashi, M.; Matsugasako, R.; Tawarayama, H.; Nagasawa, H.; Tsuru, T. Pore size tuning of sol-gel-derived triethoxysilane (TRIES) membranes for gas separation. *J. Membr. Sci.* **2017**, *524*, 64–72. [CrossRef]
18. van Gestel, T.; Velterop, F.; Meulenberg, W.A. Zirconia-supported hybrid organosilica microporous membranes for CO_2 separation and pervaporation. *Sep. Purif. Technol.* **2021**, *259*, 118114. [CrossRef]
19. de Lange, R.; Keizer, K.; Burggraaf, A.J. Analysis and theory of gas transport in microporous sol-gel derived ceramic membranes. *J. Membr. Sci.* **1995**, *104*, 81–100. [CrossRef]
20. Yang, R.T. Rate Processes in Adsorbers. In *Gas Separation by Adsorption Processes*, 1st ed.; Yang, R.T., Ed.; Butterworths: London, UK, 1987; pp. 108–124.
21. Ram Reddy, M.K.; Xu, Z.P.; Lu, G.Q.; Diniz da Costa, J.C. Layered Double Hydroxides for CO_2 Capture: Structure Evolution and Regeneration. *Ind. Eng. Chem. Res.* **2006**, *45*, 7504–7509. [CrossRef]
22. León, M.; Díaz, E.; Bennici, S.; Vega, A.; Ordóñez, S.; Auroux, A. Adsorption of CO_2 on Hydrotalcite-Derived Mixed Oxides: Sorption Mechanisms and Consequences for Adsorption Irreversibility. *Ind. Eng. Chem. Res.* **2010**, *49*, 3663–3671. [CrossRef]
23. Wang, Q.; Tay, H.H.; Ng, D.J.W.; Chen, L.; Liu, Y.; Chang, J.; Zhong, Z.; Luo, J.; Borgna, A. The effect of trivalent cations on the performance of Mg-M-CO_3 layered double hydroxides for high-temperature CO_2 capture. *ChemSusChem* **2010**, *3*, 965–973. [CrossRef]
24. Bublinski, M. CO_2-Abtrennung aus Synthesegasen mit Hydrotalciten unter Hochtemperatur-Hochdruckbedingungen. Ph.D. Thesis, Universität Stuttgart, Stuttgart, Germany, 2016.
25. Zhang, J.; Xu, Y.F.; Qian, G.; Xu, Z.P.; Chen, C.; Liu, Q. Reinvestigation of Dehydration and Dehydroxylation of Hydrotalcite-like Compounds through Combined TG-DTA-MS Analyses. *J. Phys. Chem. C* **2010**, *114*, 10768–10774. [CrossRef]
26. Kim, T.W.; Sahimi, M.; Tsotsis, T.T. The Preparation and Characterization of Hydrotalcite Thin Films. *Ind. Eng. Chem. Res.* **2009**, *48*, 5794–5801. [CrossRef]
27. Wook Kim, T.; Sahimi, M.; Tsotsis, T.T. The preparation and characterization of hydrotalcite micromembranes. *Chem. Eng. Sci.* **2009**, *64*, 1585–1590. [CrossRef]
28. Vinoba, M.; Bhagiyalakshmi, M.; Alqaheem, Y.; Alomair, A.A.; Pérez, A.; Rana, M.S. Recent progress of fillers in mixed matrix membranes for CO_2 separation: A review. *Sep. Purif. Technol.* **2017**, *188*, 431–450. [CrossRef]
29. Fajrina, N.; Yusof, N.; Ismail, A.F.; Jaafar, J.; Aziz, F.; Salleh, W.; Nordin, N. MgAl-CO_3 layered double hydroxide as potential filler in substrate layer of composite membrane for enhanced carbon dioxide separation. *J. Environ. Chem. Eng.* **2021**, *9*, 106164. [CrossRef]
30. Wiheeb, A.D.; Ahmad, M.A.; Murat, M.N.; Kim, J.; Othman, M. Identification of Molecular Transport Mechanisms in Micro-Porous Hydrotalcite–Silica Membrane. *Transp. Porous Med.* **2014**, *104*, 133–144. [CrossRef]
31. Wiheeb, A.D.; Ahmad, M.A.; Murat, M.N.; Kim, J.; Othman, M. The declining affinity of microporous hydrotalcite-silica membrane for carbon dioxide. *J. Porous Media* **2014**, *17*, 159–167. [CrossRef]
32. Wiheeb, A.D.; Ahmad, M.A.; Murat, M.N.; Kim, J.; Othman, M. The Effect of Hydrotalcite Content in Microporous Composite Membrane on Gas Permeability and Permselectivity. *Sep. Sci. Technol.* **2014**, *49*, 1309–1316. [CrossRef]
33. Wiheeb, A.D.; Kim, J.; Othman, M. Highly Perm-Selective Micro-Porous Hydrotalcite-Silica Membrane for Improved Carbon Dioxide-Methane Separation. *Sep. Sci. Technol.* **2015**, *50*, 1701–1708. [CrossRef]
34. Wiheeb, A.D.; Mohammed, T.E.; Abdel-Rahman, Z.A.; Othman, M. Flow dynamics of gases inside hydrotalcite-silica micropores. *Microporous Microporous Mater.* **2017**, *246*, 37–42. [CrossRef]
35. Wiheeb, A.D.; Ahmad, M.A.; Murat, M.N.; Kim, J.; Othman, M. Predominant Gas Transport in Microporous Hydrotalcite–Silica Membrane. *Transp. Porous Med.* **2014**, *102*, 59–70. [CrossRef]
36. Yoldas, B.E. Alumina Sol Preparation from Alkoxides. *Am. Ceram. Soc. Bull.* **1975**, *54*, 289–290.
37. Gardner, E.; Huntoon, K.M.; Pinnavaia, T.J. Direct Synthesis of Alkoxide-Intercalated Derivatives of Hydrocalcite-like Layered Double Hydroxides: Precursors for the Formation of Colloidal Layered Double Hydroxide Suspensions and Transparent Thin Films. *Adv. Mater.* **2001**, *13*, 1263. [CrossRef]
38. Gursky, J.A.; Blough, S.D.; Luna, C.; Gomez, C.; Luevano, A.N.; Gardner, E.A. Particle-particle interactions between layered double hydroxide nanoparticles. *J. Am. Chem. Soc.* **2006**, *128*, 8376–8377. [CrossRef] [PubMed]
39. Castricum, H.L.; Sah, A.; Kreiter, R.; Blank, D.H.A.; Vente, J.F.; Elshof, J.E. ten. Hydrothermally stable molecular separation membranes from organically linked silica. *J. Mater. Chem.* **2008**, *18*, 2150. [CrossRef]
40. Kanezashi, M.; Yada, K.; Yoshioka, T.; Tsuru, T. Organic–inorganic hybrid silica membranes with controlled silica network size: Preparation and gas permeation characteristics. *J. Membr. Sci.* **2010**, *348*, 310–318. [CrossRef]
41. Iruretagoyena Ferrer, D. *Supported Layered Double Hydroxides as CO_2 Adsorbents for Sorption-Enhanced H_2 Production*; Springer International Publishing: Cham, Switzerland, 2016.
42. Radha, S.; Navrotsky, A. Energetics of CO_2 Adsorption on Mg–Al Layered Double Hydroxides and Related Mixed Metal Oxides. *J. Phys. Chem. C* **2014**, *118*, 29836–29844. [CrossRef]
43. Alsyouri, H.M.; Lin, J.Y.S. Gas diffusion and microstructural properties of ordered mesoporous silica fibers. *J. Phys. Chem. B* **2005**, *109*, 13623–13629. [CrossRef]

44. Goh, P.S.; Ismail, A.F.; Sanip, S.M.; Ng, B.C.; Aziz, M. Recent advances of inorganic fillers in mixed matrix membrane for gas separation. *Sep. Purif. Technol.* **2011**, *81*, 243–264. [CrossRef]
45. Lu, P.; Liu, Y.; Zhou, T.; Wang, Q.; Li, Y. Recent advances in layered double hydroxides (LDHs) as two-dimensional membrane materials for gas and liquid separations. *J. Membr. Sci.* **2018**, *567*, 89–103. [CrossRef]
46. Xu, X.; Wang, J.; Zhou, A.; Dong, S.; Shi, K.; Li, B.; Han, J.; O'Hare, D. High-efficiency CO_2 separation using hybrid LDH-polymer membranes. *Nat. Commun.* **2021**, *12*, 3069. [CrossRef] [PubMed]
47. Monsalve-Bravo, G.; Bhatia, S. Modeling Permeation through Mixed-Matrix Membranes: A Review. *Processes* **2018**, *6*, 172. [CrossRef]
48. Zimmerman, C.M.; Singh, A.; Koros, W.J. Tailoring mixed matrix composite membranes for gas separations. *J. Membr. Sci.* **1997**, *137*, 145–154. [CrossRef]
49. Guo, M.; Zhang, Y.; Xu, R.; Ren, X.; Huang, W.; Zhong, J.; Tsuru, T.; Kanezashi, M. Ultrahigh permeation of CO_2 capture using composite organosilica membranes. *Sep. Purif. Technol.* **2022**, *282*, 120061. [CrossRef]

membranes

MDPI

Article

Cationic Imidazolium-Urethane-Based Poly(Ionic Liquids) Membranes for Enhanced CO_2/CH_4 Separation: Synthesis, Characterization, and Performance Evaluation

Guilherme Dias [1,2], Laura Rocca [1], Henrique Z. Ferrari [1,2], Franciele L. Bernard [1], Fernando G. Brandão [3], Leonardo Pereira [3] and Sandra Einloft [1,*]

1 School of Technology, Pontifical Catholic University of Rio Grande do Sul (PUCRS), Porto Alegre 90619-900, RS, Brazil; guilherme.dias@acad.pucrs.br (G.D.); l.rocca@edu.pucrs.br (L.R.); henrique.zucchetti@edu.pucrs.br (H.Z.F.); franciele.bernard@pucrs.br (F.L.B.)
2 Post-Graduation Program in Materials Engineering and Technology, Pontifical Catholic University of Rio Grande do Sul (PUCRS), Porto Alegre 90619-900, RS, Brazil
3 Petrobras/CENPES, Ilha do Fundão Qd. 07, Rio de Janeiro 21941-915, RJ, Brazil; fernando.brandao@petrobras.com.br (F.G.B.); lspereira@petrobras.com.br (L.P.)
* Correspondence: einloft@pucrs.br

Abstract: The escalating emissions of CO_2 into the atmosphere require the urgent development of technologies aimed at mitigating environmental impacts. Among these, aqueous amine solutions and polymeric membranes, such as cellulose acetate and polyimide are commercial technologies requiring improvement or substitution to enhance the economic and energetic efficiency of CO_2 separation processes. Ionic liquids and poly(ionic liquids) (PILs) are candidates to replace conventional CO_2 separation technologies. PILs are a class of materials capable of combining the favorable gas affinity exhibited by ionic liquids (ILs) with the processability inherent in polymeric materials. In this context, the synthesis of the IL GLYMIM[Cl] was performed, followed by ion exchange processes to achieve GLYMIM variants with diverse counter anions (NTf_2^-, PF_6^-, and BF_4). Subsequently, PIL membranes were fabricated from these tailored ILs and subjected to characterization, employing techniques such as SEC, FTIR, DSC, TGA, DMA, FEG-SEM, and CO_2 sorption analysis using the pressure decay method. Furthermore, permeability and ideal selectivity assessments of CO_2/CH_4 mixture were performed to derive the diffusion and solubility coefficients for both CO_2 and CH_4. PIL membranes exhibited adequate thermal and mechanical properties. The PIL-BF_4 demonstrated CO_2 sorption capacities of 33.5 mg CO_2/g at 1 bar and 104.8 mg CO_2/g at 10 bar. Furthermore, the PIL-BF_4 membrane exhibited permeability and ideal (CO_2/CH_4) selectivity values of 41 barrer and 44, respectively, surpassing those of a commercial cellulose acetate membrane as reported in the existing literature. This study underscores the potential of PIL-based membranes as promising candidates for enhanced CO_2 capture technologies.

Keywords: CO_2 capture; poly(ionic liquids); ionic liquids; polyurethane; permeability

1. Introduction

In the context of contemporary technological progress, a noticeable increase in atmospheric CO_2 emissions originating from the utilization of fossil resources has emerged [1,2]. In response, the imperative to develop new technologies or refine existing ones to improve environmental impacts has become increasingly evident. A particularly promising avenue of investigation lies in CO_2 capture. This technological approach garners attention due to its applicability to major emission sources and its potential for the recovery and repurposing of CO_2 for other industrial processes [2,3]. Currently, diverse techniques for CO_2 capture exists, among which amine solutions and polymeric membranes stand out as exemplary materials employed for this purpose [4]. The amine-based capture method, which involves the chemical absorption of CO_2, is widely used in emission plants for handling combustion gases and in natural gas sweetening [5–9]. Despite its cost effectiveness, this approach presents challenges such as the generation of undesirable by-products, the formation of corrosive acids leading to the degradation of the internals of the reactor, volatility, and the high energy costs associated with CO_2 recovery, thus diminishing its attractiveness over time [10]. On the other hand, the utilization of membranes in CO_2 separation demonstrates significant potential due to their adaptability to high CO_2 concentrations and their more compact structural requirements compared to amine solutions [11]. Ionic liquids (ILs) constitute a class of organic salts composed of a cation and an anion, being an organic cation and an organic or inorganic anion [4,12–14]. Characterized by a melting temperature below 100 °C, ILs possess several properties expected of effective CO_2 capture materials, including low vapor pressure, non-flammability, thermal stability, low corrosiveness, minimal decomposition rates, lower recovery costs, and environmental compatibility vis-à-vis organic solvents [4,12,15,16]. ILs have attracted attention in the domain of combustion gas capture due to their enhanced CO_2 affinity relative to gas mixtures containing N_2 and CH_4. However, ILs present specific attributes that impede their immediate adoption in CO_2 capture plants, such as high production costs and complexity compared to amine solutions, as well as high viscosity resulting in suboptimal CO_2 sorption/desorption rates [4]. Polymerized ionic liquids or poly(ionic liquids) (PILs) represent a distinct class of materials with broad applicability across various scientific domains [17–21]. Typically composed of a polymeric structure hosting an ionic liquid unit for each polymer repeat unit, PILs offer superior CO_2 sorption/desorption kinetics, thus conferring a significant advantage over ILs [12,22–25]. Condensation polymerization stands out as a common method for PIL synthesis. Researchers, including Bernard 2019, da Luz 2021, and Morozova 2020 [26–28] have successfully synthesized PILs and assessed their properties. Materials such as PTMG (polytetramethylene glycol), PCD (polycarbonate diol), and PCL (polycaprolactone) serve as illustrative examples of polyols used as starting reagents along with diisocyanates for PIL synthesis [26,27,29]. In the case of polyurethane (PU)-based cationic PILs, the respective ILs in the form of diols are employed, thus incorporating the IL cation into the polymer chain as documented in the literature [26,29,30]. This study endeavors to synthesize and evaluate the CO_2/CH_4 separation potential of PU-based cationic PIL membranes. The IL GLYMIM[Cl] was synthesized, followed by ion exchange processes to obtain GLYMIM with different counter anions (NTf_2^-, PF_6^- and BF_4^-). These ILs were used as diols in the synthesis of PILs, aiming to tailor polymeric chains with high affinity for CO_2, thereby producing membranes with improved selectivity.

2. Materials and Methods

2.1. Materials

For the synthesis of hydroxyl-functionalized ionic liquids (ILs), N-methylimidazole (Sigma-Aldrich, Burlington, MA, USA), 3-chloro-1,2-propanediol (Sigma-Aldrich), ethyl acetate (Neon Química, São Paulo, Brazil), lithium bis(trifluoromethanesulfonyl)imide ($LiNT_2F$, Sigma-Aldrich, USA), lithium tetrafluoroborate ($LiBF_4$, Sigma-Aldrich, USA), and sodium hexafluorophosphate ($NaPF_6$, Sigma-Aldrich, USA) were used. To obtain the PILs, polycarbonate diol (PCD, Mn = 2000 g/mol, Bayer, Berlin, Germany), hexamethylene diisocyanate (HDI, 99%, Merck, Darmstadt, Germany), dibutyltin dilaurate (DBTDL, Miracema Nuodex, Campinas, Brazil), methyl ethylketone (MEK, 99%, Mallinckrodt, Hazelwood, MO, USA) and the obtained ILs glyceryl-N-methylimidazolium chloride [GLYMIM][Cl], and derivatives ([GLYMIM][Cl], [GLYMIM][NT_2F], ([GLYMIM][BF_4] and ([GLYMIM][PF_6]) were used.

2.2. Synthesis of Hydroxyl-Functionalized Ionic Liquids

The synthesis of glyceryl-N-methylimidazolium chloride was conducted as described in the literature [25,31,32]. The reaction was carried out continuously in a glycerin bath with constant magnetic stirring at a temperature of 70 °C. During the first hour of the reaction, 10 mL (0.11 mol) of 3-chloro-1,2-propanediol was slowly added dropwise to a solution containing 12 mL (0.15 mol) of N-methylimidazole. The reaction conditions were maintained for 72 h.

Subsequently, the reaction mixture was removed from the heat and washed six times with ethyl acetate. The resulting material was then placed in a glycerin bath to maintain a constant temperature of 70 °C and dried under vacuum for 24 h.

The product was stored under a nitrogen atmosphere, resulting in a yellow oil [25,31,32]. Ion exchanges were performed in acetonitrile by reacting [GLYMIM]Cl with the respective salts: $LiNT_2F$ to obtain $[GLYMIM]NT_2F$, $LiBF_4$ to form $[GLYMIM]BF_4$, and $NaPF_6$ to form $[GLYMIM]PF_6$. It should be noted that the reagents used in the ion exchange process ([GLYMIM]Cl, $LiBF_4$, $LiNT_2F$, and $NaPF_6$) are soluble in acetonitrile, while the by-products, LiCl and NaCl, are not soluble. This allowed for easy separation of the by-products through simple filtration [25,32–34]. The obtained structures can be observed in Figure 1.

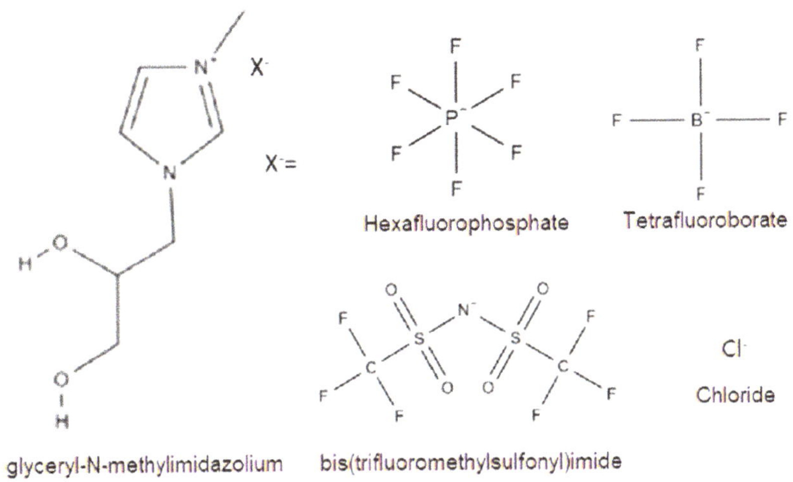

Figure 1. ILs structure.

$AgNO_3$ tests were carried out until no significant amount of white precipitate was observed, confirming satisfactory removal of the by-product. Vacuum drying was then performed to remove any excess solvent and moisture.

2.3. Cationic Poly(Ionic Liquids)

Initially, under constant stirring, the polyol PCD (0.04 mol) was melted in a five-neck reactor, the DBTDL catalyst (0.1% by weight) was added, and 50 mL of methyl ethyl ketone was added. After reaching the reaction temperature (70 °C), the diisocyanate (0.047 mol) was added to form the prepolymer. Then, dialcohol ([GLYMIM]Cl) (0.11 mol) and diisocyanate (0.11 mol) were added to form the PIL. The completion of the reaction is indicated by the disappearance of the free NCO band (around 2270 cm^{-1}) in the infrared spectrum. The synthesized PILs were labeled as PLIX, where X represents the counteranion (Cl, NT_2F, PF_6, or BF_4). Figure 2 shows the structural formula of a cationic PIL.

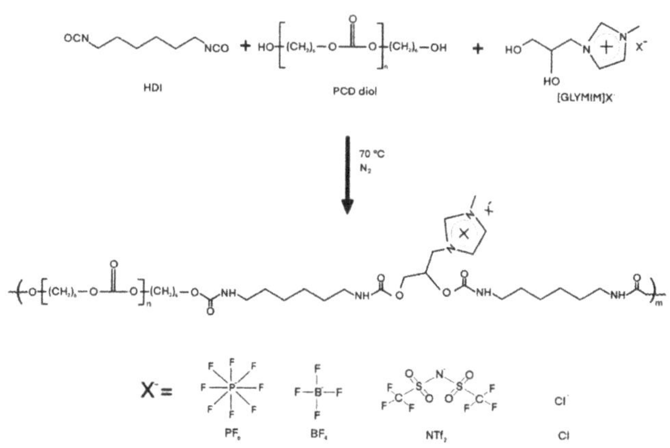

Figure 2. Structures of the cationic PILs.

2.4. Syntheses of PU

The PU preparation process was the same as the method used for the synthesis of cationic poly(ionic liquid); however, [GLYMIM]X, was not added, using an NCO/OH molar ratio of 1.1.

2.5. Preparation of Dense Membranes

Membranes were prepared using the casting or solvent evaporation method with thickness between 150 and 250 μm. Initially, a 20% *w*/*w* solution of PU or PIL was prepared by dissolving 2 g of PIL in 10 mL of methyl ethylketone by magnetic stirring and heating (50 °C) until the PIL was completely dissolved. The solution was then applied to a flat surface, such as a glass plate or a Petri dish, followed by solvent evaporation at room temperature for five days.

2.6. Characterization

Hydroxyl-functionalized ILs and synthesized PILs were characterized by Fourier transform infrared spectroscopy (FTIR, PerkinElmer Spectrum 100 spectrometer, PerkinElmer, Waltham, MA, USA) in transmission mode in the range of 4000–650 cm^{-1} to verify their structures. The synthesized ILs were also characterized by nuclear magnetic resonance (NMR) using a Bruker Ascend 400 model NMR spectrometer, Bruker, Billerica, MA, USA.

The molar masses of the PIL membranes and their distribution were determined by gel permeation chromatography (GPC), with chromatograms obtained using the isocratic HPLC pump-1515 chromatograph (LabMakelaar Benelux, Zevenhuizen, The Netherlands) coupled with a Waters Instruments 2412 (Waters Corp, Milford, CT, USA) refractive index detector and THF as eluent.

Scanning electron microscopy with field emission (FEI Inspect F50, FEI Company, Hillsboro, OR, USA) analyses were performed in secondary electrons (SE) mode to evaluate the surface of the composite membranes.

Differential scanning calorimetry (DSC, TA Instruments Q20, TA Instruments, New Castle, DE, USA) was used to determine the glass transition temperature (Tg), melting temperature (Tm), and crystallization temperature (Tc) of the PILs. Tests were performed with two heating ramps and one cooling ramp in the range of −90 to 200 °C at 10 °C/min under an inert nitrogen atmosphere.

The thermal stability of the PILs was assessed through thermogravimetric analysis (TGA) using a TA Instruments Q600 model instrument, ranging from room temperature to 600 °C with a heating rate of 10 °C/min under a nitrogen atmosphere.

Mechanical analyses were performed in triplicate according to ASTM D822 standard [35] on a TA Instruments Q800 (TA Instruments, New Castle, DE, USA) apparatus to determine Young's moduli and obtain stress–strain curves.

2.7. CO_2 Sorption Capacity

The pressure decay technique was used to determine the CO_2 sorption capacity. A detailed description of the sorption apparatus and measurement procedure can be found in our previous works [26,27]. Samples (1.0 g) were placed in the sorption chamber and degassed under vacuum (10^{-3} mbar) for 1 h at room temperature before the test began. CO_2 sorption experiments were carried out at 30 °C at different equilibrium pressures (1 bar and 10 bar).

2.8. CO_2 Permeability and Ideal CO_2/CH_4 Selectivity

The permeation and selectivity for CO_2 and CH_4 of the PLI membranes were evaluated in a system consisting of two plates joined in a 4 cm diameter hole (Figure S1), where the membrane is inserted. Vacuum was applied to the membrane and the system before gas feed. CO_2 or CH_4 was fed at a pressure of 4 bar. At the bottom, a pressure transducer computed the amount of gas passing through the membrane over time (dp/dt). The permeate gas was assumed to exhibit ideal behavior; therefore, normal temperature and pressure conditions were used. Permeability was determined from the slope (dp/dt) of the linear portion of pressure versus time using Equation (1):

$$\frac{P}{l} = \frac{dp}{dt}\left(\frac{V_{System}}{A\,\Delta p}\right)\left(\frac{T_{CNTP}}{T\,P_{CNTP}}\right) \tag{1}$$

where l is the membrane thickness, P is permeability, Δp is the pressure difference across the membrane, A corresponds to the membrane area, V_{System} is the volume of the permeation cell, T is the ambient temperature, and T_{CNTP} and P_{CNTP} are, respectively, the temperature and pressure at normal conditions [36]. The ideal selectivity was calculated from Equation (2) by dividing the CO_2 permeability by the CH_4 permeability.

$$\propto_{CO_2/CH_4} = \frac{P_{CO_2}}{P_{CH_4}} \tag{2}$$

The solution-diffusion mechanism is widely accepted as the primary transport mechanism for gas permeation through a dense membrane [37,38]. In this mechanism, the gas solubility coefficient (cm^3(STP)/(cm^3 cmHg)) is calculated using Equation (3):

$$S = \frac{P}{D} \tag{3}$$

where P represents permeability and D represents the gas diffusion coefficient (cm^2/s). The gas diffusion coefficient was determined using the time-lag method described by Equation (4) [37,39]:

$$D = \frac{l^2}{6\theta} \tag{4}$$

where D is the diffusion coefficient (cm^2/s), l is the membrane thickness (cm), and θ is the diffusion time lag (s) determined by the linear portion of the curve intercepting the time axis.

3. Results

3.1. Characterization of the ILs

Figure S2 shows the FTIR spectra of the ILs. The IL [GLYMIM]Cl exhibited the expected bands confirming the desired product formation. The OH band between 3281 and 3239 cm^{-1} was visible, as well as bands related to the imidazolium ring near 1670 cm^{-1}

(C=C), 1604 cm^{-1} (N-H), and 1371 cm^{-1} (aromatic C-N) [25,40]. The substitution of the Cl anion by NT_2F resulted in the appearance of bands between 1200 cm^{-1} and 1400 cm^{-1}, characteristic for this ion, as well as others near 1060 cm^{-1} corresponding to S=O bonds, 846 cm^{-1} corresponding to N-S bonds, 789 cm^{-1} corresponding to C-S bonds, and 751 cm^{-1} corresponding to C-F bonds. Substitution of Cl by BF_4 and PF_6 anions also caused changes in the spectra, resulting in the appearance of bands at 815 cm^{-1} for PF_6 P-F bonds and at 1050 cm^{-1} for BF_4 B-F bonds [41–43].

For illustrative purposes, the NMR spectra of the chloride-functionalized N-glyceryl-N-methylimidazole [GLYMIM]Cl (Figure S3) and [GLYMIM]NT_2F (Figure S4) ILs, obtained after the exchange of the chloride anion (Cl^-) for NT_2F, are shown. The 1H NMR spectrum (DMSO-d6) of [GLYMIM]Cl (Figure S2) exhibited characteristic peaks: 3.4 ppm (2H, CH2), 3.7 ppm (1H, CH), 3.9 ppm (3H, CH3), 4.4 ppm (2H, CH2), 5.3 ppm (1H, OH), 5.6 ppm (1H, OH), 7.7 ppm (1H, CH), 7.8 ppm (1H, CH), and 9.2 ppm (1H, CH) [25,32,44]. The 13C spectrum (DMSO-d6) of [GLYMIM]NT_2F (Figure S3) showed characteristic peaks: δ 36.1 (CH_3); 52.5 (-N+CH_2); 63.1 (CH_2OH); 70.07 (CHOH); 123.5 (CH); 123.6 (CH); 137.5 (CH); and 117.4 (2C, NT_2F) [25,32,44].

3.2. Characterization of Cationic Poly(Ionic Liquids)

The molecular weight distributions (PD, polydispersity) and the weighted molecular weights (Mw) of the PILs are presented in Table S1. The Mw values ranged from 43,264 $gmol^{-1}$ to 70,655 $gmol^{-1}$, while the polydispersity (PD) ranged from 1.2 to 1.4. In Figure 3, spectra obtained for the PILs are shown, and Figure 4 shows the enlargements of the spectra in the regions of 3000–3700 cm^{-1}; 1600–1800 cm^{-1}; and 650–1350 cm^{-1}. FTIR spectra (Figure 3) showed characteristic bands found for polyurethanes: 2936–2871 cm^{-1} (CH_2 and CH_3 stretching), 1536 cm^{-1} (N–H bending groups), 1246 cm^{-1} (vibration of C-N and C-O groups of urethane), 1041 cm^{-1} (stretching of C–O–C groups of urethane), and 955 cm^{-1} (stretching of C–O–C groups of polycarbonate diol) [45]. The bands characteristic of H bonds present in polyurethanes were also evaluated: the region between 3200 and 3500 cm^{-1} corresponding to the stretching vibration of the N-H group of urethane (Figure 4A) and the region between 1700 and 1730 cm^{-1} characteristic of bonded and non-bonded carbonyl groups, respectively (Figure 4B). Shifts in characteristic bands (N-H and C=O) evidence possible H bonds. Figure 4A shows the characteristic bands of the stretching vibration of urethane N-H groups, where the region around 3400 cm^{-1} corresponds to non-bonded N-H groups and the band in the region of 3300 cm^{-1} characterizes the vibration of bonded N-H groups. All PILs showed an increase in intensity in the regions of bonded and non-bonded N-H groups with the insertion of IL into the polymeric chain. Also, concerning the structure of polyurethane, Figure 4B contains an enlargement of this area between 1600 and 1800 cm^{-1}. In this range, it is possible to observe bands related to non-bonded carbonyl groups near 1730 cm^{-1} and bonded carbonyl groups near 1700 cm^{-1} [27]. In Figure 4B, changes in the spectrum were also noticed only when IL insertion is performed, such as the displacement of the band ~1730 cm^{-1} to values close to ~1710 cm^{-1}. The appearance of characteristic bands of bonded carbonyl groups in the regions of 1700 cm^{-1} for samples with Cl, PF_6 and BF_4, and 1650 cm^{-1} for samples with NTf_2, PF_6, and BF_4 counterions was observed, indicating the formation of H bonds [26,27]. The absence of the characteristic band for the stretching vibration of the N=C=O group in the region of 2270 cm^{-1} confirms the absence of free NCO groups in the polymeric material, indicating the formation of PU [27,29,46]. Bands related to the counter anions present in the samples can also be visualized (Figure 3). The PIL-BF4 sample presents a broad band near 1060 cm^{-1}, indicating the presence of the B-F bond [41–43]. The PIL-PF_6 presents a band near 835 cm^{-1} related to P-F bonds. For PIL-NTf_2, a band near 1346 cm^{-1} related to SO_2 bond vibrations was observed, as well as bands near 1136 cm^{-1} and 1189 cm^{-1} related to C-F bond vibrations [41–43].

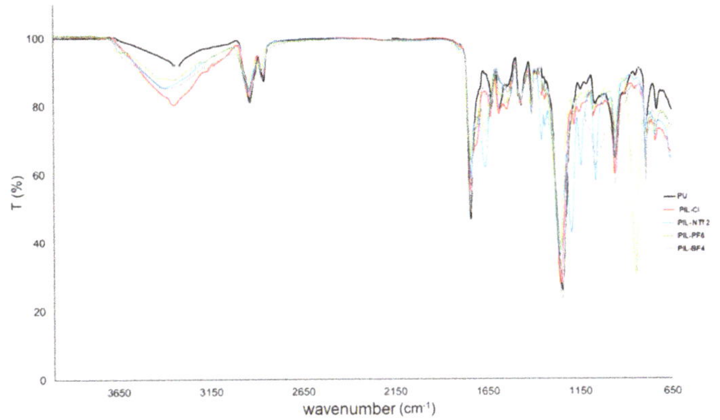

Figure 3. Infrared spectra of PU and cationic PILs.

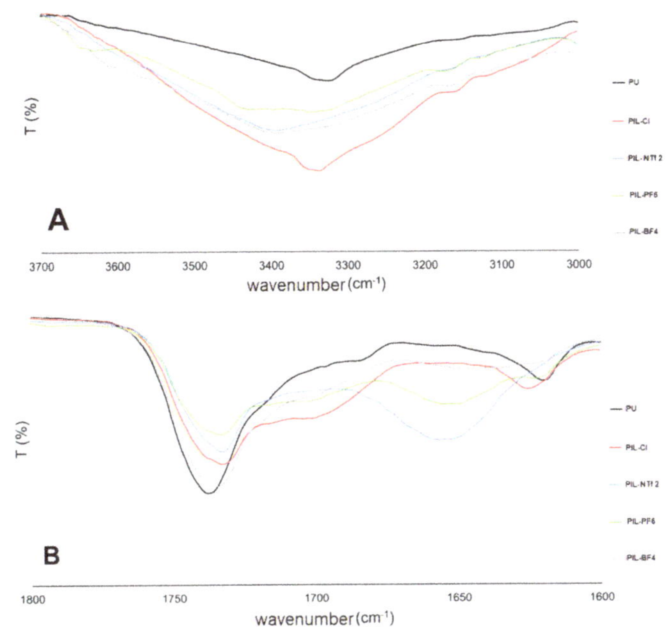

Figure 4. Infrared spectra of PU and cationic PILs. (**A**) 3000–3700 cm^{-1}; (**B**) 1600–1800 cm^{-1}.

3.2.1. DSC

The DSC curves obtained for the polymeric samples are presented in Figure 5. The glass transition temperatures (Tg) values of −44.7, −47.9, −43.2, and −42.7 °C were found for the samples PIL-Cl, PIL-NTf_2, PIL-PF_6, and PIL-BF_4, respectively. By comparing the values obtained for the PILs with the result obtained for pure PU (−42 °C), it is possible to perceive a decreased tendency in the Tg values for all PILs in relation to the non-ionic PU, which could mean that the addition of ILs in the polymeric chain increases the separation of the polymer microphases. This separation of microphases facilitates the mobility of the chains [26,27]. From the analysis of the thermograms, it is possible to identify the existence of an endothermic peak which can be attributed to the melting of a crystalline microphase (Tmf). This endothermic peak is characteristic for PU-based polymers that use polyol PCD with a molecular weight of 2000 g/mol in the polymer chain [26,27]. The results found for the PU, PIL-Cl, PIL-NTf_2, PIL-PF_6, and PIL-BF_4 samples were, respectively, 40.1 °C, 42.2 °C, 46.4 °C, 42.8 °C, and 43.5 °C. Finally, the crystallization temperature (Tc) was obtained for the PIL-Cl sample (−12.7 °C).

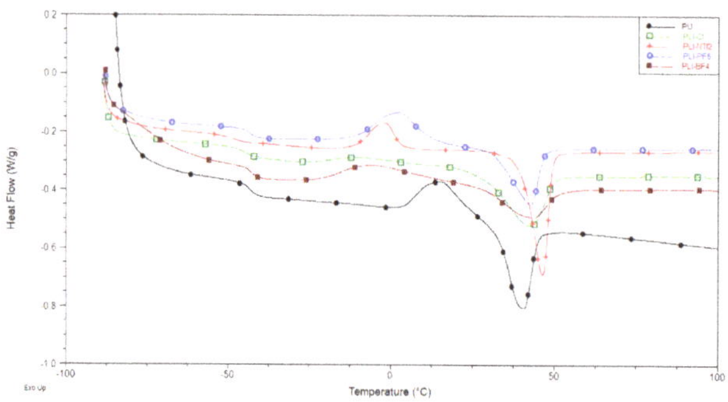

Figure 5. DSC curves of the PILs.

3.2.2. TGA

Table 1 and Figure S5 present the results of the degradation temperature obtained by TGA. PIL samples showed their first degradation temperature between 279 °C and 325 °C. This initial decomposition temperature corresponds to the rigid segments, where the degradation of the polyurethane bonds occurs. The second stage occurs around 440–469 °C, which refers to the breakdown of the flexible segments, leading to the degradation of the polyol along with the imidazole molecules. The incorporation of ILs led to an increase in the decomposition temperature, which may be related to the hydrogen bonds that may occur between PU and IL. Thermal decomposition of ILs typically occurs between 200 and 350 °C, suggesting that the decomposition of ILs may be occurring in the first stage of PILs degradation [47–49].

Table 1. Results of degradation temperatures obtained by TGA.

Sample	T_{onset1} (°C)	T_{onset2} (°C)
PU	276	428
PIL-Cl	279	469
PIL-NTf_2	318	462
PIL-PF_6	300	440
PIL-BF_4	325	445

3.2.3. DMA

Figure 6 shows the stress–strain curves referring to the PILs samples. In terms of Young's modulus, the increasing order is PIL-NTf_2 < PIL-Cl < PIL-PF_6 < PIL-BF_4 < PU. The Young's modulus values obtained for the PILs samples were 7 MPa (PIL-NTf_2), 16.4 MPa (PIL-Cl), 29 MPa (PIL-PF_6), 47 MPa (PIL-BF_4), and 53 MPa (PU). The insertion of an IL into the polymeric chain can act as a plasticizer, interfering with the flexibility and mobility of the chains, affecting the Young's modulus and resulting in its decrease [27,47]. Another possibility for this material behavior is the increased separation of the polymer microphases, which can also lead to a decrease in the Young's modulus and elongation of the curve [50,51]. Except for PIL-NTf_2, it can be observed that, besides the decrease in Young's modulus, which indicates greater elasticity of the material, the curve becomes more elongated when comparing the PIL samples with the non-ionic PU sample.

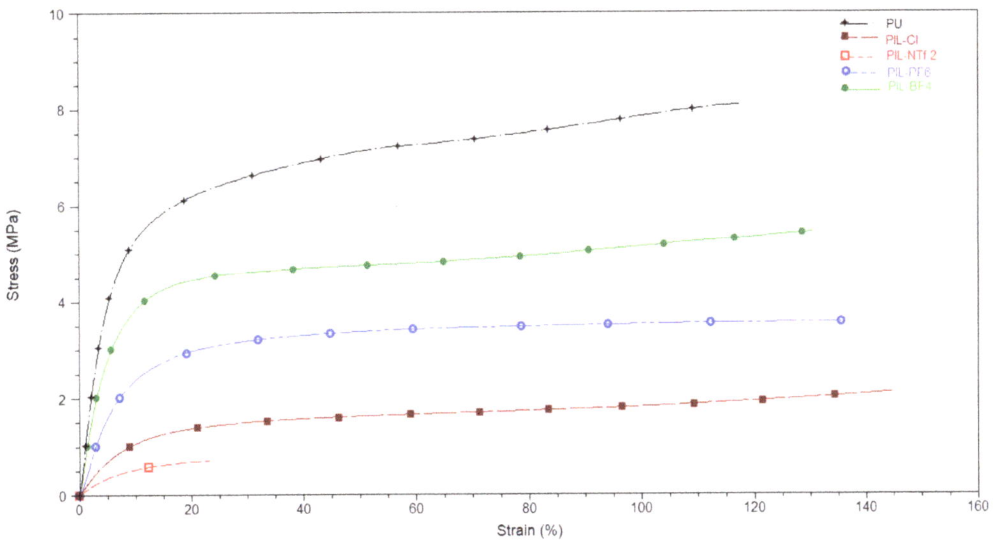

Figure 6. Stress–strain curves of PU and cationic PILs.

3.3. CO_2 Sorption Capacity

The CO_2 capture results can be observed in Figure 7. From the values found, it can be noticed that increasing pressure also increases the amount of captured CO_2. This behavior is typical of samples that undergo CO_2 sorption through physical interaction, as is the case with the PILs obtained in this work. The results obtained for the non-ionic PU sample were 24.7 mg CO_2/g (1 bar) and 83.1 mg CO_2/g (10 bar). These CO_2 capture results before the addition of the ionic liquid can be attributed to the interactions between CO_2 and the polar groups composed of nitrogen and oxygen present in the structure of the PU polymer chain. Additionally, Figure 7 shows that the insertion of the IL leads to a slight increase in CO_2 capture values for all PIL samples. The results obtained for 1 bar were 27.5 mg CO_2/g for PIL-PF_6, 26.3 mg CO_2/g for PIL-Cl, 27.6 mg CO_2/g for PIL-NTf_2, and 33.5 mg CO_2/g for PIL-BF_4. For the same samples at 10 bar, the results were 93.7 mg CO_2/g for PLI-PC-Cl, 95.4 mg CO_2/g for PLI-PC-PF6, 98.1 mg CO_2/g for PIL-NTf_2, and 104.8 mg CO_2/g for PIL-BF_4. The sorption capacity increases in the following order: $[Cl]^-$ < $[PF_6]^-$ < $[NTf_2]^-$ < $[BF_4]^-$ for 1 and 10 bar. As expected, fluorinated anions showed a higher affinity for CO_2. However, in this case, the size of the anion also appears to influence the CO_2 sorption capacity. The structures of the NTF_2^- and PF_6^- anions are larger when compared to the BF_4^- [41,52]. This may block some important CO_2 interaction sites in the PIL cation, reducing the sorption capacity [53]. In the tests conducted, the CO_2 sorption capacity of PIL-BF_4 was superior compared to PU-based PILs reported in the literature using the BF_4 anion. For example, Morozova 2017 [29] and Morozova 2020 [28] obtained 24.7 mgCO_2/g for the PIL-8.1.BF_4 and 24.8 mgCO_2/g for PUR2.BF_4 at 1 bar and 273 K.

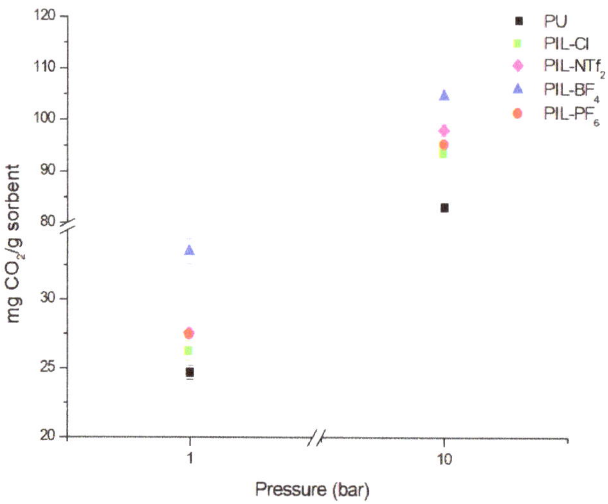

Figure 7. Sorption results for PU and cationic PILs at 1 and 10 bar with a temperature of 303.15 K.

3.4. Permeability and CO_2 Selectivity

Table 2 presents the results of the CO_2 permeability and ideal (CO_2/CH_4) selectivity obtained for dense membranes of cationic PILs and pristine PU, compared to the results of cellulose acetate and some PILs found in the literature. The incorporation of ILs into the polymer chains produced membranes with superior performance when compared to neat PU, as seen in Table 2. It is observed that PIL-CL and PIL-BF_4 samples tested exhibited higher CO_2/CH_4 selectivity than cellulose acetate [54,55]. Furthermore, it can be highlighted that the PIL-BF_4 sample showed the best permeability and selectivity results among all samples presented in Table 2. The PIL-NTf_2 sample presents poor mechanical properties, breaking during the test. Better CO_2 permeability is expected for PILs, as the diffusion and solubility of CO_2 molecules are generally high in ionic liquids [56]. The influence of PIL anions on CO_2 affinity is well documented in the literature [53] and corroborated in our experiments. Permeability and selectivity increased when the Cl^- anion was replaced by BF_4^-, indicating that the presence of fluorine in PILs can enhance CO_2 affinity. However, interestingly, the PF_6^- anion showed lower permeability and selectivity compared to the other anions. This behavior may be related to the size of the anion. The sizes of the anions used in this work follow the general trend $Cl^- < BF_4^- < PF_6^- < NTF_2^-$ [41,52]. Therefore, although BF_4^- and PF_6^- have similar structures, PF_6 has a larger size. Bulky anion structures can reduce the free volume of the PIL, hindering CO_2 penetration towards the cation, which is primarily responsible for sorption [53]. According to Vollas et al. [41], the presence of smaller anions such as $BF4^-$ facilitates polymer packing through chain interactions, leading to denser structures and consequently lower methane permeabilities. Therefore, there appears to be an optimal size for fluorinated anions (BF_4^-) to achieve high permeability and selectivity.

Table 2. Permeability and ideal selectivity results for cationic PILs (25 °C and 4 bar), compared with results for cellulose acetate and PILs from the literature.

Membrane	Perm. CO_2 (Barrer)	Perm. CH_4 (Barrer)	Sel. Ideal CO_2/CH_4
PU	2.0 ± 0.1	0.78 ± 0.02	2.5
PIL-Cl	17 ± 2.2	0.6 ± 0.1	30
PIL-NTf_2	-	-	-
PIL-PF_6	9 ± 1.7	0.6 ± 0.2	14
PIL-BF_4	41 ± 3.2	0.9 ± 0.3	44
1—CTA (25 °C, 5 bar)	18	1.98	9
2—CA (25 °C, 3 bar)	15	1.45	10
3—CA (25 °C, 1 bar)	4	0.2	17
4—CA (35° C, 3 bar)	4.3	0.21	21
5—Styrene-Based Poly(RTILS) (2 atm; RT)	9.2 ± 0.5	0.24 ± 0.01	39
5—Acrylate-Based Poly(RTILS) (2 atm; RT)	7.0 ± 0.4	0.19 ± 0.02	37
6—PIL-TFSI (1 atm; RT)	4.1 ± 0.1	-	41
7—OEG_1 (2 atm; 295 K)	16 ± 1	0.48 ± 0.01	33
7—OEG_2 (2 atm; 295 K)	22 ± 1	0.74 ± 0.02	29
8—PIL-i-propyl (3 atm; 20 °C)	10.4 ± 0.2	0.35 ± 0.01	30
9—Poly[DPyDBzPBI-BuI][Tf2N] (20 atm; 35 °C)	36.2	1.25	29

1—Raza, A. et al., 2021 [54]; 2—Mubashir, M. et al., 2018 [55]; 3—Saijan, P. et al., 2020 [57]; 4—Akbarzadeh, E. et al., 2021 [58]; 5—Bara, J. et al., 2007 [59]; 6—Vollas, A. et al., 2018 [41]; 7—Bara, J. et al., 2008 [60]; 8—Horne, W. et al., 2015 [61]; 9—Shaligram, S. et al., 2016 [62].

To obtain more information about the gas transport mechanism through the membranes, the diffusion and solubility coefficients for CO_2 were determined. These coefficients can provide a better understanding of the variations in permeability obtained for the different PILs, as presented in Table 2. Furthermore, these coefficients can offer a better understanding of the permeability variations of PIL membranes with different counter anions, as presented in Table 3. The values of the diffusion and solubility coefficients showed that the obtained PIL membranes exhibited higher diffusivity and solubility to CO_2 than to CH_4, which contributed to the achieved CO_2 permeabilities. Additionally, this behavior may be related to the high affinity of ILs with the polar groups of CO_2. It can be highlighted that the PIL-BF_4 sample showed higher diffusion and solubility coefficients, corroborating with the sorption capacity results. It is known that CO_2 solubility is influenced by various factors, such as the size of the anions, which can lead to an increase in the free volume of PIL, thus enhancing gas solubility since there is more space available for sorption [63].

Table 3. Diffusion and solubility coefficients for PIL membranes at 4 bar and 25 °C.

Membrane	D (10^{-8} cm²/s)		S (10^{-2} cm³(STP)/(cm³ cmHg))		$D_{CO2/CH4}$	$S_{CO2/CH4}$
	CO_2	CH_4	CO_2	CH_4		
PIL-Cl	21.60	2.32	0.78	0.26	9.32	3.04
PIL-NTf_2	-	-	-	-	-	-
PIL-PF_6	22.04	1.61	0.41	0.37	13.68	1.10
PIL-BF_4	32.67	1.79	1.26	0.50	18.22	2.52
CA (35 °C, 3 bar) [57]	2.05	0.55	2.11	0.38	3.72	5.55

3.5. Comparison with Robeson Upper Bound

The separation performance of dense cationic PIL membranes in this study was compared with Robeson's curves for CO_2/CH_4 [64,65], as shown in Figure 8. It can be observed that the membranes achieved performance below the Robeson limit; however, the PIL-BF_4 sample is better positioned when compared to samples with Cl^- and PF_6^- anions, being very close to the upper limit of 1991. A likely explanation lies in the relatively low CO_2 permeability of the PILs with Cl^- and PF_6^- anions, although they show superior results compared to commercial polymer cellulose acetate. The highlight of the PIL-BF_4

sample is the ideal selectivity for CO_2/CH_4, which is higher than in other studies reported in the literature [66].

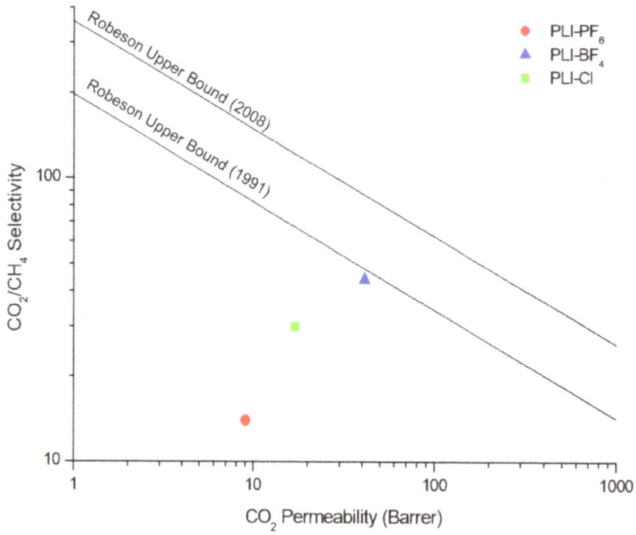

Figure 8. Performance of cationic PIL membranes at the Robeson upper bound for CO_2/CH_4 separation.

In this context, the obtained permeability and ideal selectivity results demonstrate that the synthesized PIL membranes are extremely promising for CO_2/CH_4 separation, especially because there is an intention to explore structural modification of the polymer chain to enhance the permeability and selectivity of these membranes [41]. It can be observed that the samples of cationic PILs prepared with different counter anions exhibited good permeability and high selectivity for CO_2. From the SEM images shown in Figure 9, it can be seen that the PIL membranes produced for the permeability tests exhibited a surface without pores and defects.

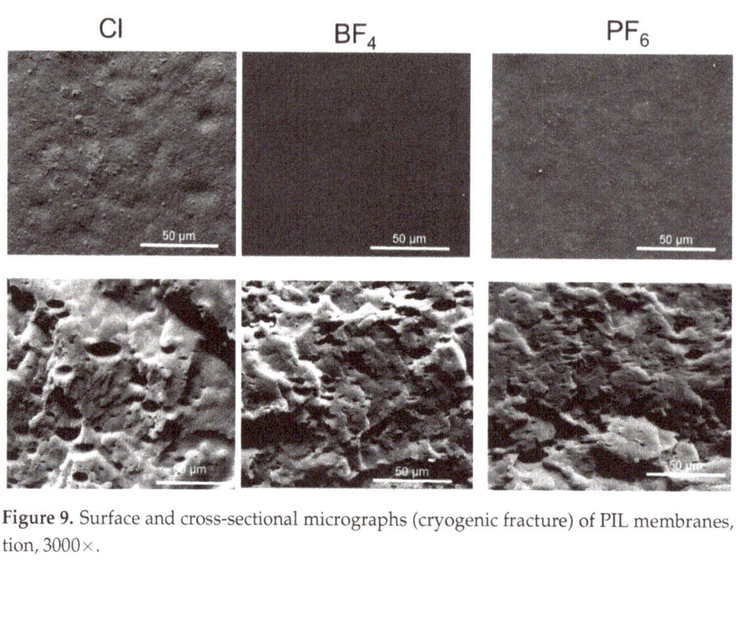

Figure 9. Surface and cross-sectional micrographs (cryogenic fracture) of PIL membranes, magnification, 3000×.

4. Conclusions

In this study, we successfully developed PIL membranes that exhibited adequate thermal and mechanical properties, high CO_2 permeability, and ideal CO_2/CH_4 selectivity. The IL GLYMIM [Cl] was synthesized and thoroughly characterized, and different counter anions' chemical structures (NTf_2, PF_6, and BF_4) were evaluated to identify the optimal combination for physicochemical properties, sorption, and permeability. FTIR analysis confirmed the presence of the characteristic bands for both ILs and PILs. The incorporation of IL into the polymer chain resulted in a decrease in the Tg values, indicating improved membrane flexibility. Among the samples tested, PIL-BF_4 demonstrated the best performance in terms of sorption, permeability, and ideal CO_2/CH_4 selectivity. This suggests that appropriately sized fluorinated anions significantly enhance CO_2 permeability and ideal CO_2/CH_4 selectivity, outperforming the commercial cellulose acetate membrane.

Supplementary Materials: The following supporting information can be downloaded at: https://www.mdpi.com/article/10.3390/membranes14070151/s1, Figure S1: Illustration of permeability system set up; Figure S2: Synthesized ionic liquids spectra; Figure S3: IL ([GLYMIM]) Cl RMN spectra, 1H (DMSO-d6); Figure S4: IL [GLYMIM]NT2F 13C RMN spectra; Figure S5: PILs TGA curves; Table S1: PILs GPC results.

Author Contributions: Conceptualization, S.E., G.D. and F.L.B.; methodology, G.D., L.R., H.Z.F. and F.L.B.; validation, G.D., L.R., H.Z.F. and F.L.B.; formal analysis, S.E., F.L.B., L.P., G.D, L.R., H.Z.F. and F.G.B.; investigation, S.E., F.L.B., L.P., G.D., L.R., H.Z.F. and F.B; resources, S.E., F.L.B., F.G.B. and L.P.; data curation, S.E., F.L.B., L.P., G.D., L.R., H.Z.F. and F.G.B.; writing—original draft preparation, G.D., F.L.B., H.Z.F. and S.E. writing—review and editing, G.D., F.L.B., H.Z.F. and S.E.; supervision, S.E.; project administration, S.E. and F.G.B.; funding acquisition, S.E., F.L.B. and F.G.B. All authors have read and agreed to the published version of the manuscript.

Funding: This research was founded by Petrobras Project Number 4600677123 and CNPq Project Number 316580/2021-0.

Institutional Review Board Statement: Not applicable.

Data Availability Statement: The original contributions presented in the study are included in the article, further inquiries can be directed to the corresponding author.

Acknowledgments: The authors would like to thank PETROBRAS for financial support; Sandra Einloft thanks CNPq for research scholarship.

Conflicts of Interest: Author Fernando Brandão and Leonardo Pereira were employed by the company Petrobras. The remaining authors declare that the research was conducted in the absence of any commercial or financial relationships that could be construed as a potential conflict of interest.

Nomenclature

PILs	poly(ionic liquids) or polymerized ionic liquids
ILs	ionic liquids
GLYMIM	glyceryl-N-methylimidazolium
Cl^-	chloride anion
NTf_2^-	bis(trifluoromethanesulfonyl)imide anion
PF_6^-	hexafluorophosphate
BF_4^-	tetrafluoroborate
PU	polyurethane
PTMG	polytetramethylene glycol
PCD	polycarbonate diol
PCL	polycaprolactone
HDI	hexamethylene diisocyanate
DBTDL	dibutyltin dilaurate
MEK	methyl ethylketone
PD	polydispersity

Mw	molecular weights
Tg	glass transition temperatures

References

1. Supasitmongkol, S.; Styring, P. High CO_2 Solubility in Ionic Liquids and a Tetraalkylammonium-Based Poly(Ionic Liquid). *Energy Environ. Sci.* **2010**, *3*, 1961–1972. [CrossRef]
2. Privalova, E.I.; Mäki-Arvela, P.; Murzin, D.Y.; Mikkhola, J.P. Capturing CO_2: Conventional versus Ionic-Liquid Based Technologies. *Russ. Chem. Rev.* **2012**, *81*, 435–457. [CrossRef]
3. Rockett, G.C. Associação de Fontes Emissoras e Reservatórios Potenciais para Armazenamento Geológico de CO_2 na Bacia de Campos, Brasil. Bachelor's Thesis, Pontifical Catholic University of Rio Grande do Sul, Porto Alegre, Brazil, 2010.
4. Dunn, C.A.; Denning, S.; Crawford, J.M.; Zhou, R.; Dwulet, G.E.; Carreon, M.A.; Gin, D.L.; Noble, R.D. CO_2/CH_4 Separation Characteristics of Poly(RTIL)-RTIL-Zeolite Mixed-Matrix Membranes Evaluated under Binary Feeds up to 40 Bar and 50 °C. *J. Memb. Sci.* **2021**, *621*, 118979. [CrossRef]
5. Aghaie, M.; Rezaei, N.; Zendehboudi, S. A Systematic Review on CO_2 Capture with Ionic Liquids: Current Status and Future Prospects. *Renew. Sustain. Energy Rev.* **2018**, *96*, 502–525. [CrossRef]
6. Privalova, E.I.; Karjalainen, E.; Nurmi, M.; Mäki-Arvela, P.; Eränen, K.; Tenhu, H.; Murzin, D.Y.; Mikkola, J.P. Imidazolium-Based Poly(Ionic Liquid)s as New Alternatives for CO_2 Capture. *ChemSusChem* **2013**, *6*, 1500–1509. [CrossRef] [PubMed]
7. Luis, P. Use of Monoethanolamine (MEA) for CO_2 Capture in a Global Scenario: Consequences and Alternatives. *Desalination* **2016**, *380*, 93–99. [CrossRef]
8. Kwak, N.S.; Lee, J.H.; Lee, I.Y.; Jang, K.R.; Shim, J.G. A Study of the CO_2 Capture Pilot Plant by Amine Absorption. *Energy* **2012**, *47*, 41–46. [CrossRef]
9. Panahi, M.; Skogestad, S. Economically Efficient Operation of CO_2 Capturing Process Part I: Self-Optimizing Procedure for Selecting the Best Controlled Variables. *Chem. Eng. Process. Process Intensif.* **2011**, *50*, 247–253. [CrossRef]
10. Kausar, A.; Nulwala, H.; Mirjafari, A.; Zhou, X.; Duan, N.; Sun, Z.; Ren, Y.; Liu, Z.; Liu, L.; Yan, F.; et al. Enhanced CO_2 Absorption of Poly(Ionic Liquid)s. *Macromolecules* **2021**, *38*, 5477–5489. [CrossRef]
11. Idem, R.; Wilson, M.; Tontiwachwuthikul, P.; Chakma, A.; Veawab, A.; Aroonwilas, A.; Gelowitz, D. Pilot Plant Studies of the CO_2 Capture Performance of Aqueous MEA and Mixed MEA/MDEA Solvents at the University of Regina CO_2 Capture Technology Development Plant and the Boundary Dam CO_2 Capture Demonstration Plant. *Ind. Eng. Chem. Res.* **2006**, *45*, 2414–2420. [CrossRef]
12. Liang, Z.; Rongwong, W.; Liu, H.; Fu, K.; Gao, H.; Cao, F.; Zhang, R.; Sema, T.; Henni, A.; Sumon, K.; et al. Recent Progress and New Developments in Post-Combustion Carbon-Capture Technology with Amine Based Solvents. *Int. J. Greenh. Gas Control.* **2015**, *40*, 26–54. [CrossRef]
13. Lei, Z.; Dai, C.; Chen, B. Gas Solubility in Ionic Liquids. *Chem. Rev.* **2014**, *114*, 1289–1326. [CrossRef] [PubMed]
14. Brennecke, J.F.; Gurkan, B.E. Ionic Liquids for CO_2 Capture and Emission Reduction. *J. Phys. Chem. Lett.* **2010**, *1*, 3459–3464. [CrossRef]
15. Olivier-Bourbigou, H.; Magna, L.; Morvan, D. Ionic Liquids and Catalysis: Recent Progress from Knowledge to Applications. *Appl. Catal. A Gen.* **2010**, *373*, 1–56. [CrossRef]
16. Zhang, X.; Zhang, X.; Dong, H.; Zhao, Z.; Zhang, S.; Huang, Y. Carbon Capture with Ionic Liquids: Overview and Progress. *Energy Environ. Sci.* **2012**, *5*, 6668–6681. [CrossRef]
17. Hasib-ur-Rahman, M.; Siaj, M.; Larachi, F. Ionic Liquids for CO_2 Capture-Development and Progress. *Chem. Eng. Process. Process Intensif.* **2010**, *49*, 313–322. [CrossRef]
18. Li, X.; Ding, S.; Zhang, J.; Wei, Z. Optimizing Microstructure of Polymer Composite Membranes by Tailoring Different Ionic Liquids to Accelerate CO2 Transport. *Int. J. Greenh. Gas Control.* **2020**, *101*, 103136. [CrossRef]
19. Sodeifian, G.; Raji, M.; Asghari, M.; Rezakazemi, M.; Dashti, A. Polyurethane-SAPO-34 Mixed Matrix Membrane for CO_2/CH_4 and CO_2/N_2 Separation. *Chin. J. Chem. Eng.* **2019**, *27*, 322–334. [CrossRef]
20. Wang, Y.; Nie, J.; Lu, C.; Wang, F.; Ma, C.; Chen, Z.; Yang, G. Imidazolium-Based Polymeric Ionic Liquids for Heterogeneous Catalytic Conversion of CO_2 into Cyclic Carbonates. *Microporous Mesoporous Mater.* **2020**, *292*, 109751. [CrossRef]
21. Baker, R.W.; Low, B.T. Gas Separation Membrane Materials: A Perspective. *Macromolecules* **2014**, *47*, 6999–7013. [CrossRef]
22. Favre, E. Membrane Processes and Postcombustion Carbon Dioxide Capture: Challenges and Prospects. *Chem. Eng. J.* **2011**, *171*, 782–793. [CrossRef]
23. Lian, S.; Song, C.; Liu, Q.; Duan, E.; Ren, H.; Kitamura, Y. Recent Advances in Ionic Liquids-Based Hybrid Processes for CO_2 Capture and Utilization. *J. Environ. Sci.* **2021**, *99*, 281–295. [CrossRef] [PubMed]
24. Zhou, Y.; Chang, M.; Zang, X.; Zheng, L.; Wang, Y.; Wu, L.; Han, X.; Chen, Y.; Yu, Y.; Zhang, Z. The polymeric ionic liquids/mesoporous alumina composites: Synthesis, characterization and CO_2 capture performance test. *Polym. Test.* **2020**, *81*, 106109. [CrossRef]
25. Bellina, F.; Bertoli, A.; Melai, B.; Scalesse, F.; Signori, F.; Chiappe, C. Synthesis and Properties of Glycerylimidazolium Based Ionic Liquids: A Promising Class of Task-Specific Ionic Liquids. *Green Chem.* **2009**, *11*, 622–662. [CrossRef]
26. Bernard, F.L.; dos Santos, L.M.; Schwab, M.B.; Polesso, B.B.; do Nascimento, J.F.; Einloft, S. Polyurethane-Based Poly(Ionic Liquid)s for CO_2 Removal from Natural Gas. *J. Appl. Polym. Sci.* **2019**, *136*, 4–11. [CrossRef]

27. da Luz, M.; Dias, G.; Zimmer, H.; Bernard, F.L.; do Nascimento, J.F.; Einloft, S. Poly(Ionic Liquid)s-Based Polyurethane Blends: Effect of Polyols Structure and ILs Counter Cations in CO_2 Sorption Performance of PILs Physical Blends. *Polym. Bull.* **2022**, *79*, 6123–6139. [CrossRef]
28. Morozova, S.M.; Lozinskaya, E.I.; Sardon, H.; Su, F.; Vlasov, P.S.; Vaudemont, R.; Vygodskii, Y.S.; Shaplov, A.S. Ionic Polyureas—A Novel Subclass of Poly(Ionic Liquid) s for CO_2 Capture. *Membranes* **2020**, *10*, 240. [CrossRef]
29. Morozova, S.M.; Shaplov, A.S.; Lozinskaya, E.I.; Mecerreyes, D.; Sardon, H.; Zulfiqar, S.; Suárez-García, F.; Vygodskii, Y.S. Ionic Polyurethanes as a New Family of Poly(Ionic Liquid)s for Efficient CO_2 Capture. *Macromolecules* **2017**, *50*, 2814–2824. [CrossRef]
30. Morozova, S.M.; Shaplov, A.S.; Lozinskaya, E.I.; Vlasov, P.S.; Sardon, H.; Mecerreyes, D.; Vygodskii, Y.S. Poly(Ionic Liquid)-Based Polyurethanes Having Imidazolium, Ammonium, Morpholinium or Pyrrolidinium Cations. *High Perform. Polym.* **2017**, *29*, 691–703. [CrossRef]
31. Dong, Y.; Holm, J.; Kärkkäinen, J.; Nowicki, J.; Lassi, U. Dissolution and Hydrolysis of Fibre Sludge Using Hydroxyalkylimidazolium Hydrogensulphate Ionic Liquids. *Biomass Bioenergy* **2014**, *70*, 461–467. [CrossRef]
32. Duan, N.; Sun, Z.; Ren, Y.; Liu, Z.; Liu, L.; Yan, F. Imidazolium-Based Ionic Polyurethanes with High Toughness, Tunable Healing Efficiency and Antibacterial Activities. *Polym. Chem.* **2020**, *11*, 867–875. [CrossRef]
33. Sonnenschein, M.F.; Lysenko, Z.; Brune, D.A.; Wendt, B.L.; Schrock, A.K. Enhancing Polyurethane Properties via Soft Segment Crystallization. *Polymer* **2005**, *46*, 10158–10166. [CrossRef]
34. Tsai, T.H.; Maes, A.M.; Vandiver, M.A.; Versek, C.; Seifert, S.; Tuominen, M.; Liberatore, M.W.; Herring, A.M.; Coughlin, E.B. Synthesis and Structure-Conductivity Relationship of Polystyrene-Block- Poly(Vinyl Benzyl Trimethylammonium) for Alkaline Anion Exchange Membrane Fuel Cells. *J. Polym. Sci. B Polym. Phys.* **2013**, *51*, 1751–1760. [CrossRef]
35. ASTM D822 Standard. Available online: https://industrialphysics.com/standards/astm-d822/ (accessed on 2 July 2024).
36. Barboza, E.M.; Delpech, M.C.; Garcia, M.E.F.; Pimenta, F.D. Avaliação Das Propriedades de Barreira de Membranas Obtidas a Partir de Dispersões Aquosas à Base de Poliuretanos e Argila. *Polimeros* **2014**, *24*, 94–100. [CrossRef]
37. Ferrari, H.Z.; Rodrigues, D.M.; Bernard, F.L.; dos Santos, L.M.; Le Roux, C.; Micoud, P.; Martin, F.; Einloft, S. A New Class of Fillers in Mixed Matrix Membranes: Use of Synthetic Silico-Metallic Mineral Particles (SSMMP) as a Highly Selective Component for CO_2/N_2 Separation. *Chem. Eng. J. Adv.* **2023**, *14*, 100488. [CrossRef]
38. Waheed, N.; Mushtaq, A.; Tabassum, S.; Gilani, M.A.; Ilyas, A.; Ashraf, F.; Jamal, Y.; Bilad, M.R.; Khan, A.U.; Khan, A.L. Mixed Matrix Membranes Based on Polysulfone and Rice Husk Extracted Silica for CO_2 Separation. *Sep. Purif. Technol.* **2016**, *170*, 122–129. [CrossRef]
39. Wu, H.; Thibault, J.; Kruczek, B. The Validity of the Time-Lag Method for the Characterization of Mixed-Matrix Membranes. *J. Memb. Sci.* **2021**, *618*, 118715. [CrossRef]
40. Jacoby, C.G. Síntese de Compostos Imidazol-Tiazolidina e Sua Aplicação como Organocatalisadores em Reações Aldólicas Estereosseletivas. Master's Thesis, The Federal University of Rio Grande do Sul, Porto Alegre, Brazil, 2016.
41. Vollas, A.; Chouliaras, T.; Deimede, V.; Ioannides, T. New Pyridinium Type Poly(Ionic Liquids) as Membranes for CO_2 Separation. *Polymers* **2018**, *10*, 912. [CrossRef]
42. Yu, G.; Man, Z.; Li, Q.; Li, N.; Wu, X.; Asumana, C.; Chen, X. New Crosslinked-Porous Poly-Ammonium Microparticles as CO_2 Adsorbents. *React. Funct. Polym.* **2013**, *73*, 1058–1064. [CrossRef]
43. Zhang, Z.; Gai, H.; Li, Q.; Feng, B.; Xiao, M.; Huang, T.; Song, H. Effect Anions on the Hydrogenation of Nitrobenzene over N-Rich Poly(Ionic Liquid) Supported Pd Catalyst. *Chem. Eng. J.* **2022**, *429*, 132224. [CrossRef]
44. Chiappe, C.; Pomelli, C.S.; Rajamani, S. Influence of Structural Variations in Cationic and Anionic Moieties on the Polarity of Ionic Liquids. *J. Phys. Chem. B* **2011**, *115*, 9653–9661. [CrossRef] [PubMed]
45. Guo, J.; Zhao, M.; Ti, Y.; Wang, B. Study on Structure and Performance of Polycarbonate Urethane Synthesized via Different Copolymerization Methods. *J. Mater. Sci.* **2007**, *42*, 5508–5515. [CrossRef]
46. Rogulska, M. Polycarbonate-Based Thermoplastic Polyurethane Elastomers Modified by DMPA. *Polym. Bull.* **2019**, *76*, 4719–4733. [CrossRef]
47. Cakić, S.M.; Špírková, M.; Ristić, I.S.; B-Simendić, J.K.; M-Cincović, M.; Poręba, R. The Waterborne Polyurethane Dispersions Based on Polycarbonate Diol: Effect of Ionic Content. *Mater. Chem. Phys.* **2013**, *138*, 277–285. [CrossRef]
48. Zhang, Y.; Sunarso, J.; Liu, S.; Wang, R. Current Status and Development of Membranes for CO2/CH4 Separation: A Review. *Int. J. Greenh. Gas Control.* **2013**, *12*, 84–107. [CrossRef]
49. Yuan, J.; Mecerreyes, D.; Antonietti, M. Poly(Ionic Liquid)s: An Update. *Prog. Polym. Sci.* **2013**, *38*, 1009–1036. [CrossRef]
50. Zhang, M.; Hemp, S.T.; Zhang, M.; Allen, M.H.; Carmean, R.N.; Moore, R.B.; Long, T.E. Water-Dispersible Cationic Polyurethanes Containing Pendant Trialkylphosphoniums. *Polym. Chem.* **2014**, *5*, 3795–3803. [CrossRef]
51. Behera, P.K.; Usha, K.M.; Guchhait, P.K.; Jehnichen, D.; Das, A.; Voit, B.; Singha, N.K. A Novel Ionomeric Polyurethane Elastomer Based on Ionic Liquid as Crosslinker. *RSC Adv.* **2016**, *6*, 99404–99413. [CrossRef]
52. Vila, J.; Varela, L.M.; Cabeza, O. Cation and Anion Sizes Influence in the Temperature Dependence of the Electrical Conductivity in Nine Imidazolium Based Ionic Liquids. *Electrochim. Acta* **2007**, *52*, 7413–7417. [CrossRef]
53. Zulfiqar, S.; Sarwar, M.I.; Mecerreyes, D. Polymeric Ionic Liquids for CO_2 Capture and Separation: Potential, Progress and Challenges. *Polym. Chem.* **2015**, *6*, 6435–6451. [CrossRef]
54. Raza, A.; Farrukh, S.; Hussain, A.; Khan, I.; Othman, M.H.D.; Ahsan, M. Performance Analysis of Blended Membranes of Cellulose Acetate with Variable Degree of Acetylation for CO_2/CH_4 Separation. *Membranes* **2021**, *11*, 245. [CrossRef]

55. Mubashir, M.; Yeong, Y.F.; Lau, K.K.; Chew, T.L.; Norwahyu, J. Efficient CO_2/N_2 and CO_2/CH_4 Separation Using NH_2-MIL-53(Al)/Cellulose Acetate (CA) Mixed Matrix Membranes. *Sep. Purif. Technol.* **2018**, *199*, 140–151. [CrossRef]
56. Farrokhara, M.; Dorosti, F. New High Permeable Polysulfone/Ionic Liquid Membrane for Gas Separation. *Chin. J. Chem. Eng.* **2020**, *28*, 2301–2311. [CrossRef]
57. Sajjan, P.; Nayak, V.; Padaki, M.; Zadorozhnyy, V.Y.; Klyamkin, S.N.; Konik, P.A. Fabrication of Cellulose Acetate Film through Blending Technique with Palladium Acetate for Hydrogen Gas Separation. *Energy Fuels* **2020**, *34*, 11699–11707. [CrossRef]
58. Akbarzadeh, E.; Shockravi, A.; Vatanpour, V. High Performance Compatible Thiazole-Based Polymeric Blend Cellulose Acetate Membrane as Selective CO_2 Absorbent and Molecular Sieve. *Carbohydr. Polym.* **2021**, *252*, 117215. [CrossRef]
59. Bara, J.E.; Lessmann, S.; Gabriel, C.J.; Hatakeyama, E.S.; Noble, R.D.; Gin, D.L. Synthesis and Performance of Polymerizable Room-Temperature Ionic Liquids as Gas Separation Membranes. *Ind. Eng. Chem. Res.* **2007**, *46*, 5397–5404. [CrossRef]
60. Bara, J.E.; Gabriel, C.J.; Hatakeyama, E.S.; Carlisle, T.K.; Lessmann, S.; Noble, R.D.; Gin, D.L. Improving CO2 Selectivity in Polymerized Room-Temperature Ionic Liquid Gas Separation Membranes through Incorporation of Polar Substituents. *J. Memb. Sci.* **2008**, *321*, 3–7. [CrossRef]
61. Horne, W.J.; Andrews, M.A.; Shannon, M.S.; Terrill, K.L.; Moon, J.D.; Hayward, S.S.; Bara, J.E. Effect of Branched and Cycloalkyl Functionalities on CO_2 Separation Performance of Poly(IL) Membranes. *Sep. Purif. Technol.* **2015**, *155*, 89–95. [CrossRef]
62. Shaligram, S.V.; Rewar, A.S.; Wadgaonkar, P.P.; Kharul, U.K. Incorporation of Rigid Polyaromatic Groups in Polybenzimidazole-Based Polymeric Ionic Liquids: Assertive Effects on Gas Permeation Properties. *Polymer* **2016**, *93*, 30–36. [CrossRef]
63. Fang, W.; Luo, Z.; Jiang, J. CO_2 Capture in Poly(Ionic Liquid) Membranes: Atomistic Insight into the Role of Anions. *Phys. Chem. Chem. Phys.* **2013**, *15*, 651–658. [CrossRef]
64. Robeson, L.M. The Upper Bound Revisited. *J. Memb. Sci.* **2008**, *320*, 390–400. [CrossRef]
65. Robeson, L.M. Correlation of Separation Factor versus Permeability for Polymeric Membranes. *J. Membr. Sci.* **1991**, *62*, 165–185. [CrossRef]
66. Zhang, M.; Semiat, R.; He, X. Recent Advances in Poly(Ionic Liquids) Membranes for CO_2 Separation. *Sep. Purif. Technol.* **2022**, *299*, 121784. [CrossRef]

membranes

Article

3D-CFD Modeling of Hollow-Fiber Membrane Contactor for CO_2 Absorption Using MEA Solution

Alexandru-Constantin Bozonc, Vlad-Cristian Sandu, Calin-Cristian Cormos and Ana-Maria Cormos *

Faculty of Chemistry and Chemical Engineering, Babeș-Bolyai University, Arany Janos 11, 400028 Cluj-Napoca, Romania; alexandru.bozonc@stud.ubbcluj.ro (A.-C.B.); vlad.sandu@ubbcluj.ro (V.-C.S.); calin.cormos@ubbcluj.ro (C.-C.C.)

* Correspondence: ana.cormos@ubbcluj.ro

Abstract: Membrane technology is considered an innovative and promising approach due to its flexibility and low energy consumption. In this work, a comprehensive 3D-CFD model of the Hollow-Fiber Membrane Contactor (HFMC) system for CO_2 capture into aqueous MEA solution, considering a counter-current fluid flow, was developed and validated with experimental data. Two different flow arrangements were considered for the gas mixture and liquid solution inside the HFMC module. The simulation results showed that the CO_2 absorption efficiency was considerably higher when the gas mixture was channeled through the membranes and the liquid phase flowed externally between the membranes, across a wide range of gas and liquid flow rates. Sensitivity studies were performed in order to determine the optimal CO_2 capture process parameters under different operating conditions (flow rates/flow velocities and concentrations) and HFMC geometrical characteristics (e.g., porosity, diameter, and thickness of membranes). It was found that increasing the membrane radius, while maintaining a constant thickness, positively influenced the efficiency of CO_2 absorption due to the higher mass transfer area and residence time. Conversely, higher membrane thickness resulted in higher mass transfer resistance. The optimal membrane thickness was also investigated for various inner fiber diameters, resulting in a thickness of 0.2 mm as optimal for a fiber inner radius of 0.225 mm. Additionally, a significant improvement in CO_2 capture efficiency was observed when increasing membrane porosity to values below 0.2, at which point the increase dampened considerably. The best HFMC configuration involved a combination of low porosity, moderate thickness, and large fiber inner diameter, with gas flow occurring within the fiber membranes.

Keywords: CO_2 capture process; HFMC; absorption; 3D-CFD modeling

1. Introduction

Global warming has become a major issue in recent years due to the steady rise in greenhouse gas emissions. Carbon dioxide (CO_2) accounts for about 76% of total greenhouse gas emissions [1]. The rapid rate of growth of atmospheric CO_2 concentrations, from 280 ppm in 1850 to 418 ppm in 2022 [2], has driven a substantial increase in research concerning mitigation of CO_2 emissions. Carbon capture, utilization, and storage (CCUS) is a technology used to reduce CO_2 emissions and diminish the global warming phenomenon [3,4]. Among CCUS technologies, post-combustion technologies are the most convenient to implement on an industrial scale. The CO_2 separation process takes place after the combustion of fossil fuels in power plants or after the generation of products for other industrial processes (i.e., steel and iron production). As such, a complete overhaul of the existing industrial plant is not necessary and only the installation of an additional unit for CO_2 capture is required. There are various techniques applied in CO_2 capture, such as absorption, adsorption, and membrane-based and cryogenic separation [5–7]. Considering the advantages and disadvantages of these CO_2 capture methods, the post-combustion gas–liquid absorption process demonstrates enhanced viability due to its high absorption capacity and low energy consumption [8,9], as well its ability to be integrated into existing power plants by introducing the CO_2 capture stage.

In the gas–liquid absorption process, the absorption solution represents a crucial component as the nature of the absorbent dictates the speed of absorption and also plays an important role in the dimensions of the absorber, with a faster reaction leading to a reduction in unit size. The most commonly used chemical solvents in the absorption of CO_2 are monoethanolamine (MEA), diethanolamine (DEA), methyldiethanolamine (MDEA), ammonia, and sodium hydroxide [2]. Among all alkanolamines, MEA is the most widely used in industrial processes, primarily due to its very fast reaction with CO_2 and high CO_2 absorption capacity. These properties enable the minimization of the absorption column dimensions and reduction of the liquid flow rate, inherently lowering the energy consumption associated with transportation, heating, and cooling. In addition, MEA has a low price and high water solubility. The disadvantages of MEA include high viscosity at high concentrations, high corrosion, and high regeneration energy [2,10].

Compared to conventional packed-bed columns, the use of hollow-fiber membrane contactors (HFMC) for CO_2 capture provides a considerable number of advantages, such as: (i) a significantly larger mass transfer area per unit volume, resulting in a smaller absorber; (ii) membrane systems are highly modular, so scale-up is simpler; (iii) the mass transfer area is constant and independent of hydrodynamic conditions; (iv) membrane systems are easier to replace or repair; (v) more environmental friendly; (vi) due to the physical separation between the two phases, the operating problems related to foaming, flooding, channeling, entrainment, and formation of emulsions are avoided; (vii) reduced loss of solvent; (viii) lower operating and investment costs; (ix) fluid flow rates are not limited to each other [10–14]. Given these advantages, the use of HFMC shows great potential for intensification of the gas–liquid absorption CO_2 capture process. However, there are also drawbacks to using HFMC for carbon capture, such as: (i) membranes introduce an additional mass transfer resistance for CO_2; (ii) transfer resistance increases significantly over time due to membrane watering; (iii) due to the short lifetime, the membranes need to be replaced periodically [2,10–12].

A schematic representation of the HFMC is presented in Figure 1, along with a representation of the CO_2 transfer from the gas phase into the liquid phase through the membrane pores.

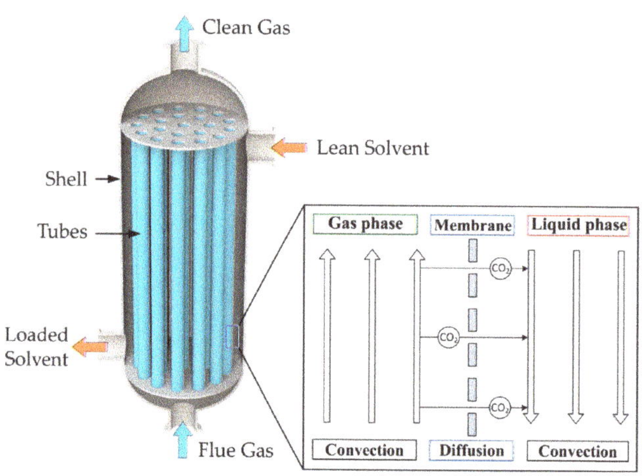

Figure 1. Schematic representation of the HFMC for CO_2 absorption.

The HFMC module consists of a large number of hollow cylindrical membranes inside the shell compartment. There are three distinct regions in the HFMC: the space inside the membranes—the tube side, the hollow-fiber porous membranes, and the space between the membranes—the shell compartment. As the gas mixture flows inside the membrane tubes and the liquid absorbent solution flows between them in a counter-current arrangement, the membranes physically separate the two phases. The CO_2 diffuses from the gas phase through the membrane pores and is absorbed into the liquid solution, where it reacts with the MEA to form a stable compound. The resulting product will then be transported to the desorption column in order to obtain a high purity stream of CO_2 and regenerate the MEA solution (which is subsequently recycled to the absorption stage).

Several models for the capture of CO_2 using HFMC are presented in the literature, divided into one-dimensional (1D), two-dimensional (2D), and three-dimensional (3D) spatial models, based on the process-simplifying assumptions that have been made [15]. For the 1D models, the angular and axial variations of the process parameters are not accounted for, and only the variation over one dimension is considered. For these models, the radial direction is the one of interest, considering a series resistance for CO_2 mass transfer (i.e., resistance of gas, porous membrane, and liquid solution). Based on the two-film theory, the CO_2 will diffuse through the gas film, through the membrane pores and into the liquid solution. The mass transfer for each diffusion step can be represented with partial mass transfer coefficients, which can then be used to determine a global mass transfer coefficient [16]. In addition, the chemical reaction can also be integrated by using the enhancement factor brought by the reaction [17]. The 1D models available in the literature also consider different solvents for the absorption process and whether the membranes are wetted or not [18–23]. The 2D models suppose a single fiber inside the HFMC, considering the variation of parameters over the axial and radial directions occurring within the diffusive and convective transport mechanisms. The CFD modeling consists of a 2D axisymmetric spatial component, thus allowing the representation of a 3D geometry for one hollow-fiber membrane by revolving the 2D results around the axis of symmetry [24–32]. The 2D models provide a better understanding of the CO_2 absorption process using HFMC compared to the 1D models. However, the 2D models are limited to obtaining results for only a single fiber. The development of a 3D model is necessary to achieve higher accuracy and consider phenomena that are not captured in the 1D and 2D models, such as: non-uniform distribution of the fluids, non-uniform properties of the fluids and membranes, interaction between the fibers, etc. The main limitation of 2D models, compared to the 3D model developed in this work, is that they are axisymmetric and use the Happel's approximation model to determine the effective radius of the shell around one membrane. Additionally, the 3D model also considers the fluid distribution around the shell wall, an effect that is also overlooked in the available 2D models. A drawback to the increased complexity enabled by the 3D models is the requirement to use more computational resources in order to solve the governing equations. There is a limited number of 3D models currently available in the literature [33–35], indicating substantial potential for the development of 3D models for membrane-based CO_2 capture [15].

In this work, a comprehensive computational fluid dynamics (CFD) 3D model of the HFMC system for CO_2 capture considering an aqueous MEA solution and counter-current fluid flow is developed and validated with experimental data. The novelty of this work lies in the modeling and simulation of the full 3D HFMC geometry for CO_2 capture. The effect of different operating conditions (flow rates/flow velocities and concentrations) and HFMC geometrical characteristics (porosity, number, inner diameter, thickness, and length of fiber membranes) on CO_2 capture process parameters is investigated. In addition, the impact of how the fluids flow inside the fiber membrane module is evaluated.

2. Mathematical Model Approach

The CFD model for CO_2 absorption was developed using COMSOL Multiphysics, implementing the HFMC geometry represented in Figure 2. In Figure 2a, the entire HFMC geometry is represented, containing 510 membranes arranged symmetrically under a hexagonal distribution inside the shell. In order to reduce the simulation time and the computational resources needed, the complete HFMC module is sliced into four symmetrical parts (Figure 2b), resulting in only a quarter of the HFMC module being simulated. A 2D horizontal section of the simulated geometry is represented in Figure 2c, where the hexagonal symmetrical distribution can be better observed, as well as the distinct sections of the geometry (i.e., tubes—inside the membranes, membranes, and the shell compartment). A section through one of the membranes is represented in Figure 2d. The gas mixture flows inside the membranes (i.e., tubes) and the liquid MEA solution flows between the membranes (i.e., shell) in a counter-current arrangement.

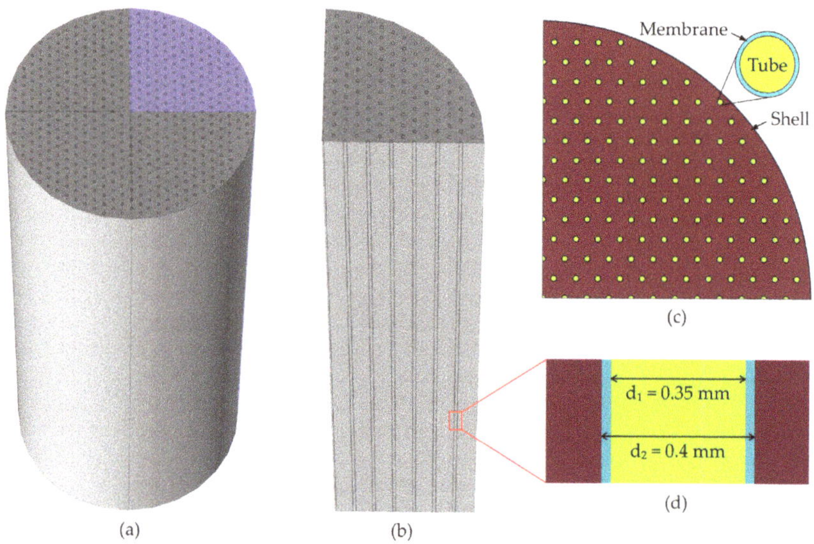

Figure 2. (**a**) 3D HFMC geometry representation, (**b**) quarter of the HFMC 3D representation, (**c**) 2D horizontal section representation of the distribution of membranes in the shell of the HFMC, (**d**) 2D vertical section representation of a membrane.

The mesh used to spatially discretize the HFMC in the CFD model is represented in Figure 3.

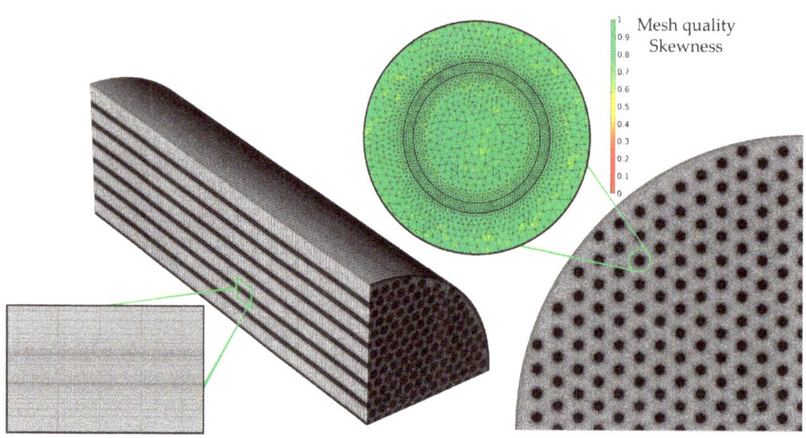

Figure 3. Mesh of the HFMC and quality of the mesh.

The geometry was sliced in 100 equal pieces. The mesh for each slice (presented in the right corner of Figure 3) consists of a triangular mesh, with an increase in element density in the regions of interest: inside and around the membranes, as this is where CO_2 from the gas phase diffuses through the membrane, is absorbed in the liquid, and reacts with the MEA. In addition, the mesh quality is presented in terms of element skewness, with green representing the highest quality. The mesh was sufficiently refined until mesh convergence was reached and the model predicted mesh-independent results. The number of mesh elements for a single fiber membrane was 130,800 (the elements for all membranes in the implemented geometry was 16,731,100), 166,000 for discretization of a tube (all tubes required 20,979,300 elements), and 34,107,600 for the shell, with a total average quality of about 0.85.

The HFMC characteristics are presented in Table 1, along with the process operating conditions, which are represented in Table 2.

Table 1. HFMC Polypropylene (PP) module characteristics [31].

Parameters	Value	Unit
Membrane inner radius (r_1)	0.175	mm
Membrane outer radius (r_2)	0.2	mm
Number of membranes (n)	510	-
Module inner radius (r_{mod})	1.75	cm
Membrane length (L)	27	cm
Membrane porosity (ε_{mem})	0.17	-
Membrane tortuosity (τ_{mem})	19.7	-
Average pores radius (r_p)	0.05	µm

Table 2. Process operating conditions [31].

Parameters	Value	Unit
Gas flow rate (Q_G)	1–3	L/min
Liquid flow rate (Q_L)	10–30	L/h
CO_2 gas concentration (C_{CO_2})	10	vol%
MEA liquid concentration (C_{MEA})	5	wt%
Temperature (T)	298	K
Pressure (P)	1	bar

In order to develop the CFD model for CO_2 absorption using HFMC, the following assumptions have been made:

- Isothermal conditions for the fluids;
- The membranes are operating in non-wet condition;
- The CO_2 solubility in aqueous MEA solution is calculated with Henry's law;
- Fully developed laminar flow for both fluid phases (i.e., gas and liquid).

The 3D CFD model of the HFMC was implemented in COMSOL Multiphysics 6.1, using a 64-bit operating system, with two Intel(R) Xeon(R) Platinum 8168 CPU @ 2.70 GHz and 512 GB of RAM.

2.1. Continuity Equation

The equations are solved in dynamic conditions, therefore the continuity equations (i.e., conservation of mass) for both phases are:

$$\frac{\partial}{\partial t}(\rho_i) + \nabla \cdot (\rho_i u_i) = 0 \tag{1}$$

2.2. Navier–Stokes Equations—Tubes and Shell

The flow regimes for the gas mixture inside the tubes and the liquid solution inside the shell are determined by calculating the Reynolds numbers with values of 18.4 and 44, respectively, indicating laminar flow types for both phases. In order to solve the laminar flows inside the membranes and surrounding them, the model used the continuity equations (Equation (1)) and the Navier–Stokes equations (i.e., conservation of momentum) (Equation (2)):

$$\rho_i \frac{\partial u_i}{\partial t} + \rho_i (u_i \cdot \nabla) u_i = \nabla \cdot (-p_i I + \tau_i) \tag{2}$$

where τ_i is the viscous stress tensor, calculated using (Equation (3)):

$$\tau_i = \mu_i \left(\nabla u_i + (\nabla u_i)^T \right) - \frac{2}{3} \mu_i (\nabla \cdot u_i) I \tag{3}$$

2.3. Species Transport—Tubes

In the gas phase, there is no chemical reaction, therefore the mass transfer of species inside the membranes is described only by convection and diffusion, considering an isotropic diffusion (Equation (4)):

$$\frac{\partial c_j}{\partial t} + \nabla \cdot \left(-D_{j,G} \nabla c_j \right) + u \cdot \nabla c_j = 0 \tag{4}$$

Equation (5), known as the Graetz–Lévêque solution [10], is used to predict the partial mass transfer coefficient of CO_2 inside the membranes (i.e., tube side):

$$Sh_{tube} = \frac{k_{CO_2,g} d_1}{D_{CO_2,g}} = 1.62 \left(\frac{d_1}{L} Re \right)^{0.33} \tag{5}$$

2.4. Species Transport—Membranes

The mass transfer in the hydrophobic microporous membranes is represented by the convection and diffusion of gas components through the membrane pores, which are considered to be filled only with gas as the membrane functions in a non-wet state (Equation (6)):

$$\varepsilon_{mem} \frac{\partial c_j}{\partial t} + \nabla \cdot \left(-D_{j,mem} \nabla c_j \right) + u \cdot \nabla c_j = 0 \tag{6}$$

The effective diffusion coefficient of component j through the membrane pores $D_{j,mem}$ is calculated considering the membrane porosity ε_{mem} and tortuosity τ_{mem} (Equation (7)), assuming an isotropic diffusion:

$$D_{j,mem} = \frac{\varepsilon_{mem}}{\tau_{mem}} D_{j,G} \tag{7}$$

The partial mass transfer coefficient of CO_2 through the membrane is determined using Equation (8) [26]:

$$k_{CO_2,mem} = \frac{D_{CO_2,mem}}{\delta_{mem}} \tag{8}$$

2.5. Species Transport—Shell

Inside the shell compartment, the CO_2 absorbed from the gas phase into the liquid solution reacts with MEA. Therefore, the equation that describes mass transfer inside the shell (Equation (9)) also contains the chemical reaction rate R_j, together with terms for diffusion and convection:

$$\frac{\partial c_j}{\partial t} + \nabla \cdot \left(-D_{j,L} \nabla c_j \right) + u \cdot \nabla c_j = R_j \tag{9}$$

According to Yang and Cussler in their work on gas absorption and stripping, the partial mass transfer coefficient of CO_2 inside the shell side, where fluid flow is parallel to the fibers, can be calculated with Equation (10) ($0.5 < Re < 500$) [36]:

$$Sh_{shell} = \frac{k_{CO_2,l} d_h}{D_{CO_2,l}} = 1.25 \left(Re \frac{d_h}{L} \right)^{0.93} Sc^{0.33} \tag{10}$$

According to the double-film theory, the overall resistance of mass transfer is the sum of the resistances of all layers through which the CO_2 diffuses. The resistance of each layer can be measured by calculating the inverse of the partial mass transfer coefficient of CO_2, indicating the difficulty for the CO_2 molecules to diffuse through that particular layer (fluid or material). Therefore, the overall mass transfer coefficient of CO_2 based on the liquid phase can be calculated using the sum of the three mass transfer resistances inside of the HFMC: liquid ($1/k_{CO_2,l}$), membrane ($1/k_{CO_2,m}$), and gas ($1/k_{CO_2,g}$), while also accounting for the dimensions of the system and the CO_2 solubility in the absorbent solution, as shown in Equation (11) [37]:

$$\frac{1}{K_L d_2} = \frac{1}{k_{CO_2,l} d_2} + \frac{1}{k_{CO_2,m} H_{CO_2} d_{lm}} + \frac{1}{k_{CO_2,g} H_{CO_2} d_1} \tag{11}$$

The CO_2 mass transfer flux from the gas phase to the liquid phase is calculated based on the overall mass transfer coefficient of CO_2, the concentration gradient at the gas–liquid interface, and the absorption enhancement factor E given by the chemical reaction (Equation (12)) [23]:

$$J_{CO_2} = E \cdot K_L \cdot \left(\frac{C^g_{CO_2}}{H_{CO_2}} - C^L_{CO_2} \right) \tag{12}$$

The reaction of CO_2 and MEA, which takes place within the liquid phase located in the shell compartment of the HFMC, is described by the following reaction rate expression [26]:

$$R_{CO_2-MEA} = \frac{10^{(10.99-2152/T_L)}}{1000} C^L_{CO_2} C^L_{MEA} \tag{13}$$

The efficiency of the absorption process of CO_2 using HFMC in an MEA solution is calculated using Equation (14), considering the variation in CO_2 concentration in the flue gas and the clean gas:

$$CO_2 \text{ capture rate} = \left(1 - \frac{C_{CO_2,outlet}}{C_{CO_2,inlet}} \right) \cdot 100 \tag{14}$$

3. Results and Discussion

The CFD model for the CO_2 absorption process in an MEA solution using HFMC was implemented with the membrane module characteristics presented in Table 1 at different operating conditions (Table 2).

3.1. Model Validation

The CFD model of the HFMC for CO_2 absorption was validated with experimental data published in the literature [31]. The validation was performed by comparing the experimental data with the simulation results in terms of CO_2 capture rate (Equation (14)), at various liquid flow rates (Figure 4), gas flow rates (Figure 5), and CO_2 gas inflow concentrations (Figure 6). Overall, the simulation results showed excellent agreement with the experimental data, with $R^2 > 0.922$, demonstrating that the developed CFD model represented the process of CO_2 absorption in an MEA solution using HFMC with high accuracy.

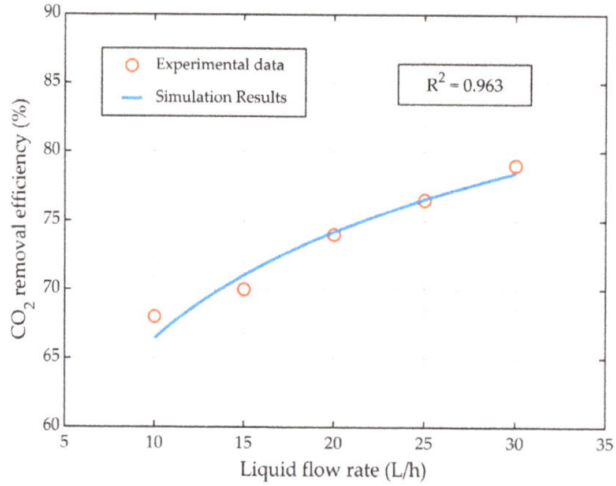

Figure 4. CO_2 capture rate profile at different liquid flow rates, at a constant gas flow rate of 2 L/min.

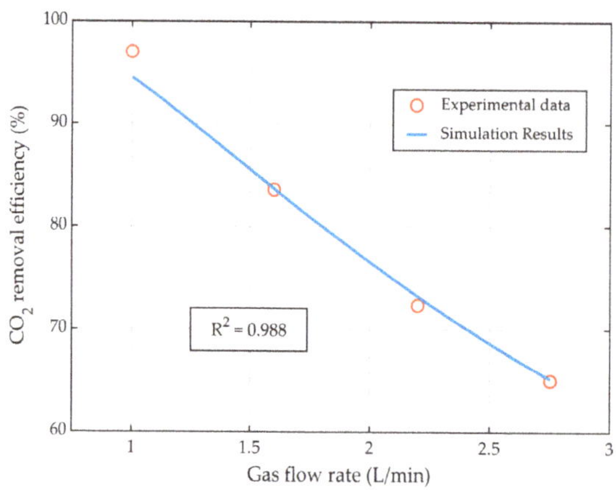

Figure 5. CO_2 capture rate profile at different gas flow rates, at a constant liquid flow rate of 25 L/h.

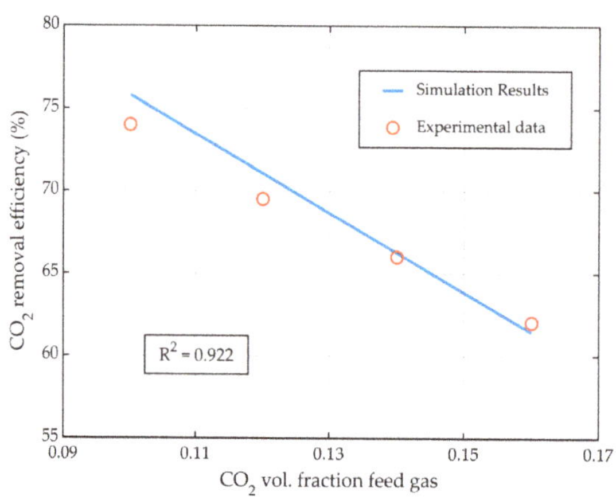

Figure 6. CO_2 capture rate profile at different inflow CO_2 concentration in gas phase, at constant gas (2 L/min) and liquid flow rates (25 L/h).

By increasing the liquid flow rate and maintaining a constant gas flow rate, the absorption efficiency of CO_2 showed an increase (Figure 4) due to the rise in the amount of MEA that flowed inside the HFMC and reacted with the absorbed CO_2. Comparing the experimental data with the simulation results, a very good correlation was observed, with an $R^2 = 0.963$. The CO_2 removal efficiency was increased from about 66.5% to nearly 79% for values of the liquid flow rate ranging from 10 L/h to 30 L/h, respectively.

While maintaining the liquid flow rate constant at 25 L/h, the increase in the gas flow rate determined a decrease in CO_2 capture rate (Figure 5) due to the reduction in the CO_2 residence time inside the HFMC. The experimental data and the simulation results exhibited a very good correlation, with an R^2 value of nearly 0.99. By increasing the gas flow rate from 1 L/min to 2.75 L/min, the CO_2 absorption efficiency decreased from 94.5% to nearly 65%, respectively.

The increase in the concentration of CO_2 in the inflow gas led to a decrease in the absorption efficiency of CO_2 (Figure 6), while maintaining constant gas and liquid flow rates, due to the increase in the amount of CO_2 in the gas phase. The simulation results were in good agreement with the experimental data, with an $R^2 = 0.922$. Changing the CO_2 concentration in the inflow gas from 0.1 to 0.16 vol. fraction, the CO_2 removal efficiency was decreased from nearly 76% to 61.5%, respectively.

3.2. Velocity Profiles

The gas velocity inside the membranes, within the tubes, is represented in Figure 7a. The velocity profile predicted by the model showed a typical laminar shape for fluid flow inside a tube.

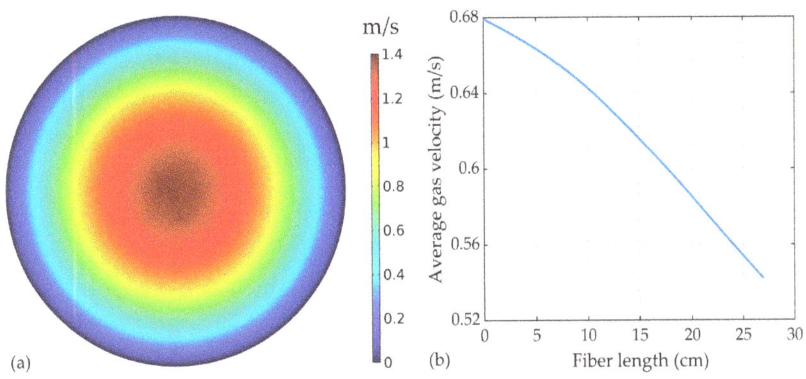

Figure 7. (**a**) Gas mixture velocity profile inside the tubes, (**b**) average gas velocity profile along the HFMC length. Gas flow rate of 2 L/min.

The maximum velocity was observed in the center of the tubes, at around 1.37 m/s (*Re* = 14), while a decreasing velocity was seen closer to the inside membrane wall, with values of zero at the contact with the membrane, due to the friction with the membrane wall and the no-slip boundary conditions assumed at the wall. The average gas velocity profile along the HFMC module is presented in Figure 7b. A reduction in the gas velocity from around 0.68 m/s at the inlet to 0.542 m/s at the outlet of the HFMC module.

The solution for the liquid phase velocity profile surrounding the membrane tubes is represented in Figure 8, with the maximum velocity of the liquid seen between the membranes, at around 0.0132 m/s (*Re* = 37). The liquid velocity was decreasing closer to the membranes and module walls, with a velocity near zero where the friction between the fluid and the walls was highest at the edges.

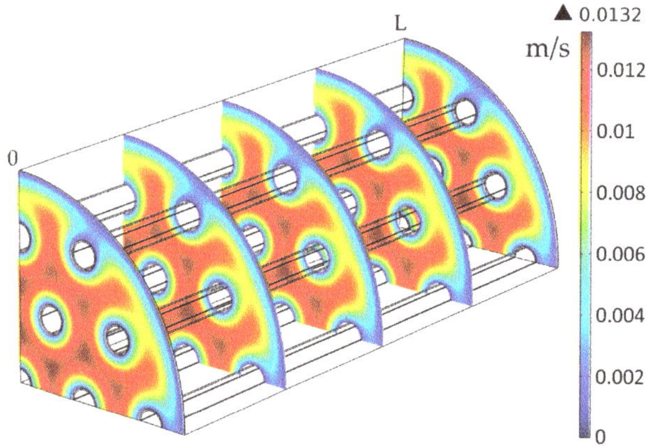

Figure 8. Liquid velocity profile in the shell compartment, Q_L = 25 L/h.

3.3. Concentration Profiles

The CO_2 concentration in the gas phase inside the membranes (tubes) decreased over the length of the membranes, from about 4.06 mol/m^3 at the tube inlet to nearly 0.62 mol/m^3 at the outlet, resulting in a CO_2 absorption efficiency of about 84.7% due to the absorption in the liquid phase (Figure 9). The concentration was maximum in the middle of the tubes and was decreasing in the radial direction of the tubes and along the membrane thickness due to diffusion.

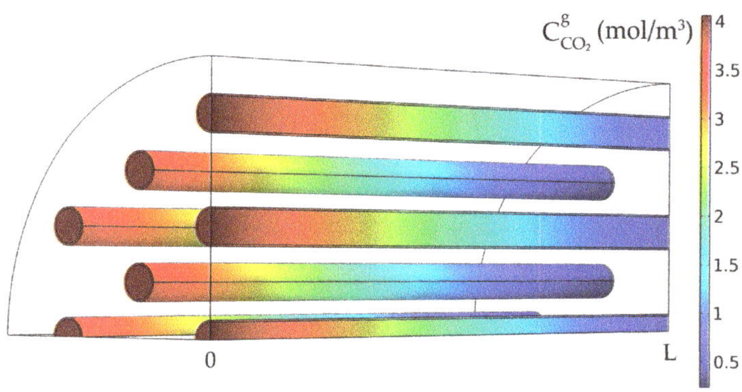

Figure 9. CO_2 gas concentration in the tubes. $C_{MEA,0}$ = 818 mol/m^3 (5 wt%), $C_{CO_2,0}$ = 4.06 mol/m^3 (10 vol%), Q_g = 2 L/min, Q_L = 25 L/h.

For the absorbent solution, which flows between the membranes in the shell, the MEA concentration was also decreasing along the membrane length and around the membranes due to the reaction with the absorbed CO_2 (Figure 10). The MEA concentration decreased along the length of the membranes from 818 mol/m^3 at the inlet to about 720 mol/m^3 at the exit of the HFMC. The concentration decreased around the membranes and was lower closer to the membrane walls due to the reaction with the absorbed CO_2 from the gas phase, while a higher concentration was seen further in the liquid phase, far from the membranes. The absorbed CO_2 reacted first with the MEA close to the membrane walls, then the CO_2 diffused inside the liquid phase further into the liquid solution, which resulted in the concentration gradient around the membranes.

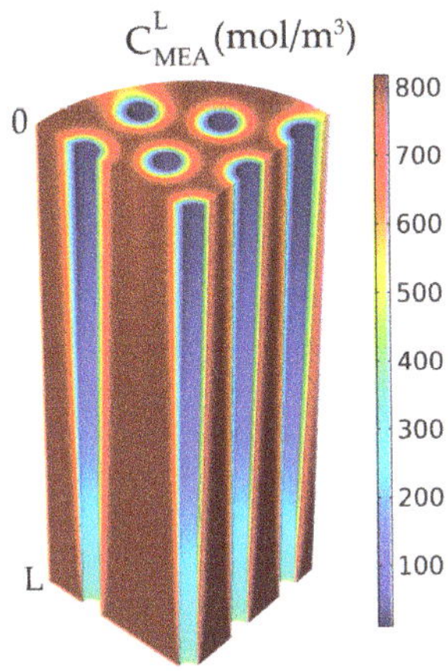

Figure 10. MEA liquid concentration in the shell compartment. $C_{MEA,0}$ = 818 mol/m^3 (5 wt%), $C_{CO_2,0}$ = 4.06 mol/m^3 (10 vol%), Q_g = 2 L/min, Q_L = 25 L/h.

3.4. Tubes vs. Shell

Two different ways for the flow of the gas mixture and liquid solution were considered inside the HFMC module.

In the first case, the gas mixture flowed inside the membranes and the liquid MEA solution between them (Figure 11a), while in the second case, the liquid flow was considered inside the membrane and the gas between them (Figure 11b). In both cases, counter-current flow was assumed, with constant flow rates and compositions.

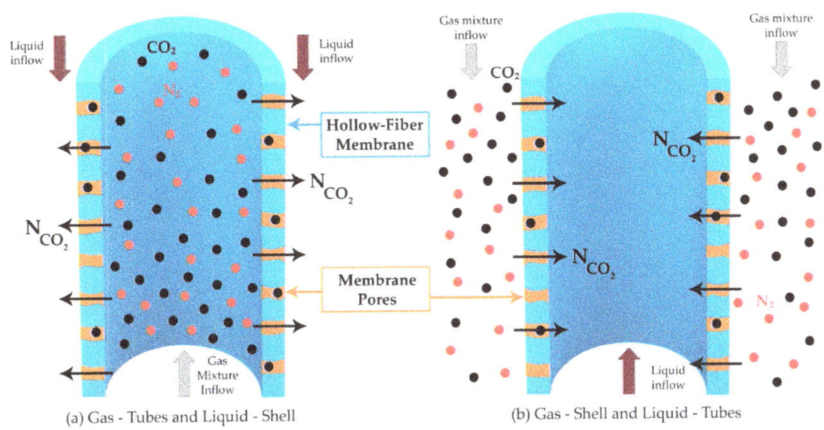

Figure 11. (**a**) Gas mixture inside the membranes (Tubes) and Liquid solution between membranes (Shell), (**b**) Gas mixture between membranes (Shell) and Liquid solution inside the membranes (Tubes).

Regardless of flow configuration (Figure 11), increasing the gas flow rate showed a decrease in the absorption rate of CO_2 (Figure 12) due to the decrease in the overall residence time for the gas phase. When the gas mixture flowed inside the membranes and the liquid aqueous MEA solution between them (Figure 11a), the absorption efficiency decreased from about 94.5% at a gas flow rate of 1 L/min to 62.8% at a gas flow rate of 3 L/min. In the other case, when the gas mixture passed between the membranes and the liquid absorption solution ran inside them (Figure 11b), the CO_2 capture rate decreased from 81% at a gas flow rate of 1 L/min to 43.3% at a gas flow rate of 3 L/min. Over the entire range of gas flow variation, the absorption efficiency was considerably higher for the case when the gas mixture flow inside the membranes and the liquid solution between them (Figure 11a). At a gas flow rate of 1 L/min, the difference in absorption efficiency between the two configurations was 13.5%, while at a gas flow rate of 3 L/min, the difference increased to 19.5%.

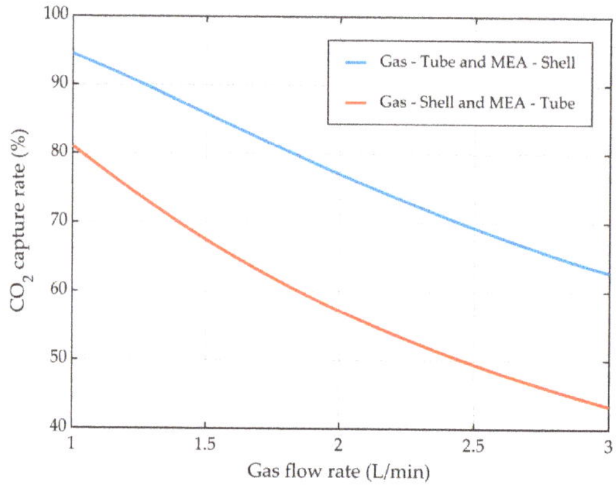

Figure 12. CO_2 capture rate at different gas flow rates. $C_{MEA,0}$ = 818 mol/m^3 (5 wt%), $C_{CO_2,0}$ = 4.06 mol/m^3 (10 vol%), Q_L = 25 L/h.

The effect of the liquid flow rate considering the two different flow configurations inside the HFMC is represented in Figure 13. The increase in the liquid flow rate resulted in an increase in the CO_2 capture rate. When the gas mixture flowed inside the membranes and the liquid aqueous MEA solution between them, the absorption efficiency increased from about 66.5% at a liquid flow rate of 10 L/h to nearly 78.5% at a liquid flow rate of 30 L/h. In the other case, when the gas mixture passed between the membranes and the liquid absorption solution inside them, the CO_2 capture rate increased from 46.5% at a liquid flow rate of 10 L/h to 58.8% at a liquid flow rate of 30 L/h. Along the entire range of investigated liquid flow rates, the absorption efficiency was considerably higher in the case when the gas mixture ran inside the membranes and the liquid solution between them, with a near constant difference between the two cases of around 12.2%.

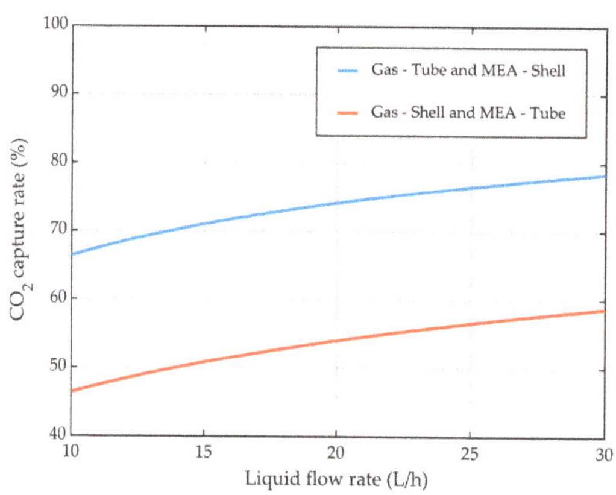

Figure 13. CO_2 capture rate at different liquid flow rates. $C_{MEA,0}$ = 818 mol/m^3 (5 wt%), $C_{CO_2,0}$ = 4.06 mol/m^3 (10 vol%), Q_g = 2 L/min.

3.5. Membrane Dimensions

An important factor in the absorption process of CO_2 using HFMC is the dimensions of the membranes, as the mass transfer area between gas and liquid is primarily dictated by the radius of the membrane and the mass transfer resistance through the pores is determined by the thickness of the membrane.

3.5.1. Constant Membrane Thickness

The CO_2 mass transfer resistance through the membrane pores was kept consistent by maintaining a constant membrane thickness. By increasing the membrane radius, the mass transfer area between the gas mixture and liquid solution increased and the gas velocity decreased, resulting in a higher residence time for the gas phase and an increase in the CO_2 removal efficiency (Figure 14), from around 58.5% for the membrane inner radius of 0.1 mm to nearly 99.5% for a membrane radius of 0.5 mm.

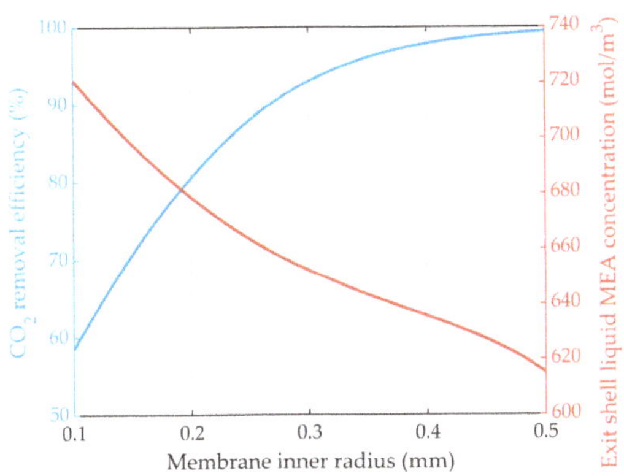

Figure 14. CO_2 removal efficiency and liquid MEA shell exit concentration profiles at different membrane inner radii and a constant membranes thickness of 0.025 mm. $C_{MEA,0}$ = 818 mol/m^3 (5 wt%), $C_{CO_2,0}$ = 4.06 mol/m^3 (10 vol%), Q_g = 2 L/min, Q_L = 25 L/h.

The liquid MEA concentration decreased from about 720 mol/m^3 for a membrane inner radius of 0.1 mm to 615 mol/m^3 for a membrane inner radius of 0.5 mm.

The MEA concentration decreased along the membrane length and around them due to the reaction with the absorbed CO_2. The liquid MEA concentration 3D profiles in the liquid phase surrounding the membranes, for different membrane inner radius values but constant thickness, are represented in Figure 15. The increase in the membrane radius led to an increase in the liquid velocity inside the shell compartment of the HFMC.

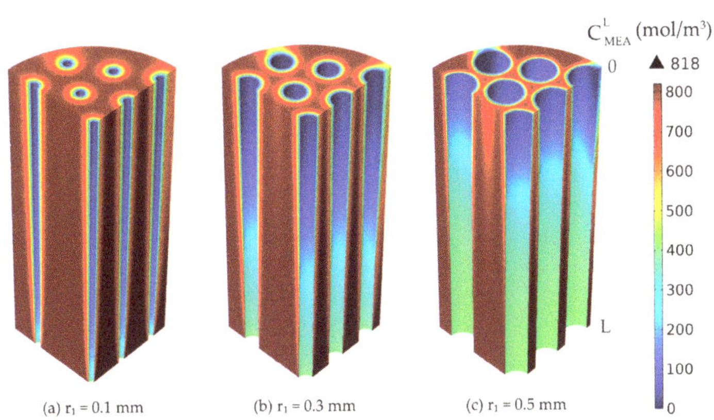

Figure 15. Liquid MEA concentration profiles in the shell compartment, constant membrane thickness 0.025 mm and variable inner radius: (**a**) 0.1 mm, (**b**) 0.3 mm, (**c**) 0.5 mm. $C_{MEA,0}$ = 818 mol/m^3 (5 wt%), $C_{CO_2,0}$ = 4.06 mol/m^3 (10 vol%), Q_g = 2 L/min, Q_L = 25 L/h.

3.5.2. Variable Membrane Thickness—r_2 Constant

By increasing the membrane thickness, the inner radius decreased, meaning that the velocity for the fluid that ran in the tubes (i.e., the gas mixture) increased, leading to a decrease in the overall residence time of CO_2 inside the HFMC. In addition, increasing the membrane thickness determined a rise in the mass transfer resistance of CO_2 through membranes pores. Accounting for both effects, when keeping the membrane outer radius constant, the increase in the membrane thickness led to a decrease in the absorption efficiency of CO_2 (Figure 16). For the outer membrane radius of 0.3 mm, the CO_2 removal efficiency was reduced from about 91% at a membrane thickness of 0.025 mm to nearly 76.5% when the membrane thickness was increased to 0.2 mm. By increasing the outer radius of the membranes from 0.3 mm to 0.4 mm, at a constant membrane thickness of 0.1 mm, the absorption efficiency of CO_2 increased from 86% to about 94.2%, respectively. The change was primarily due to the increase in the mass transfer area between the gas and liquid, but also because of a reduction in gas velocity, thus increasing the residence time of the gas.

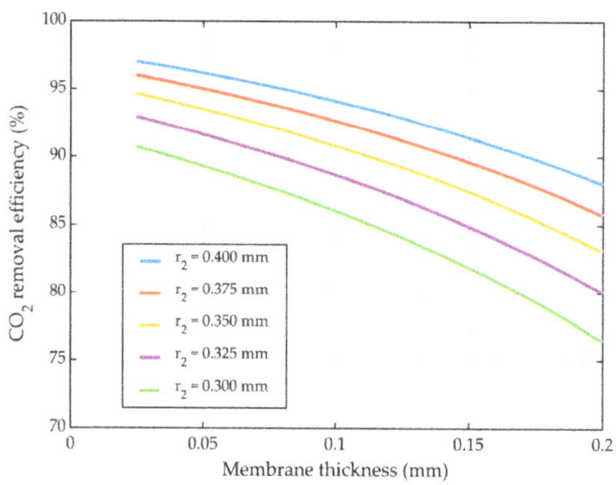

Figure 16. CO_2 removal efficiency at different membrane thicknesses, outside membrane radius constant, variable inner radius. $C_{MEA,0}$ = 818 mol/m^3 (5 wt%), $C_{CO_2,0}$ = 4.06 mol/m^3 (10 vol%), Q_g = 2 L/min, Q_L = 25 L/h.

The CO_2 gas concentration profiles in tubes and membranes at constant outer radius values are presented in Figure 17. The concentration decreased along the length of the membrane due to the absorption in the liquid phase. By increasing the membrane thickness, the mass transfer resistance of CO_2 through the membranes was increased, thus resulting in an increase in the CO_2 gradient concentration in the membranes. In addition, by reducing the inner radius of the membranes, the residence time of the gas mixture inside the tubes decreased, resulting in a higher CO_2 concentration in the exit gas and a lower absorption efficiency.

3.5.3. Variable Membrane Thickness—r_1 Constant

Maintaining the inner radius of the membrane constant, the mass transfer resistance of CO_2 through the gas phase was also kept constant. The increase in the membrane thickness led to a higher mass transfer resistance of CO_2 through the pores of the membrane, as well as an increase in the outer radius of the membranes, which determined an increase in the mass transfer area between the gas mixture and liquid solution. Considering the sum outcome of these two opposite effects on the absorption efficiency of CO_2, at lower values for the membrane thickness, the effect related to the membrane CO_2 mass transfer resistance was lower than the effect of the increase in the mass transfer area, resulting in an overall positive effect on the efficiency of absorption. By continuing to increase the membrane thickness, a maximum in CO_2 removal efficiency was observed at different values of thickness based on the inner radius of the membranes. By further increasing the thickness, the mass transfer resistance of the membranes also increased, leading to a higher overall negative effect on the absorption efficiency.

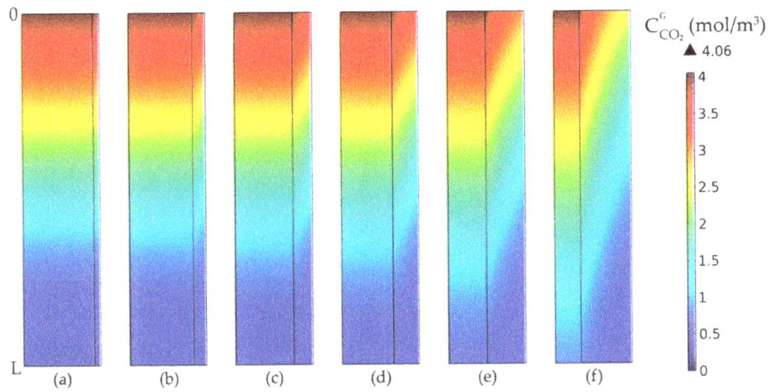

Figure 17. CO_2 gas concentration profiles in tubes and membranes at a constant outside radius of r_2 = 0.3 mm and different membrane thicknesses: (**a**) δ = 0.025 mm, (**b**) δ = 0.05 mm, (**c**) δ = 0.075 mm, (**d**) δ = 0.1 mm, (**e**) δ = 0.15 mm, (**f**) δ = 0.2 mm. $C_{MEA,0}$ = 818 mol/m^3 (5 wt%), $C_{CO_2,0}$ = 4.06 mol/m^3 (10 vol%), Q_g = 2 L/min, Q_L = 25 L/h.

The effect of the membrane thickness for a constant inner radius on the CO_2 removal efficiency is represented in Figure 18. At a membrane inner radius of 0.2 mm, by increasing the membrane thickness from 0.025 mm to 0.4 mm, the CO_2 removal efficiency was increased from around 81% to a maximum of about 88% at a thickness of 0.22 mm, after which the removal efficiency decreased to nearly 87%. At different values for the membrane's inner radius, the maximum values of absorption efficiency were seen at various values for membrane thickness. With increasing inner radius values, the absorption efficiency peaked at lower values for the membrane thickness. For an inner radius of 0.225 mm, the maximum value for the removal efficiency (i.e., nearly 90%) was observed at a thickness of 0.2 mm, while for the lowest inner radius of 0.15 mm, the maximum (i.e., about 83.5%) was noticed at a membrane thickness of around 0.26 mm.

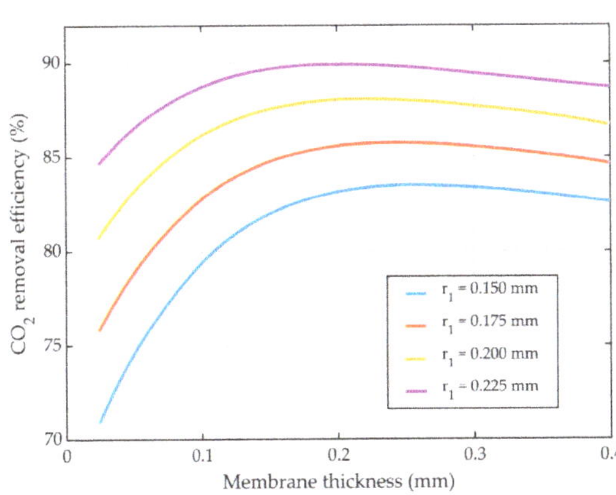

Figure 18. CO_2 removal efficiency at different membrane thicknesses, for different constant inside radii, and variable outer radii. $C_{MEA,0}$ = 818 mol/m^3 (5 wt%), $C_{CO_2,0}$ = 4.06 mol/m^3 (10 vol%), Q_g = 2 L/min, Q_L = 25 L/h.

The CO_2 gas concentration profiles in tubes and membranes, for constant inner radius values and different membrane thickness values, are represented in Figure 19. The CO_2

concentration in the gas phase decreased along the length and radius of the membranes due to the absorption process. By increasing the membrane thickness, the CO_2 mass transfer resistance was increased together with the concentration gradient inside the membrane.

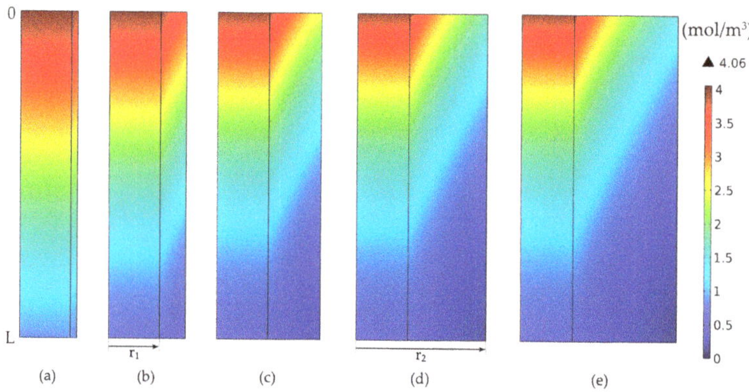

Figure 19. CO_2 gas concentration profile in tubes and membranes over the membranes' length and radius, at a constant membrane inside radius r_1 = 0.2 mm. (**a**) r_2 = 0.225 mm, (**b**) r_2 = 0.3 mm, (**c**) r_2 = 0.4 mm, (**d**) r_2 = 0.5 mm, (**e**) r_2 = 0.6 mm. $C_{MEA,0}$ = 818 mol/m^3 (5 wt%), $C_{CO_2,0}$ = 4.06 mol/m^3 (10 vol%), Q_g = 2 L/min, Q_L = 25 L/h.

3.6. Membrane Porosity

The membrane porosity is an important parameter in the absorption process of CO_2, with a direct impact on the efficiency of the process. The effect of the membrane porosity on the CO_2 capture rate is represented in Figure 20.

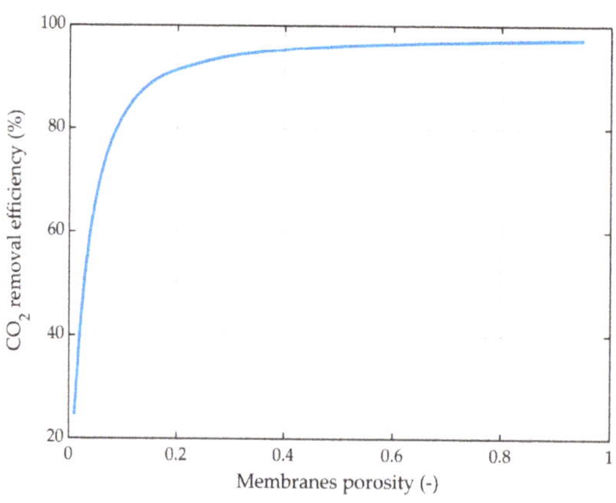

Figure 20. CO_2 removal efficiency at different membrane porosities. r_1 = 0.225 mm, δ = 0.2 mm, $C_{MEA,0}$ = 818 mol/m^3 (5 wt%), $C_{CO_2,0}$ = 4.06 mol/m^3 (10 vol%), Q_g = 2 L/min, Q_L = 25 L/h.

By increasing the porosity while maintaining constant membrane dimensions, the absorption efficiency increased from about 24.5% at a porosity of 0.01 to around 97.3% at a porosity of 0.95. However, at porosity values lower than 0.2, the increase in the absorption efficiency was considerably higher. The increase from a porosity of 0.01 to 0.2 led to an increase in absorption efficiency of around 67%, while a porosity increase from 0.2 to 1 determined an increase in absorption efficiency of only around 6%.

The CO_2 concentration profiles inside the membrane tube compartments and inside the membranes, at different porosities, are represented in Figure 21, considering the same membrane dimensions in all cases. In all cases, the CO_2 concentration in the gas phase decreased along the tubes and length and radius of the membranes due to the absorption in the aqueous MEA solution. Because of the diffusion phenomena, a concentration gradient was observed, with the gradient being more pronounced in the membranes compared to the tubes. At a lower porosity, the final concentration of CO_2 in the purified gases was higher than at a higher porosity, resulting in an improvement in the absorption efficiency with the increase in the membranes' porosity. At a porosity of 0.01, the CO_2 concentration in the gas phase at the exit reached nearly 3.07 mol/m^3 (Figure 21a), but increasing the porosity to 0.1 resulted in a decrease of the exit concentration to around 0.69 mol/m^3 (Figure 21f), and further increasing the porosity to 0.8 led to a decrease in CO_2 concentration to around 0.12 mol/m^3 (Figure 21j).

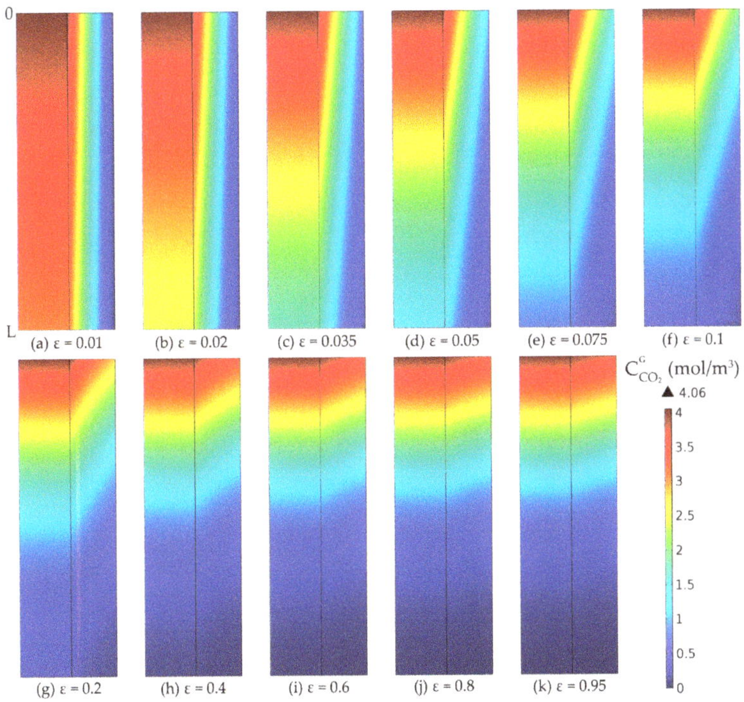

Figure 21. CO_2 gas concentration profiles in tubes and membranes wall, at a constant membrane radius (r_1 = 0.225 mm) and thickness (δ = 0.2 mm) and different porosities: (**a**) ε = 0.01, (**b**) ε = 0.02, (**c**) ε = 0.035, (**d**) ε = 0.05, (**e**) ε = 0.075, (**f**) ε = 0.1, (**g**) ε = 0.2, (**h**) ε = 0.4, (**i**) ε = 0.6, (**j**) ε = 0.8, (**k**) ε = 0.95. $C_{MEA,0}$ = 818 mol/m^3 (5 wt%), $C_{CO_2,0}$ = 4.06 mol/m^3 (10 vol%), Q_g = 2 L/min, Q_L = 25 L/h.

4. Conclusions

A 3D-CFD model of a hollow-fiber membrane contactor for the CO_2 capture process using an aqueous MEA solution was developed and implemented in COMSOL Multiphysics, considering the entire HFMC geometry. The model described the absorption process of CO_2 from the gas phase, through the pores of the membrane and into the liquid solution, followed by the chemical reaction with MEA, accounting for both convection and diffusion mechanisms. The model was validated by comparing experimental data published in literature with model predictions on CO_2 capture rates at different gas and liquid flow rates, as well as different initial concentrations of the flue gases. The simulation results demonstrated excellent correlation with the experimental data, with an R^2 coefficient value exceeding 0.922. The highest CO_2 absorption efficiency was achieved at a high liquid flow rate, low gas flow rate, and low CO_2 concentration in the inflow gas.

Additionally, the optimal mode for fluid flow inside the HFMC was studied, considering identical operating conditions and a counter-current arrangement. The simulation results indicated that superior CO_2 capture rates were achieved when the gas mixture flowed inside the membranes and the liquid absorption solution ran between them. Furthermore, the effect of the membrane dimensions (i.e., radius and thickness) on the capture process of CO_2 using HFMC was investigated. Increasing the membrane radius, while maintaining a constant thickness, positively influenced the efficiency of the process due to the higher mass transfer area and longer residence time of the gas phase. On the other hand, higher membrane thickness resulted in higher CO_2 mass transfer resistance. In order to maximize the absorption efficiency, the findings indicate a low membrane thickness, with optimal dimensions being an inner radius of 0.225 mm and a membrane thickness of 0.2 mm. Additionally, the effect of porosity on the absorption process was investigated, revealing that increasing the porosity led to an improved CO_2 capture efficiency, but with significant impact only for porosity values below 0.2 and a lesser influence for porosities above that threshold.

Author Contributions: Conceptualization, A.-C.B. and V.-C.S.; methodology, V.-C.S. and A.-M.C.; software, V.-C.S.; validation, A.-C.B., V.-C.S. and A.-M.C.; formal analysis, A.-C.B.; investigation, A.-C.B. and V.-C.S.; resources, C.-C.C.; data curation, V.-C.S., C.-C.C. and A.-M.C.; writing—original draft preparation, A.-C.B.; writing—review and editing, C.-C.C. and A.-M.C.; visualization, V.-C.S. and A.-M.C.; supervision, C.-C.C. and A.-M.C.; project administration, C.-C.C. and A.-M.C.; funding acquisition, C.-C.C. All authors have read and agreed to the published version of the manuscript.

Funding: This work was supported by a scholarship no. 36.574/25.11.2022 awarded by Babes-Bolyai University, Cluj-Napoca, Romania, and NO Grants 2014–2021, under project contract no. 13/2020.

Institutional Review Board Statement: Not applicable.

Data Availability Statement: The original contributions presented in the study are included in the article. Further inquiries can be directed to the corresponding author.

Conflicts of Interest: The authors declare no conflicts of interest.

Nomenclature

c	concentration of species (mol/m^3)
d_1	membrane inside diameter (m)
d_h	hydraulic diameter (m)
d_{lm}	logarithmic diameter (m)
$D_{j,i}$, $D_{j,mem}$	diffusion coefficient of species j in phase i and membrane (m^2/s)
E	enhancement factor (-)
H_{CO_2}	Henry's coefficient of CO_2 in the liquid MEA solution (-)
J_{CO_2}	CO_2 mass transfer flux from gas to liquid (mol/m^2 s)
$k_{CO_2,i}$, $k_{CO_2,mem}$	partial CO_2 mass transfer coefficient in the i phase and through membrane pores (m/s)
K_L	global CO_2 mass transfer coefficient (m/s)
L	membranes length (m)
n	number of membranes (-)
P	pressure (bar)
Q	volumetric flow rate (m^3/s)
r_1	membrane inside radius (m)
r_2	membrane outside radius (m)
r_{mod}	module inner radius (m)
r_p	average membranes pores radius (m)
R_{CO_2}	reaction rate (mol/m^3 s)
Re	Reynolds number (-)
Sc	Schmidt number (-)
Sh	Sherwood number (-)
T	temperature (K)
u	superficial fluid velocity (m/s)

Superscripts/subscripts	
i	gas (g) and liquid (L) phase
j	system components
Greek letters	
δ_{mem}	membrane thickness (m)
ε_{mem}	membrane porosity (-)
μ	dynamic viscosity (Pa s)
ρ	fluid density (kg/m^3)
τ_{mem}	membrane tortuosity (-)

References

1. Nocito, F.; Dibenedetto, A. Atmospheric CO_2 Mitigation Technologies: Carbon Capture Utilization and Storage. *Curr. Opin. Green Sustain. Chem.* **2020**, *21*, 34–43. [CrossRef]
2. Eskandari, M.; Khaksar, S.A.N.; Keshavarz, P. CO_2 Absorption Using Benzylamine as Absorbent and Promoter in a Hollow Fiber Membrane Contactor: A Numerical Study. *J. CO2 Util.* **2022**, *66*, 102287. [CrossRef]
3. Storrs, K.; Lyhne, I.; Drustrup, R. A Comprehensive Framework for Feasibility of CCUS Deployment: A Meta-Review of Literature on Factors Impacting CCUS Deployment. *Int. J. Greenh. Gas Control* **2023**, *125*, 103878. [CrossRef]
4. Gkotsis, P.; Peleka, E.; Zouboulis, A. Membrane-Based Technologies for Post-Combustion CO_2 Capture from Flue Gases: Recent Progress in Commonly Employed Membrane Materials. *Membranes* **2023**, *13*, 898. [CrossRef]
5. Wang, Z.; Liu, L.; Ren, T.; Gao, P.; Sang, H.; Yang, X.; Li, Z.; Wang, Z.; Li, S.; Li, G.K. Carbon Capture from Humid Gases Using Alkaline-Promoted Polypyrrole by a Vacuum Swing Adsorption Process. *Sep. Purif. Technol.* **2023**, *323*, 124399. [CrossRef]
6. Entesari, N.; Divsalar, A.; Tsotsis, T.T. Modeling and Simulation of a Reactive Separation System for Carbon Capture and Utilization in Biogas Streams. *Chem. Eng. Process.-Process Intensif.* **2020**, *156*, 108093. [CrossRef]
7. Vadillo, J.M.; Díaz-Sainz, G.; Gómez-Coma, L.; Garea, A.; Irabien, A. Chemical and Physical Ionic Liquids in CO_2 Capture System Using Membrane Vacuum Regeneration. *Membranes* **2022**, *12*, 785. [CrossRef]
8. Fu, L.; Ren, Z.; Si, W.; Ma, Q.; Huang, W.; Liao, K.; Huang, Z.; Wang, Y.; Li, J.; Xu, P. Research Progress on CO_2 Capture and Utilization Technology. *J. CO2 Util.* **2022**, *66*, 102260. [CrossRef]
9. Ding, H.; Zhang, Y.; Dong, Y.; Wen, C.; Yang, Y. High-Pressure Supersonic Carbon Dioxide (CO_2) Separation Benefiting Carbon Capture, Utilisation and Storage (CCUS) Technology. *Appl. Energy* **2023**, *339*, 120975. [CrossRef]
10. Gabelman, A.; Hwang, S.-T. Hollow Fiber Membrane Contactors. *J. Membr. Sci.* **1999**, *159*, 61–106. [CrossRef]
11. Mansourizadeh, A.; Ismail, A.F. Hollow Fiber Gas–Liquid Membrane Contactors for Acid Gas Capture: A Review. *J. Hazard. Mater.* **2009**, *171*, 38–53. [CrossRef]
12. Xia, J.; Zhang, Z.; Wang, L.; Wang, F.; Miao, H.; Zhang, H.; Xia, L.; Yuan, J. Performance Evaluation and Optimization of Hollow Fiber Membrane Contactors for Carbon Dioxide Absorption: A Comparative Study of Ammonia, Ethanolamine, and Diethanolamine Solvents. *J. Environ. Chem. Eng.* **2023**, *11*, 111354. [CrossRef]
13. Cesari, L.; Castel, C.; Favre, E. Membrane Contactors for Intensified Gas-Liquid Absorption Processes with Physical Solvents: A Critical Parametric Study. *J. Membr. Sci.* **2021**, *635*, 119377. [CrossRef]
14. Taghvaie Nakhjiri, A.; Heydarinasab, A.; Bakhtiari, O.; Mohammadi, T. Numerical Simulation of CO_2/H_2S Simultaneous Removal from Natural Gas Using Potassium Carbonate Aqueous Solution in Hollow Fiber Membrane Contactor. *J. Environ. Chem. Eng.* **2020**, *8*, 104130. [CrossRef]
15. Rivero, J.R.; Panagakos, G.; Lieber, A.; Hornbostel, K. Hollow Fiber Membrane Contactors for Post-Combustion Carbon Capture: A Review of Modeling Approaches. *Membranes* **2020**, *10*, 382. [CrossRef]
16. Mavroudi, M.; Kaldis, S.P.; Sakellaropoulos, G.P. A Study of Mass Transfer Resistance in Membrane Gas–Liquid Contacting Processes. *J. Membr. Sci.* **2006**, *272*, 103–115. [CrossRef]
17. Kumar, P.S.; Hogendoorn, J.A.; Feron, P.H.M.; Versteeg, G.F. Approximate Solution to Predict the Enhancement Factor for the Reactive Absorption of a Gas in a Liquid Flowing through a Microporous Membrane Hollow Fiber. *J. Membr. Sci.* **2003**, *213*, 231–245. [CrossRef]
18. Rode, S.; Nguyen, P.T.; Roizard, D.; Bounaceur, R.; Castel, C.; Favre, E. Evaluating the Intensification Potential of Membrane Contactors for Gas Absorption in a Chemical Solvent: A Generic One-Dimensional Methodology and Its Application to CO_2 Absorption in Monoethanolamine. *J. Membr. Sci.* **2012**, *389*, 1–16. [CrossRef]
19. Saeed, M.; Deng, L. Post-Combustion CO_2 Membrane Absorption Promoted by Mimic Enzyme. *J. Membr. Sci.* **2016**, *499*, 36–46. [CrossRef]
20. Khaisri, S.; deMontigny, D.; Tontiwachwuthikul, P.; Jiraratananon, R. A Mathematical Model for Gas Absorption Membrane Contactors That Studies the Effect of Partially Wetted Membranes. *J. Membr. Sci.* **2010**, *347*, 228–239. [CrossRef]
21. Chu, Y.; Lindbråthen, A.; Lei, L.; He, X.; Hillestad, M. Mathematical Modeling and Process Parametric Study of CO_2 Removal from Natural Gas by Hollow Fiber Membranes. *Chem. Eng. Res. Des.* **2019**, *148*, 45–55. [CrossRef]

22. Boributh, S.; Rongwong, W.; Assabumrungrat, S.; Laosiripojana, N.; Jiraratananon, R. Mathematical Modeling and Cascade Design of Hollow Fiber Membrane Contactor for CO_2 Absorption by Monoethanolamine. *J. Membr. Sci.* **2012**, *401–402*, 175–189. [CrossRef]
23. Bozonc, A.-C.; Cormos, A.-M.; Dragan, S.; Dinca, C.; Cormos, C.-C. Dynamic Modeling of CO_2 Absorption Process Using Hollow-Fiber Membrane Contactor in MEA Solution. *Energies* **2022**, *15*, 7241. [CrossRef]
24. Al-Marzouqi, M.H.; El-Naas, M.H.; Marzouk, S.A.M.; Al-Zarooni, M.A.; Abdullatif, N.; Faiz, R. Modeling of CO_2 Absorption in Membrane Contactors. *Sep. Purif. Technol.* **2008**, *59*, 286–293. [CrossRef]
25. Hosseinzadeh, A.; Hosseinzadeh, M.; Vatani, A.; Mohammadi, T. Mathematical Modeling for the Simultaneous Absorption of CO_2 and SO_2 Using MEA in Hollow Fiber Membrane Contactors. *Chem. Eng. Process. Process Intensif.* **2017**, *111*, 35–45. [CrossRef]
26. Shirazian, S.; Pishnamazi, M.; Rezakazemi, M.; Nouri, A.; Jafari, M.; Noroozi, S.; Marjani, A. Implementation of the Finite Element Method for Simulation of Mass Transfer in Membrane Contactors. *Chem. Eng. Technol.* **2012**, *35*, 1077–1084. [CrossRef]
27. Faiz, R.; Al-Marzouqi, M. CO_2 Removal from Natural Gas at High Pressure Using Membrane Contactors: Model Validation and Membrane Parametric Studies. *J. Membr. Sci.* **2010**, *365*, 232–241. [CrossRef]
28. Goyal, N.; Suman, S.; Gupta, S.K. Mathematical Modeling of CO_2 Separation from Gaseous-Mixture Using a Hollow-Fiber Membrane Module: Physical Mechanism and Influence of Partial-Wetting. *J. Membr. Sci.* **2015**, *474*, 64–82. [CrossRef]
29. Rezakazemi, M.; Niazi, Z.; Mirfendereski, M.; Shirazian, S.; Mohammadi, T.; Pak, A. CFD Simulation of Natural Gas Sweetening in a Gas–Liquid Hollow-Fiber Membrane Contactor. *Chem. Eng. J.* **2011**, *168*, 1217–1226. [CrossRef]
30. Shirazian, S.; Moghadassi, A.; Moradi, S. Numerical Simulation of Mass Transfer in Gas–Liquid Hollow Fiber Membrane Contactors for Laminar Flow Conditions. *Simul. Model. Pract. Theory* **2009**, *17*, 708–718. [CrossRef]
31. Nakhjiri, A.T.; Heydarinasab, A.; Bakhtiari, O.; Mohammadi, T. Experimental Investigation and Mathematical Modeling of CO_2 Sequestration from CO_2/CH_4 Gaseous Mixture Using MEA and TEA Aqueous Absorbents through Polypropylene Hollow Fiber Membrane Contactor. *J. Membr. Sci.* **2018**, *565*, 1–13. [CrossRef]
32. Vadillo, J.M.; Gómez-Coma, L.; Garea, A.; Irabien, A. CO_2 Desorption Performance from Imidazolium Ionic Liquids by Membrane Vacuum Regeneration Technology. *Membranes* **2020**, *10*, 234. [CrossRef] [PubMed]
33. Pozzobon, V.; Perré, P. Mass Transfer in Hollow Fiber Membrane Contactor: Computational Fluid Dynamics Determination of the Shell Side Resistance. *Sep. Purif. Technol.* **2020**, *241*, 116674. [CrossRef]
34. Cai, J.J.; Hawboldt, K.; Abdi, M.A. Analysis of the Effect of Module Design on Gas Absorption in Cross Flow Hollow Membrane Contactors via Computational Fluid Dynamics (CFD) Analysis. *J. Membr. Sci.* **2016**, *520*, 415–424. [CrossRef]
35. Boucif, N.; Corriou, J.P.; Roizard, D.; Favre, E. Carbon Dioxide Absorption by Monoethanolamine in Hollow Fiber Membrane Contactors: A Parametric Investigation. *AIChE J.* **2012**, *58*, 2843–2855. [CrossRef]
36. Yang, M.-C.; Cussler, E.L. Designing Hollow-Fiber Contactors. *AIChE J.* **1986**, *32*, 1910–1916. [CrossRef]
37. Ghobadi, J.; Ramirez, D.; Jerman, R.; Crane, M.; Khoramfar, S. CO_2 Separation Performance of Different Diameter Polytetrafluoroethylene Hollow Fiber Membranes Using Gas-Liquid Membrane Contacting System. *J. Membr. Sci.* **2018**, *549*, 75–83. [CrossRef]

membranes

Article

Permeance of Condensable Gases in Rubbery Polymer Membranes at High Pressure

Karina Schuldt [1,2], Jelena Lillepärg [1,*], Jan Pohlmann [1], Torsten Brinkmann [1] and Sergey Shishatskiy [1]

[1] Helmholtz-Zentrum Hereon, Institute of Membrane Research, Max-Planck-Str. 1, 21502 Geesthacht, Germany; karina.schuldt@camfil.com (K.S.); jan.pohlmann@hereon.de (J.P.); torsten.brinkmann@hereon.de (T.B.); sergey.shishatskiy@hereon.de (S.S.)

[2] Camfil GmbH, Feldstraße 26-32, 23585 Reinfeld, Germany

* Correspondence: jelena.lillepaerg@hereon.de; Tel.: +49-4152-87-2448

Abstract: The gas transport properties of thin film composite membranes (TFCMs) with selective layers of PolyActive™, polydimethylsiloxane (PDMS), and polyoctylmethylsiloxane (POMS) were investigated over a range of temperatures (10–34 °C; temperature increments of 2 °C) and pressures (1–65 bar abs; 38 pressure increments). The variation in the feed pressure of condensable gases CO_2 and C_2H_6 enabled the observation of peaks of permeance in dependence on the feed pressure and temperature. For PDMS and POMS, the permeance peak was reproduced at the same feed gas activity as when the feed temperature was changed. PolyActive™ TFCM showed a more complex behaviour, most probably due to a higher CO_2 affinity towards the poly(ethylene glycol) domains of this block copolymer. A significant decrease in the permeate temperature associated with the Joule–Thomson effect was observed for all TFCMs. The stepwise permeance drop was observed at a feed gas activity of $p/po \geq 1$, clearly indicating that a penetrant transfer through the selective layer occurs only according to the conditions on the feed side of the membrane. The permeate side gas temperature has no influence on the state of the selective layer or penetrant diffusing through it. The most likely cause of the observed TFCM behaviour is capillary condensation of the penetrant in the swollen selective layer material, which can be provoked by the clustering of penetrant molecules.

Keywords: thin film composite membrane; condensable gases; gas transport properties; high pressure; CO_2 permeance

1. Introduction

Thin film composite polymeric membranes (TFCMs) have been established as versatile and flexible tools to be used in gas and vapor separation processes [1–3]. This type of membrane benefits from a multi-layer structure, which allows the designer to achieve mechanical stability by choosing the appropriate support structure and optimizing membrane performance by applying a thin layer of the selective material that is most suitable for the separation in question. Carbon dioxide separation is among the most important applications, especially in natural and bio-gas purification and in the emerging carbon dioxide separation from combustion flue gas streams and industrial off-gases [4–9]. In recent years, a significant amount of membrane material has been developed aimed at the separation of CO_2/CH_4, CO_2/N_2, and CO_2/H_2 gas pairs. Detailed studies on CO_2 separation by polymer-based membranes at a wide range of temperatures and pressures have been published over the last four decades [10–13].

Some glassy polymers as materials with selective layers are very promising for CO_2 separation, but aging phenomena and plasticization affect their permeation properties in a negative way. For TFCMs with glassy separation layers, a loss in permeance of more than 25% was observed [14,15].

While the transport of gases through rubbery polymeric membranes at low pressures is well studied, only a few publications have addressed the subject of transport behavior at high pressures [16–21]. At low pressures with limited swelling, the transport behavior is well described by the Free Volume model (FVM) [22].

It was found that at an average pressure p(feed)-p(permeate) not exceeding 20 bar, the permeation of a complex gas mixture is adequately described by the FVM using model parameters derived from results of simple, single gas experiments [22–26]. However, above a certain pressure difference, the transport of CO_2 did not follow the expected tendency. This has initiated experimental work concerning the investigation of single and mixed gas permeation using, as it is considered to be well studied, PolyActive^{TM} 1500 TFCMs [2,27–29]. The results showed that a clear pressure threshold exists below which the FVM can describe the transport of multicomponent gas mixtures with an accuracy that is sufficient for practical separation process designs. Above the threshold, the CO_2 permeance was significantly higher than the values predicted by the FVM.

An investigation of condensable gas transport through thick polymer films or integral asymmetric membranes in the full range of pressures has only been conducted a few times, for example, in the work of Favre et al. [30]. This study shows no unexpected changes in the permeation curve for highly soluble penetrants such as chloroform, 2-butanol, 1-butanol, etc., in the full range of gas activity, defined as the ratio of partial and saturated vapor pressures at a given temperature. In the case of microporous materials, where condensation of penetrants is possible and capillary or selective surface flow can occur, the shape of the permeation curve is very complex [31,32].

This paper describes the experimental results of permeation behavior investigations of two single condensable gases, CO_2 and C_2H_6, in three different rubbery polymers serving as the selective layer for TFCMs in the full range of pressures up to the point of gas condensation on the feed side of the membrane. For the systematic investigation of the permeation behavior of gas separation membranes, a special laboratory-scale test facility for high pressures was developed.

The main emphasis was put on the investigation of CO_2 behavior in PolyActive^{TM} 1500 TFCM due to the envisaged membrane applications for CO_2 separation from various gas streams [2,24,33,34]. In order to find out whether the observed facts are specific to only the CO_2/PolyActive^{TM} pair or if a similar behavior of condensable gas and rubbery polymer can be observed for other combinations, ethane as an alternative condensable gas and two siloxane-based polymers, polydimethylsiloxane and polyoctylmethylsiloxane, were used in the experiments. Helium was used to prove the absence of porosity in the prepared TFCM samples and to prove that the designed experimental facility delivers adequate results. The experiments were performed in a pressure range of 1–65 bar at temperatures of 10–34 °C.

2. Materials and Methods

The TFCM selective layers were made from the rubbery polymers PolyActive^{TM} 1500 (PolyVation BV, Groningen, The Netherlands), composed of 77 wt% poly(ethylene glycol) soft segments and 23 wt% poly(butylene terephthalate) hard segments [35] (PA), polyoctylmethylsiloxane (POMS) (abcr GmbH, Karlsruhe, Germany), and polydimethylsiloxane (PDMS) (the supplier cannot be disclosed due to licensing limitations).

The gases had a purity of at least 99.95% for C_2H_6, 99.996% for He (both Air Liquide Deutschland GmbH, Stelle, Germany), and 99.995% for CO_2 (Linde GmbH, Leuna, Germany).

2.1. Preparation of Samples for Gas Sorption Experiments

The samples for gas sorption experiments were prepared as follows: The 3 wt% solution of PA was prepared in tetrahydrofuran (for analysis grade, Merck KGaA, Darmstadt, Germany) and stirred under reflux for at least 2 h until full polymer dissolution. To prepare a thick isotropic film, an Al cylinder with a polished bottom surface was placed on a leveled Teflon^{TM}-coated glass plate. The polymer solution was poured into the formed vessel; and the solvent was evaporated under a slow N_2 flow for 48 h at 25 °C, as described in [36]. The 6 wt% PDMS solution in iso-octane was stirred for 2 h under ambient conditions and poured into a Teflon™ beaker with a leveled bottom. The solvent was left to evaporate for 72 h at ambient conditions under a constant N_2 flow. All samples were treated in a vacuum oven at 70 °C for at least 3 h in order to remove the residual solvent. To complete the crosslinking reaction in the PDMS sample, it was additionally exposed to 100 °C for 2 h in a vacuum oven.

2.2. TFCM Samples for Gas Transport Experiments

The TFCMs used in the current study contained a selective layer consisting of PA, PDMS, or POMS. The PA selective layer was deposited on a gutter layer of crosslinked PDMS, which in turn was formed on a porous ultrafiltration (UF) polyacrylonitrile (PAN) membrane supported by a polyester nonwoven, which provided the mechanical strength. A protective layer of PDMS was deposited over the PA selective layer to cover possible defects (pin holes) and ensure membrane protection [37]. This membrane is known for its high CO_2 permeance, with the selectivity of the TFCM close to that of the selective layer material [2]. To compare the gas transport properties determined at various pressures and temperatures, TFCMs with PDMS and POMS selective layers on the PAN UF membrane were studied as well. The list of membranes is presented in Table 1.

Table 1. Specification of membrane samples investigated in the current study.

Sample Nomenclature	Sample Composition	Note
UF-PAN [1]	Microporous PAN membrane on polyester non-woven.	Standard support for TFCMs, as described elsewhere [38].
PDMS1280 [2] -PAN	PDMS 1280 nm on UF-PAN support (Figure S2a) [3].	Separation layer of PDMS prepared from high-concentration solution.
GL	PDMS 600 nm on UF-PAN support.	Standard PDMS gutter layer membrane [4], as described elsewhere [39].
POMS6250-GL	POMS on GL membrane (Figure S2b).	POMS layer deposited using 8 wt% casting solution.
PA77-GL	PDMS as top layer on PA selective layer-coated GL membrane (Figure S2c,d).	Standard TFCM, as described elsewhere [40].
PA186-GL	PDMS as top layer on PA selective layer-coated GL membrane.	PA layer deposited using 1 wt% casting solution.
PA film	PA 388 μm, 173 μm, and 19 μm thick films.	Isotropic films with uniform thickness.
PDMS film	PDMS 285 μm and 261 μm thick films.	Isotropic films with uniform thickness.

[1] UF-PAN support is an ultrafiltration membrane composed of a porous polyacrylonitrile layer deposited by the phase inversion method on top of a polyester non-woven. [2] The sample name is composed of the abbreviation for the selective layer material and the effective selective layer thickness, which is calculated as the ratio of the membrane permeance and the permeability coefficient of the selective material. [3] Selective layer of the PDMS1280-PAN was formed without the addition of adhesion-providing components. [4] The GL membrane is formed with PDMS layer containing adhesion-providing components (composition cannot be disclosed due to license infringement) with a thickness of ca. 150 nm deposited on a PAN UF support.

2.3. Gas Permeance Measurements up to and beyond the Saturated Vapor Pressure of a Penetrant

The single gas permeances of TFCM samples were determined using a "constant pressure/variable volume" (PI) facility developed at Helmholtz-Zentrum Hereon and described elsewhere [24,39,41]. PI measurements conducted at a constant temperature and low pressure were also used for the estimation of Knudsen-type dependence for the porous UF-PAN membranes.

For high-pressure experiments, a special experimental setup was developed that allows automatic measurements in wide pressure and temperature ranges. The setup utilizes the "single gas experiments at high pressures" method of gas transport experiments and is described elsewhere [24]. This setup was designed for experiments with high-performance flat sheet gas separation membranes. The immersion of the membrane test cell into the water tank of a cooling thermostat Huber MPC-K20 (HUBERLAB AG, Aesch, Switzerland) made it possible to carry out experiments at temperatures in the range of 5 to 60 °C. The whole setup can be used for experiments at feed pressures ranging from 1.5 to 100 bar abs with the pressure on the permeate side of the membrane close to ambient.

The conduction of automated measurements is realized using an electrically adjustable pressure reducer (B.E.S.T. Fluid Systems GmbH/Swagelok Hamburg, Brackel, Germany) that controls the feed pressure. After the pressure reducer, the gas line becomes a 20 mm in diameter test cell, which is placed in the water tank of a thermostat. The thermostat ensures that the test cell has a defined and constant temperature. Before and after the test cell, the pressure sensors LEO 3 (Keller Druckmesstechnik GmbH, Jestetten, Germany) are installed to control the feed and permeate pressures during the dead-end experiment. For the recording of the permeate flow, the experimental setup was equipped with four mass flow sensors SLA5800 Series (Brooks Instrument GmbH, Dresden, Germany) with gas flow ranges of 0–5, 0–50, 0–500, and 0–5000 cm^3 (STP) min^{-1}. Since the sensors are calibrated with N_2, the results for each gas were adjusted using a gas factor provided by Brooks Instrument GmbH. The flow sensors were combined "in series" with the "smallest" flow sensor first, and such an arrangement enabled the experiments to be conducted in the full range of pressures for every gas under study. To take into account the effect of the flow resistance of the smallest flow sensor, a pressure sensor with a pressure range of 0–4 bars was installed immediately at the permeate exit of the measurement cell, and 6 mm stainless steel tubes were used for connections on the permeate side. The chosen design allows one to work with permeate flows up to 2500 cm^3 (STP)/min, which is in the range of a laminar-type flow in the permeate side tubing. The schematic design of the test cell is given in Figure S1.

In order to control changes in membrane performance that could be caused by penetrants, highly soluble in the material of the selective layer, each measurement sequence was started and ended with the inert gas He. Before the measurement was taken, the experimental setup was flushed with the gas to be measured. The measuring program was applied in such a way that the system was first brought to and stabilized at the desired temperature for at least 10 min. Afterwards, the specified pressure curve was run through. For each measuring point, the pressure and temperature on the feed and permeate sides, as well as the permeate flow rate, were recorded automatically.

The temperature sensor Pt100 with a 500 μm diameter was installed into the porous sintered metal disc in direct contact with the permeate side of the membrane. It enabled monitoring the temperature of the gas permeating through the membrane at a location that was as close as possible to the permeate side of the selective layer. Long, 300 s pressure equilibration times and high permeate flow rates through the membrane were sufficient for temperature equilibration on the permeate side of the membrane. Since Pt100 was installed in the center of the sintered metal support, it was assumed that the measured temperature would be dominated by the temperature of the gas flowing through the membrane and that there would be a negligible influence of the heat coming from the thermostat bath through the bulky stainless steel body of the measurement cell and finally through the highly porous sintered metal disc.

Gas transport experiments were carried out for membrane samples with an effective membrane area of 1.72 cm^2. Samples of 20 mm in diameter were placed into the measurement cell on a sintered metal support with an embedded temperature sensor. The membrane was sealed with an EPDM O-ring. The measurement cell was temperature-equilibrated for 1200 s after the thermostat reached the desired temperature. The initial temperature of the experiment was always chosen to be 10 °C. After the temperature equilibration step, feed pressure was applied to the sample, starting from the lowest pressure point possible for the pressure regulator. The pressure was equilibrated for 300 s at the achieved pressure, and after this, the flow rate data were acquired from the mass flow sensors, and the next pressure value was set. In total, 38 pressure points were acquired, covering the full possible range of gas pressures. The highest possible pressure for both CO_2 and C_2H_6 was in accordance with the gas bottle temperature, which was kept in a temperature-stabilized lab. Such an arrangement allowed, in some cases, to overcome the condensation pressure of the gas under study at the measurement cell temperature of 10 °C, as will be shown later. After reaching the highest pressure setting, the experiment was continued with a stepwise decrease in the feed pressure down to the initial feed pressure setting. This allowed us to investigate possible hysteresis in the gas permeance for feed pressure, increasing and decreasing parts of the experiment. After the minimum pressure was reached, the temperature of the measurement cell increased by 2 °C and the experiment was repeated. In the current publication, experimental data mainly at 10, 20, and 30 °C are demonstrated, as well as a series of measurements at temperatures between 10 and 34 °C in 2 °C steps.

The permeate pressure was changing insignificantly during the experiment. The highest observed permeate pressure was 1.2 bar abs for the case of the highest observed permeance of 14 m^3(STP) m^{-2} h^{-1} bar^{-1}. The change in the permeate pressure is related to the resistance to the gas flow caused by the "smallest" mass flow sensor installed on the permeate side of the experimental facility. The increase in the permeate pressure by 0.2 bar when the applied feed pressure was above 50 bar was considered insignificant to cause changes in the gas flow through the membrane.

2.4. Gas Sorption over a Wide Pressure Range

Gas sorption measurements of pure gases were performed in an experimental system equipped with a magnetic suspension balance (MSB) (TA Instruments, Eschborn, Germany). The system includes a thermostat (Julabo GmbH, Seelbach, Germany), a gas supply, and a vacuum pump (Pfeiffer Vacuum GmbH, Aßlar, Germany). The MSB allows the continuous determination of the mass of sorbent materials from 10 mg to 10 g. Auxiliary equipment such as a titanium sinker permits an estimation of the density of the fluid in situ. The adsorption isotherms of pure CO_2 in thick films were acquired in a temperature range from 10 °C to 30 °C, with recordings of the sample masses, temperatures, and pressures every 5 s. The density of the fluid was estimated every 5 min. Samples were degassed at ambient temperature under a vacuum for 24 h to remove residual solvents and pre-adsorbed gases.

Densities of thick isotropic films required for the evaluation of gas sorption experiment results were estimated by Archimedes' principle using the analytical balance Excellence Plus XP105DR (Mettler-Toledo GmbH, Gießen, Germany), density determination kit, and auxiliary liquid FC-770 (3M, Saint Paul, MN, USA). For each material, five pieces of a sample were weighed for an accurate determination of the density and experimental error. A description of the density estimation is given in Appendix A.

2.5. Experimental Uncertainties

The uncertainties of the equipment used in the current study are presented in Table 2.

Table 2. Experimental uncertainties.

Parameter	Value, 95% Confidence
Membrane area	±0.5%
Volumetric flowrate	±0.9% of S.P. [1]
Feed and permeate pressure determined with LEO3 sensors	0.2% F.S. [1]
Measuring load [1]	±0.01 mg [1]
Pressure determined with sensor (0–50) bar	0.5% F.S. [1]
Huber thermostat temperature stability	±0.05 °C [1]
Temperature determined with sensor Pt100	±0.05 K at temperature below 370 K [1]

[1] Data provided by producer in calibration certificate.

3. Results and Discussion

3.1. Experimental Results

3.1.1. Quality Validation of the Experimental Setup and TFCM Stability in the Available Pressure Range

Before the experiments for the determination of membrane permeance subjected to the high activities of condensable gases, it was necessary to test the functioning of the experimental facility in a wide pressure range. To investigate the effect of pressure on gas transport through TFCMs with a rubbery polymer-based selective layer without taking into account the interaction between the penetrant and membrane material, experiments with He as a permanent, noble gas that showed allowable solubility were conducted. Figure 1a shows the results of helium transport measurements for PDMS1280-PAN TFCM in the pressure range up to 55 bar for 10 °C, 20 °C, and 30 °C. For each temperature, no significant change in the permeance with increasing pressure was observed. The same was observed for membranes with other selective layer materials. Hence, there is no link between the mechanical pressure applied to the membrane and the transport behavior of the tested rubbery membranes. No significant compaction could be detected if the experiments were carried out with a permanent gas under the chosen measurement conditions. In Figure 1b, two curves are compared, which show He permeance before and after the full set of experiments carried out for this membrane sample with condensable gases. The similarity of the two measurement curves proves that the membrane did not significantly change its properties when exposed to high activities of condensable penetrants. The experimental facility was proven to deliver reliable experimental results at the chosen pressures and temperatures.

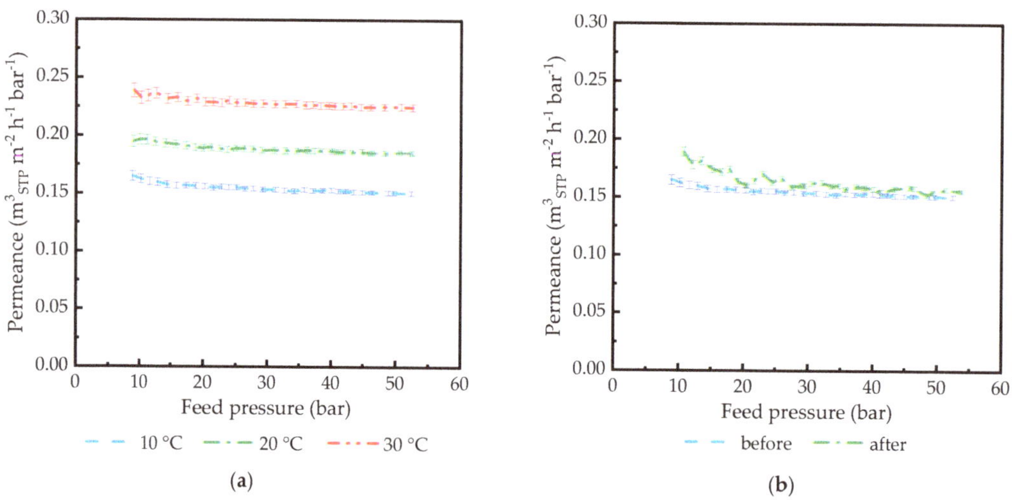

Figure 1. He permeance of PDMS1280-PAN TFCM plotted against feed pressure: (**a**) permeance in the pressure range 5–50 bar at 10, 20, and 30 °C; (**b**) comparison of membrane permeance at 10 °C before and after measurement session using condensable gases. Experimental uncertainty was calculated using linear error propagation.

3.1.2. Sorption of CO_2 in PDMS and PA Isotropic Films

Sorption of CO_2 in PDMS and PA was carried out at temperatures from 10 °C to 30 °C and at pressures up to p/po = 0.9 (Figure 2). Higher pressure was not applied to the samples due to the danger of gas condensation within the measurement equipment.

Experiments with C_2H_6 were not carried out due to technical issues. Nevertheless, both PDMS and PA were numerously investigated for gas sorption in a wide range of pressures, and all published isotherms were adequately described by mathematical models applicable to these materials [42–45].

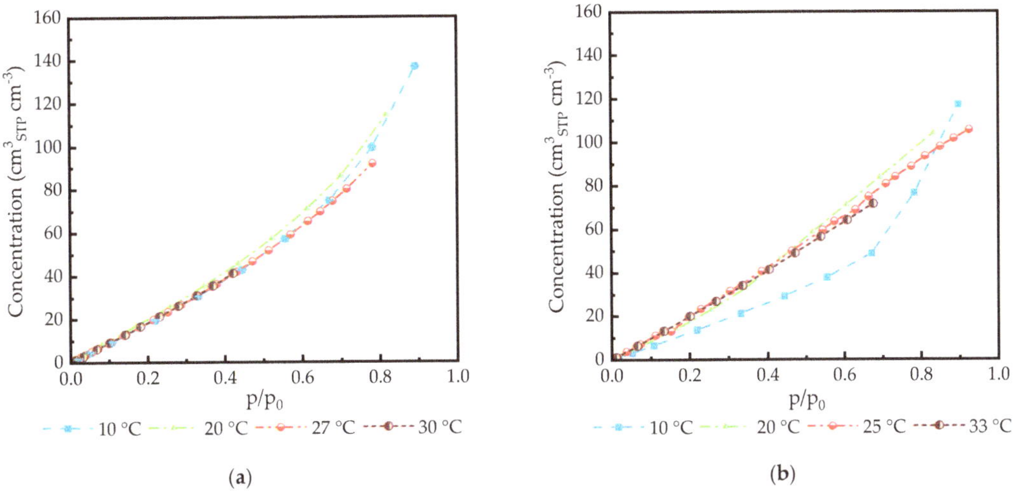

Figure 2. Sorption of CO2 in isotropic films of (**a**) PDMS films 261 μm (at 10 °C, 20 °C, and 27 °C); 283 μm (at 30 °C); (**b**) PA films 388 μm (at 10 °C, 20 °C, and 33 °C), and 173 μm (at 25 °C). The experimental uncertainty is less than 0.05%, and the standard deviation in each measurement point represents a set of more than 100 data points. Some error bars are smaller than the symbols.

The density levels of PDMS and PA, necessary for evaluating the data acquired during the gas sorption experiment, were determined using two methods utilizing the Archimedes principle. In one case, the density was determined using perfluorinated FC770 as the fluid with a large molecule size, thus reducing the probability of molecule diffusion in the polymer sample (Appendix A); in the second case, the density was determined in He within the magnetically suspended microbalance. He is an inert gas with a density that allows for good resolution of the sample weight difference between values measured in a vacuum and in the He atmosphere. The density values of isotropic films of PDMS and PA are listed in Table 3.

Table 3. Density of isotropic films of PDMS and PA.

Material	Density in FC770 [1] g cm^{-3}	Uncertainty g cm^{-3}	Density in He [2] g cm^{-3}	Uncertainty g cm^{-3}
PDMS	1.100	±0.012	1.164	$\pm 7 \times 10^{-6}$
PA	1.176	±0.005	1.188 [3]	$\pm 8 \times 10^{-6}$

[1] Density measurements performed at room temperature; [2] density measurements performed at 30 °C. [3] As reported as well by Car et al. [27].

In the pressure range up to $p/p_o = 0.6$, isotherms can be adequately described using Henry's law for both PDMS and PA. The influence of PDMS swelling in this pressure range is not significant. A swelling degree of more than 10% will be expected at pressures higher than 50 bar for a non-crosslinked polymer [46–48].

For the PA thick film, a linear increase in CO_2 uptake was observed for temperatures from 25 °C to 33 °C (Figure 2b). The non-linearity behavior was observed for the temperature of 10 °C, which is below the crystallization temperature of poly(ethylene glycol) domains of PA reported as 25–28 °C [49,50].

Numerous publications on gas sorption in rubbery polymers have always demonstrated a continuous dependence of the quantity of gas dissolved in the polymer on the applied pressure, which can be described by the Henry-type isotherm in cases of insignificant interaction of the solute with the polymer matrix or by Flory–Huggins-type isotherms in cases where the solute/polymer interaction is significant and an increased concentration of the solute in the polymer causes swelling of the polymer [30,51]. In this case, swelling can lead to changes in the diffusion coefficient, as was clearly shown by, e.g., Lin and Freeman. They studied the sorption of various condensable gases in semi-crystalline poly(ethylene oxide) [44,52].

3.1.3. Gas Flow Rate through TFCMs and Isotropic Films of PDMS and PA

The measurements of condensable gas permeances of TFCMs in the full activity range showed that the shape of the flow rate/feed pressure (activity) curve was very far from expectations based on observations reported in the literature. As shown in Figure 3a, the shape of the permeate flow curve is not monotonous, in contrast to the same parameter determined for thick films (Figure 3b).

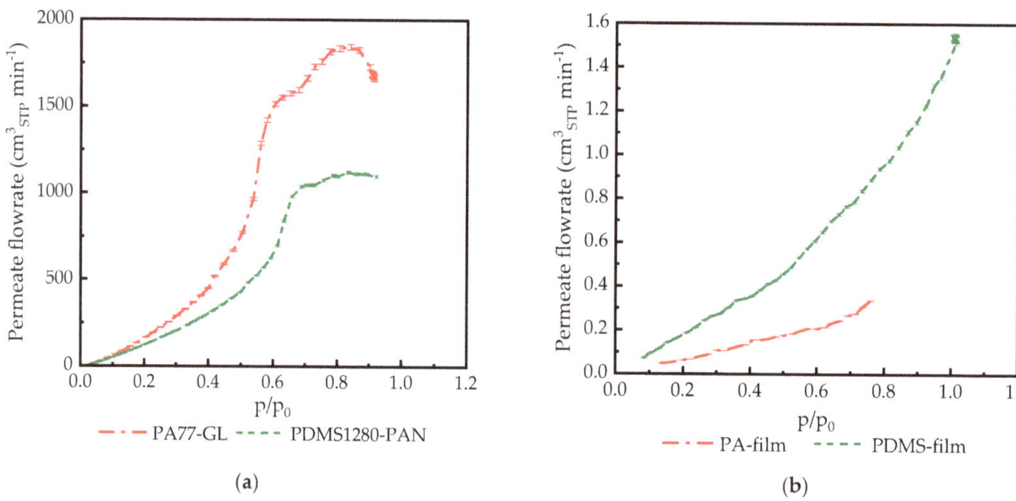

Figure 3. CO_2 permeate flow rate at 20 °C for (**a**) PA77-GL and PDMS1280-PAN; (**b**) PA 19 μm and PDMS 285 μm isotropic films.

Experiments with thick isotropic films of PA and PDMS on the experimental setup used in the current study presented certain difficulties because the setup was designed for experiments with mechanically robust TFCMs with extremely small selective layer thicknesses. Isotropic films of PA and PDMS with thicknesses of 19 μm and 285 μm, respectively, were not sufficiently mechanically stable to be compressed with the standard O-ring used for membrane sealing in high-pressure experiments. The ethylene-propylene-diene(monomer) rubber (EPDM) O-ring with the Shore 70 hardness was used for experiments with polymer films, and the measurement cell was closed with the lowest possible pressure applied to the membrane to reduce O-ring intrusion into the polymer film and thus prevent changes in the film shape within the cell due to compression. The experiments were carried out at 20 °C only and with CO_2 only. C_2H_6 caused film damage already at low feed pressures. The PDMS film withstands CO_2 in the full range of CO_2 activity when the feed pressure is gradually increased and breaks when the feed pressure begins to be reduced. PA film withstands CO_2 only up to an activity level of 0.75. Both films showed the presence of significant swelling when removed from the measurement cell, and vertically oriented wrinkles broken on the top were observed for both polymers.

It is interesting to note that for PA77-GL and PDMS films, multiple experimental points were collected at the highest pressure. The positioning of these points on top of one another indicates that the sample under investigation is in equilibrium and that it does not change permeance in time, only under changing pressures.

3.1.4. Investigation of TFCM Permeance in Relation to the Feed Pressure of Condensable Penetrants

The experiments with various TFCMs listed in Table 1 were carried out in the full possible range of pressures. At the experimental temperature of 10 °C, it was possible to overcome the saturation pressure since the gas bottle was kept in the lab with the temperature stabilized at 20 °C. This allowed for the investigation of membrane behavior under the influence of a liquid penetrant applied to the feed surface of the TFCM. Since the experiment contained pressure increasing and decreasing parts, it was possible to observe whether irreversible changes to the membrane could be introduced by the application of a high-pressure gaseous or even liquid penetrant. No significant deviation between the two curves corresponding to the pressure increasing and decreasing parts of the experiment was found, indicating that the chosen experimental conditions allowed for membrane property stabilization after the pressure was changed. Only the pressure increasing parts of the experiments are depicted in Figures 4 and 5.

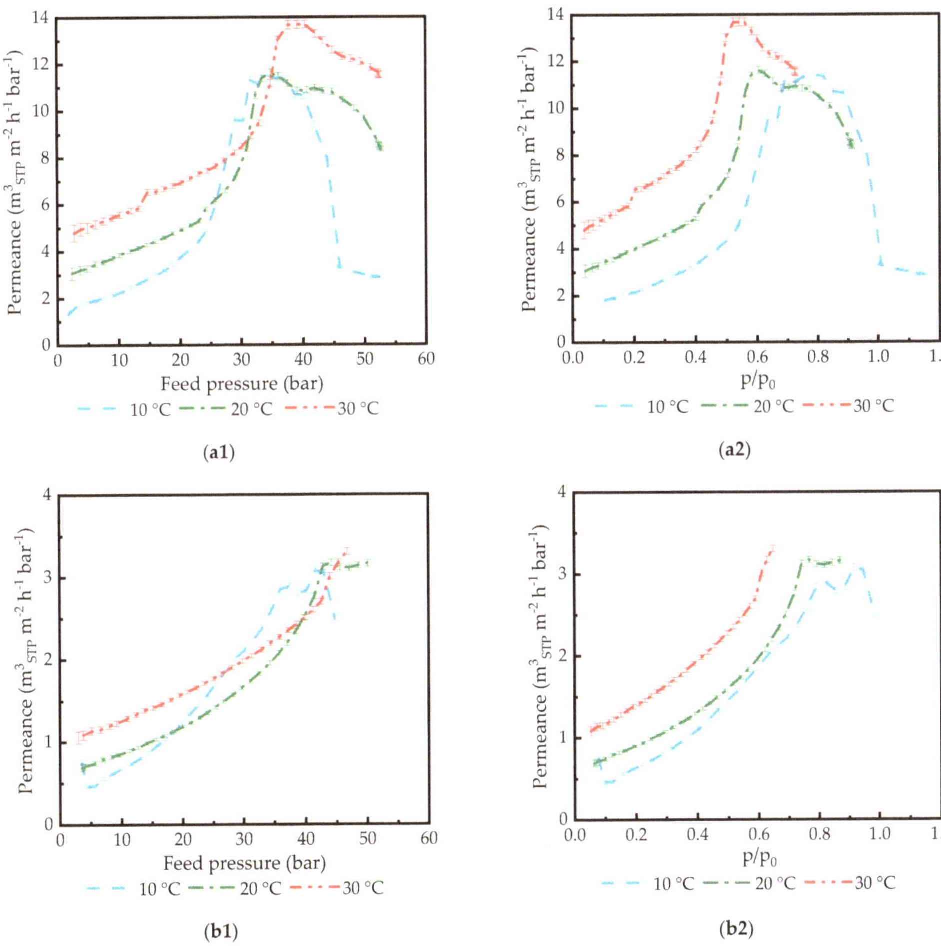

Figure 4. *Cont.*

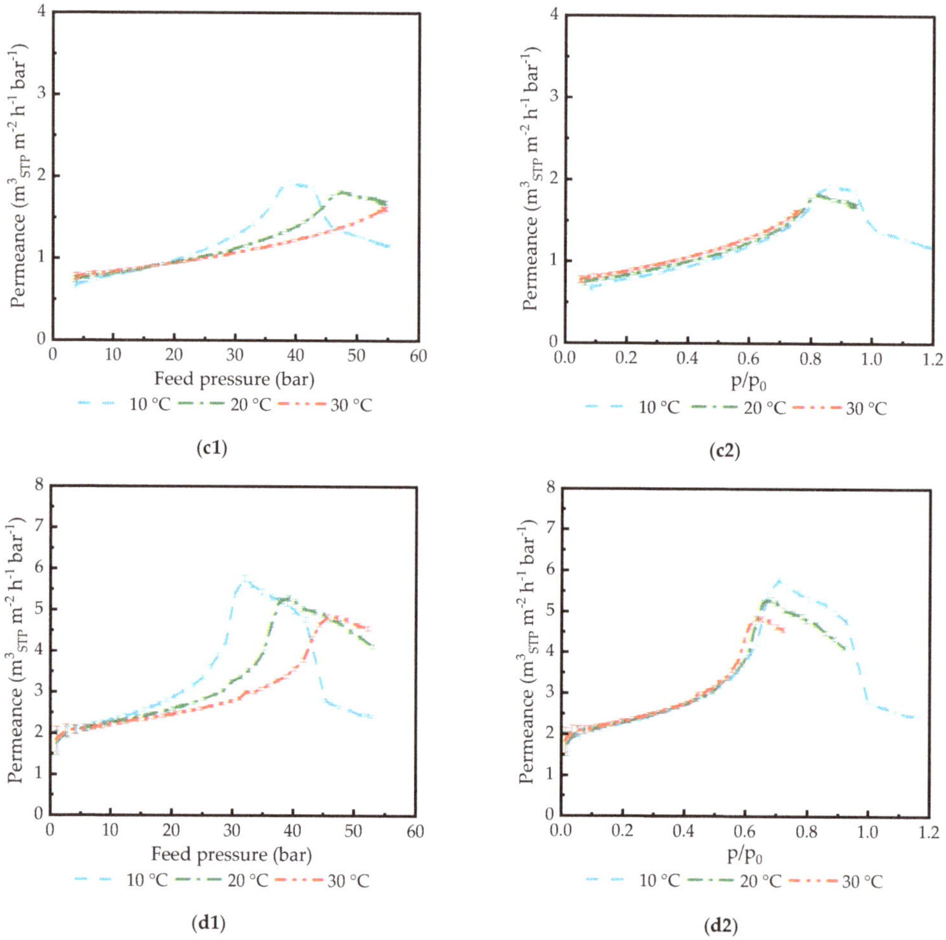

Figure 4. CO_2 permeance of various TFCMs measured at 10 °C, 20 °C, and 30 °C: (**a1**) PA77-GL vs. feed pressure; (**a2**) vs. activity; (**b1**) PA186-GL vs. feed pressure; (**b2**) vs. activity; (**c1**) POMS6250-GL vs. feed pressure; (**c2**) vs. activity; (**d1**) PDMS1280-PAN vs. feed pressure; (**d2**) vs. activity. The feed pressure increasing part of the experiment is shown for better figure clarity.

3.1.5. Gas Transport Properties of GL and UF-PAN Membranes Used as Supports for TFCMs

Membranes used as supports for the formation of selective TFCMs were tested for gas transport properties in order to investigate a possible influence of the support on the resulting gas transport properties of TFCMs. Both GL (Figure 6) and UF-PAN (Figure 7) membranes demonstrate permeances that are significantly higher than that determined for selective TFCMs.

According to the resistance model introduced by Henis and Tripodi [53] for the analysis of gas separation membrane performance, the selective layer of TFCM governs the gas transport properties of TFCM.

The measurement of UF-PAN was carried out only at as small a feed pressure as possible due to the very high membrane permeance. The gas transport properties of the UF-PAN membrane can be described by the Knudsen-type gas flow through porous media (Figure 1) [54].

The GL membrane demonstrates the presence of the peak in the CO_2 permeance curve already at ca. 0.5 feed CO_2 activity with a significant permeance decrease in the activity range 0.5–1.0.

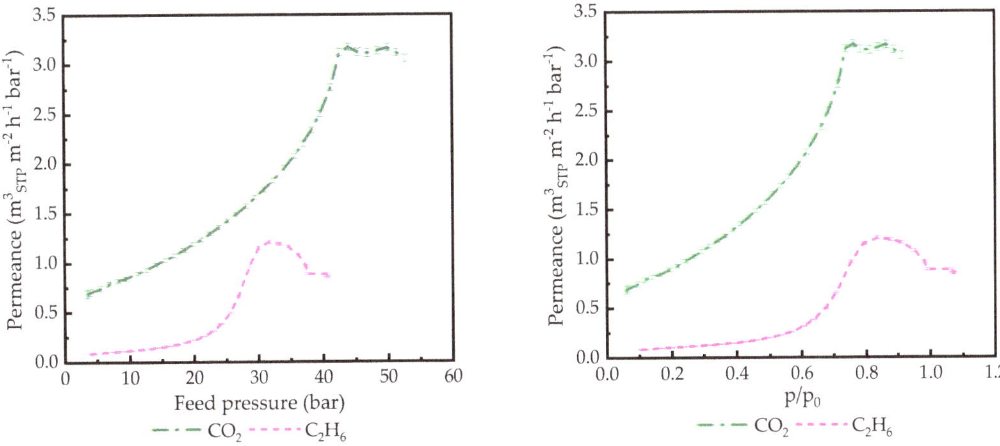

(**a**) (**b**)

Figure 5. CO_2 and C_2H_6 permeance in PA186-GL TFCM plotted: (**a**) vs. feed pressure measured at 20 °C; (**b**) vs. penetrant activity at 20 °C. Feed pressure increasing part of the experiment is shown for better figure clarity.

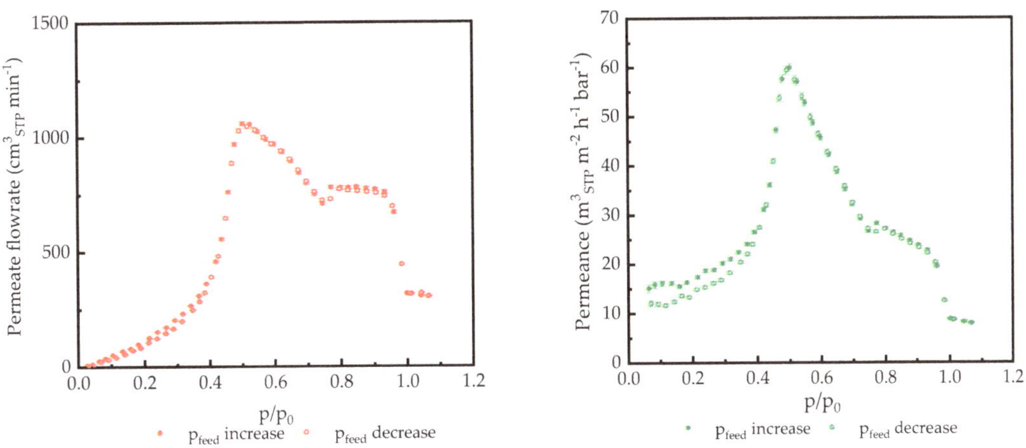

(**a**) (**b**)

Figure 6. CO_2 Transport parameters at 20 °C plotted against feed CO_2 activity for GL used as a standard gutter layer support: (**a**) permeate flowrate; (**b**) permeance. Data points related to the feed pressure p_{feed} increasing and decreasing parts of the experiment are demonstrated.

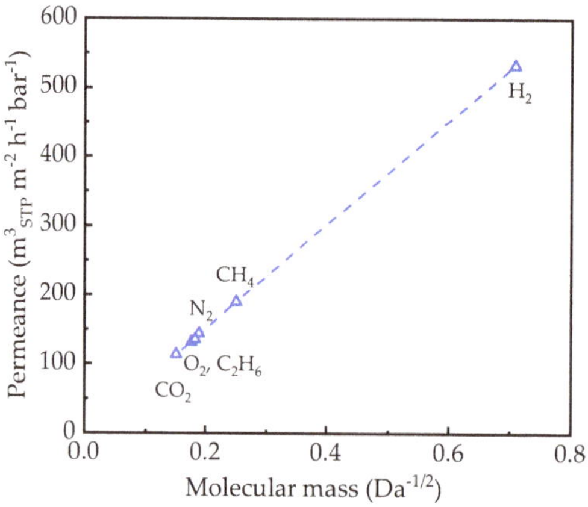

Figure 7. Knudsen-type dependence of gas permeance determined for the UF-PAN membrane at 25 °C for H_2, CH_4, N_2, O_2, CO_2, and C_2H_6 at a feed pressure of 170 mbar abs.

3.1.6. Changes in Permeate Temperature during Experiments in the Full Range of CO_2 Activity

Permeate temperature was controlled using temperature sensor placed immediately at the permeate surface of the membrane under study. Figures 8 and 9 demonstrate clear relation between peak in permeance and change of the permeating gas temperature.

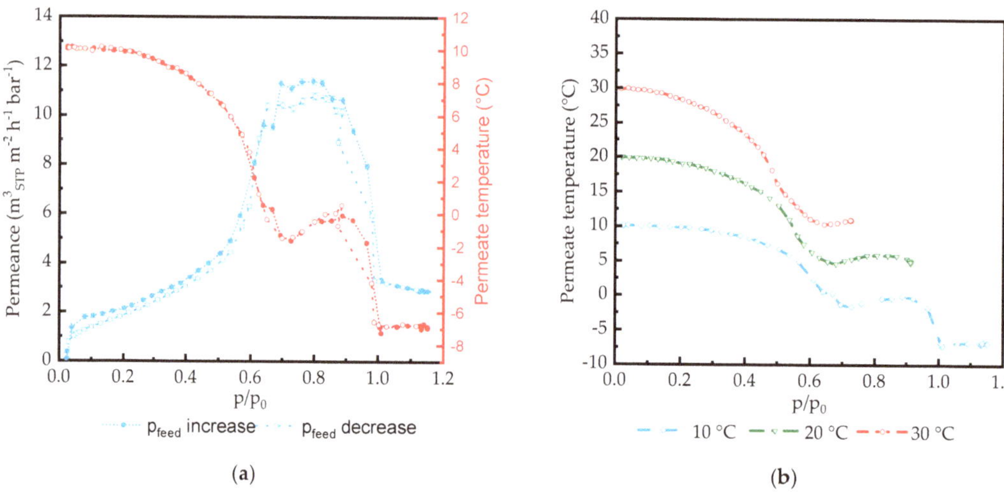

Figure 8. (**a**) Permeance (blue line) and permeate temperature (red line) vs. CO_2 feed activity for PA77-GL TFCM acquired for feed temperature 10 °C, and filled and open symbols show feed pressure increasing and decreasing parts of the experiment, respectively. (**b**) Permeate temperature at feed pressure increased in part of the experiment carried out at feed temperatures of 10 °C, 20 °C, and 30 °C.

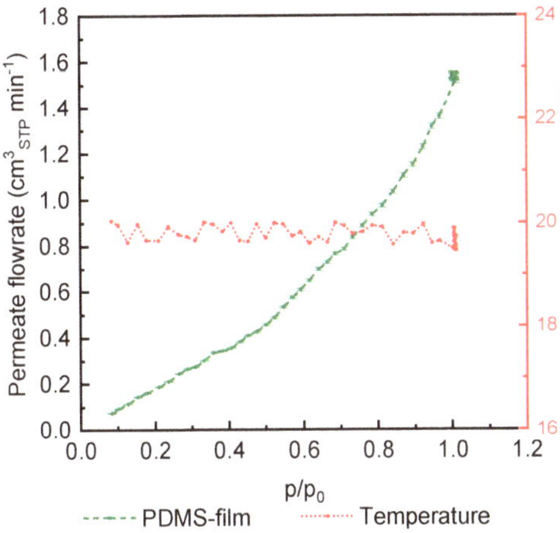

Figure 9. CO_2 permeate flowrate and permeate temperature obtained at 20 °C feed temperature for PDMS film.

3.2. Analysis of the Experimental Results

3.2.1. Experiments with Thick Isotropic Films

The experimental results obtained for thick isotropic PDMS and PA films are presented in Figures 2, 3 and 9.

CO_2 sorption data obtained in the CO_2 activity range 0–0.95 show no unusual behavior of the polymer/gas pair. Isotherms for different temperatures are very similar for one polymer. While CO_2 isotherms for PDMS follow the Flory–Huggins behavior typical for rubbery polymers [45], PA isotherms can be described as simple Henry isotherms at temperatures over the melting point of poly(ethylene glycol) at 25 °C. The beginning of the crystallization process can be seen at 20 °C. At 10 °C, poly(ethylene glycol) blocks of the PA block copolymer are in a highly crystalline state, and CO_2 starts to act as a plasticizer, reducing the melting temperature and promoting crystallites melting at activity levels of 0.7 and above [50]. At the highest activity, CO_2 solubility in PA is the highest, as presented in Figure 2b, and this is in clear accordance with the theory of gas transport, though polymers were introduced first by Barrer and Rideal in 1939 [55,56]. Only small differences in the isotherms acquired at different temperatures are in agreement with the low values of CO_2/polymer partial enthalpies of sorption that have been numerously reported elsewhere [57–59].

Gas transport experiments carried out in the full possible range of condensable gas pressures did not show deviations from the permeance/pressure relationship expected for rubbery polymers. The slightly non-linear shape of the CO_2 flow curve can be explained by the limited swelling of the polymer by the penetrant, leading to an increase in permeance at higher pressures. Unfortunately, the PDMS film did not withstand the decrease in the permeate pressure, and the permeance data were not collected for the pressure decreasing part of the experiment. The PA film, as stated above, did not withstand an increase in the CO_2 pressure above an activity of $p/p_0 = 0.75$. After the experiment, the measurement cell was opened and the states of both the PDMS and the PA films were investigated. It was found that vertical wrinkles were formed with cracks on the highest point of the wrinkle. Such a change in the film's geometry can only be attributed to the swelling of the polymer by the penetrant, as no other influence capable of causing a change in the film's geometry can be identified.

3.2.2. TFCM Permeance Peak at High Pressures

As mentioned above, during the work on membrane gas separation of complex gas mixtures using PA TFCM [24], it was found that the FVM cannot adequately predict CO_2 transport at elevated partial pressures. The model gave a CO_2 permeance lower than the experimental value, and the deviation increased with a partial pressure increase.

To investigate this phenomenon in the full range of partial pressures up to and over the saturation pressure of a penetrant, the transport of condensable gases through rubbery polymer-based TFCMs was studied for membranes with selective layers made of different materials, as presented in Table 1. The PA TFCM was developed for the separation of CO_2 containing gas mixtures, with the main emphasis on the separation of CO_2 from flue gas of various origins. This consideration led us to the decision to conduct extensive experimental work with CO_2, while C_2H_6 was used for comparison purposes. The results of the experiments with different membranes for pressures up to 65 bar are shown in Figures 3–6 and 9. All the tested rubbery TFCMs show a strong increase in permeance with increasing feed pressure. Up to about 20 bar, the observed increase is as expected for membrane material swollen by the penetrant, but above this pressure, the curve shape becomes more complex, as with an exponential increase. The curve shows a peak followed by a small plateau, followed by a steady permeance decrease up to the point of gas condensation. The peak position, plateau width, and overall curve shape depend on the selective layer material and membrane morphology.

Basically, the transport of small molecules through rubbery membranes is driven by the fugacity difference in accordance with the solution/diffusion mechanism [60]. Hence, the gas flow rate through the membrane is determined by several factors: the adsorption of gas on the feed membrane surface; the diffusion of dissolved gas molecules through the bulk of the selective layer; and finally, the desorption on the permeate side of the selective layer. The transfer of the penetrant from the selective layer into supporting layers depends on the resistance of these supporting layers to the penetrant flow; it can either be the porous surface of PAN with surface porosity not more than 13% or a gutter layer formed of highly permeable and low selective PDMS, which provides a smooth surface for the selective layer deposition and effective drainage of the penetrant from the whole permeate surface of the selective layer into the porous structure of the UF-PAN membrane.

The quantity of diffusing gas molecules Is determined by the solubility coefficient of the gas under study in the selective layer material. This parameter can be directly acquired from the results of the gas sorption experiment. The corresponding isotherms are presented in Figure 2.

While the respective membrane material directly influences the dissolution and desorption of gas molecules during the gas transport process, the diffusion within the selective layer can be affected by the degree of swelling caused by the penetrant interaction with the selective material. The sharp increase in permeance beyond that described by the FVM indicates that as the concentration of the dissolved molecules in the selective layer increases, either the solubility or the diffusivity of the penetrant in the polymer increases significantly. One can speculate that in addition to polymer swelling as a possible cause of the permeance increase, the effect of the penetrant–penetrant molecules' interaction confined in the free volume voids of the polymer matrix starts to be significant, even overcoming the effects of the penetrant–polymer interactions [61,62]. The penetrant–penetrant interaction can also lead to a negative, flow-reducing effect since penetrant molecule clustering becomes more probable, resulting in an increase in the effective diameter of the penetrant and a consequent mobility decrease in the sorbed molecules [63]. The cluster formation leads to a decrease in the diffusion coefficient, and it has already been detected for carbon dioxide [64]. The effects of swelling and clustering have opposite effects regarding the permeate flow rate. Due to the different interactions of CO_2 with the selective membrane materials under study, the shape of the curves presented in Figure 4 varies for each material.

Figure 4 presents the data collected for CO_2 permeance in TFCMs with selective layers made of different polymers (PA, PDMS, and POMS) and different selective layer thicknesses. For each gas/TFCM pair, the following two plots are presented: one for permeance plotted against feed pressure, and another for penetrant activity. As one can see, plots with penetrant activity give more generalized information. For two polymers with no specific interaction between the polymer and CO_2, the peak value of permeance for all three temperatures is reached at the same penetrant activity, ca. 0.83 for POMS6250-GL and ca. 0.66 for PDMS1280-GL. PA77-GL and PA186-GL give a different picture: the position of the permeance peak shifts to a lower CO_2 activity value with increasing temperature. It is important to mention that the maximum permeance value at 10 °C in the case of PA77-GL is very similar to that at 20 °C, indicating the melting of poly(ethylene glycol) crystallites under the influence of high CO_2 pressure.

The difference in the peak position between two PA-based membranes indicates that the thickness of the selective layer is important for the observed effect. At smaller selective layer thicknesses, peak permeance is reached at lower penetrant activity.

After the peak, the permeance of PA77-GL started to decrease in terms of the gas flowrate through the membrane (as in Figure 3a), which became mostly independent of pressure. One can presume that at this state of the polymer/penetrant system, the maximum possible amount of penetrant is dissolved in the polymer matrix, further swelling is not possible anymore, and increases in the pressure (activity) are not followed by a proportional increase in the penetrant flowrate. Another possibility is changes in the penetrant state in the polymer matrix, e.g., clustering, as discussed above.

The curve corresponding to the PA77-GL permeance at 10 °C shows a decreasing trend of CO_2 activity above 0.8 (Figure 4(a2)) and follows it to the point of feed gas condensation. At a CO_2 feed activity of 1, the trend changes to a mostly horizontal line. At this moment, the experiment on gas transport turns to the experiment of liquid CO_2 pervaporation through the polymeric membrane. The flowrate of the penetrant is not dependent anymore on the pressure. As will be discussed later, the presence of liquid CO_2 on the feed membrane surface does not influence membrane integrity, and experiments involving gas condensation were repeated numerous times and clearly demonstrated that the investigated polymeric membranes are not altered by the liquid penetrant.

In order to find out whether the observed effect is only related to CO_2 or if it is common for condensable gases, experiments with C_2H_6 were carried out with PA186-GL. As follows from Figure 5, the system PA/C_2H_6 demonstrates the same behavior as PA/CO_2 in the full range of C_2H_6 activity. The C_2H_6 permeance reached its peak value at 0.85 activity of C_2H_6. The peak permeance of C_2H_6 was significantly lower than that of CO_2. After C_2H_6 condensation on the feed membrane surface, the permeance did not show dependence on pressure. As in the case of CO_2, the membrane successfully survived exposure to the liquid penetrant on its surface.

As mentioned above, the experiment was carried out with small steps in pressure at the same temperature and 2 °C steps in temperature between the collection of permeance data at stable feed temperature conditions. So far, only data collected at 10 °C, 20 °C, and 30 °C are presented in the Figures and analyzed. For a detailed analysis of the position of the permeance peak at different temperatures, Figure 10 demonstrates the data for all of the available temperature points for the PA77-GL TFCM membrane. For each feed temperature, the maximum values of permeance at the peak, the corresponding feed pressure, and thus, the CO_2 feed activity were collected and presented as a function on the temperature (Figure 10). In the temperature range 10–14 °C, when poly(ethylene glycol) domains are in a semicrystalline state, the position of the permeance peak is at ca. 0.8 feed CO_2 activity, and the permeance decreases as the temperature increases. At temperatures above 14 °C, the permeance peak was reached at a significantly lower CO_2 feed activity, finally lowering down to a value of 0.55 at 34 °C, and the permeance gradually increased in an asymptotic manner. For the membranes under investigation, the highest achieved CO_2 permeance was 13.8 m^3 (STP) m^{-2} h^{-1} bar^{-1} at the highest experimental temperature of 34 °C. The melt peak temperature for this membrane material, as reported above, is around 28 °C. Figure 10 indicates the stabilization of the maximum value for the permeance at this temperature.

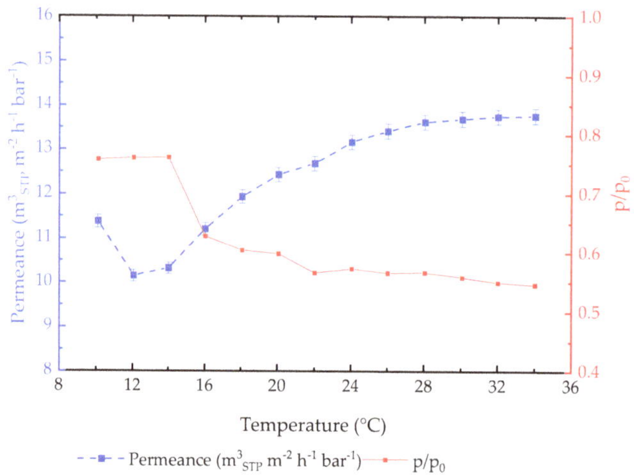

Figure 10. Analysis of PA77-GL TFCM peak permeance and corresponding CO_2 feed activity in dependence on experimental feed temperature.

For both activity and permeance curves, the change in behavior occurs in the range 14–16 °C, which is in good agreement with the observations reported earlier, where during the gas transport of the PA, TFCM was investigated at low feed pressures, and change in the Arrhenius permeance/temperature dependence associated with semicrystalline parts of the polymer was observed at temperatures of 18–20 °C [49,58]. Earlier, it was reported that the melting temperature of the poly(ethylene glycol) in the PA is 27 °C in the case of isotropic films with a ca. 100 nm thickness [27]. The difference in the melting temperatures of the bulk polymer and the polymer in the selective layer of TFCM is associated with the state of the thin layer of the polymer in the selective layer, leading to significant changes in thermal properties. Differences between the melting point observation of the current study (14–16 °C) and previous work (18–20 °C) conducted for the PA TFCM arise from the fact that, in the current study, gas transport properties were determined at much higher pressures and thus feed activities of the CO_2, which is a plasticizing agent for the PA, can induce a decrease in the melting temperature.

3.2.3. Influence of Support on TFCM Properties

The TFCMs under study have complex morphologies and consist of multiple layers. An ultrafiltration PAN layer is deposited on top of the polyester non-woven; a gutter layer of adhesive PDMS is deposited on top of the UF-PAN, followed by the selective and sometimes protective layers. The drainage of the penetrant transported through the selective layer occurs through the gutter and UF layers. It is important to investigate the possible resistance of these layers to the penetrant flow.

Figures 6 and 7 show that the gas transport in UF-PAN and GL membranes is faster than in the selective membranes discussed above. The UF-PAN membrane shows a clear Knudsen-type gas flow mechanism with a CO_2 permeance above 100 m^3(STP) m^{-2} h^{-1} bar^{-1}.

The gas transport of CO_2 through the GL membrane shows the same behavior as for PA77-GL, with a permeance peak at ca. 0.5 CO_2 activity. The peak permeance of 60 m^3(STP) m^{-2} h^{-1} bar^{-1} was observed at 20 °C for the GL membrane, followed by a sharp decrease in both the flowrate and permeance. It is interesting that in the CO_2 activity range of 0.75–0.95, the penetrant flowrate through the membrane is independent of pressure, and above this range, it drops significantly to the point of penetrant condensation. Both pressure increase and decrease curves are shown in Figure 6. A very good agreement between these curves clearly indicates that at chosen experimental conditions (pressure step and equilibration time between experimental points), the polymer of the gutter layer reaches an equilibrium state. The exact positioning of the two curves in the activity range of 0.5–1.1 gives one proof that the observed effect of polymer/penetrant interaction in TFCMs under study is not an artefact but a real physical fact.

3.2.4. Permeate Temperature Drop during Gas Transport Experiments

A gas separation membrane can be thought of as an orifice with a very small opening used for gas throttling. The gas is expanded from the high pressure side to the low-pressure side, and there is a significant drop in temperature, known as the Joule–Thomson effect [65]. The expansion of condensable gases through high-performance gas separation membranes under study should result in a significant temperature drop on the permeate side of the membrane. This drop can theoretically influence the temperature of the feed side of the selective layer.

The experimental setup enables the investigation of the temperature change in the permeate flow. The temperature sensor was immediately placed on the permeate surface of the membrane within the sintered metal support and as far away from measurement cell walls as possible to minimize the influence of the temperature outside of the cell, as shown in Figure S1 of the Supporting Information. Figure 8a shows that in the case of the PA77-GL, simultaneously with the strong variation in permeance, the temperature of the permeate stream changes. With the increase in CO_2 permeance in the range of CO_2 feed activity of 0–0.7, the permeate temperature drops, reaching a local minimum at the peak of permeance. In the activity range of 0.75–0.9, both the permeance and temperature change insignificantly, and in the activity range up to 1.0, both the permeance and temperature decrease. This simultaneous drop is unexpected since a reduction in the penetrant flowrate through the membrane should be accompanied by an increase in the permeate pressure in the system, where heat flow from external space is not prohibited. At an activity of one, another strong drop occurs, and afterwards, the temperature remains stable. This relationship results from the desorption process, or more precisely, the desorption enthalpy. As the amount of substance permeating through the membrane increases, the desorption of gas molecules on the permeate side of the membrane also rises. The change in temperature on the permeate side of the membrane is therefore related to the desorption rate of molecules passing through the membrane. Figure 8b shows the permeate temperature in dependence of the activity for the measurement series obtained at 10 °C, 20 °C, and 30 °C. For all measurement series, the course of the curves is similar. For higher temperatures, no such high activity was achieved in the measurement series as that at 10 °C, since the vapor pressure increases with the temperature while the maximum operating pressure is the same for all measurement series. Nevertheless, the data show that the higher the feed temperature, the more pronounced the decrease in the permeate temperature during the strong rise of the permeance. This is due to the higher permeance at higher temperatures and, therefore, the increased desorption rate.

To verify whether the observed change in the permeate temperature is a real behavior, a measurement was carried out using a PDMS thick film under the same experimental conditions as the measurements performed with TFCMs. Figure 9 shows that in this investigation, at a feed temperature of 20 °C, the permeate temperature fluctuates from the feed temperature of 0.57 °C with a linearly increasing permeate flow rate. Therefore, a technically induced temperature fluctuation can be excluded for the applied measuring range. The results show that for the measurement of the thick film membrane, no effect like the one observed for thin film membranes is observed for the measurement conditions applied.

Furthermore, Figure 8a shows the results for measurements with an increasing and decreasing feed pressure in one series of measurements. There is just a small vertical shift; it is almost negligible. The course of the curve is almost identical regardless of whether the pressure is increased or reduced during the measurement. The absence of hysteresis confirms the observation of a real relationship between the polymer and the penetrant, so measurement-based effects can be excluded. The fact that the value for permeance after pressure reduction corresponds again to the initial value at low feed pressure proves the reversible behavior and shows the stability of the membrane even at high pressures, extreme swelling, and a change in the temperature around the membrane.

Figure 8b indicates that at a feed activity of 1.0, the state of the penetrant on the feed side of the membrane changes from gaseous to liquid. The line corresponding to the feed temperature of 10 °C shows that the temperature on the permeate side of the membrane drops by 17 °C, but the gas condensation on the feed side occurs at conditions on the feed side, meaning that information on the gas state on the permeate side of the membrane is not transferred to the feed side through the selective layer. The expansion of the gas accompanied by the drop in temperature occurs on the permeate side interface of the selective layer, or, in the case of the presence of the gutter layer, on the interface between the gutter layer and porous PAN, and the penetrant flow through the selective layer is sufficiently high to keep the selective layer at feed temperature conditions. This fact can, of course, be related to the arrangement of the experimental setup used in the current study, which is similar to dead-end filtration, where there is no lateral transfer of a penetrant stream along the membrane surface on both the feed and permeate sides of the membrane. Earlier works have reported significant retentate temperature drops in comparison to the feed temperature during gas and vapor separation experiments involving full-scale membrane modules [66,67].

4. Conclusions

In light of the earlier reported inability of the FVM to predict gas permeances of highly efficient TFCMs at elevated partial pressures of condensable penetrants, experiments in the full range of penetrants, CO_2, and C_2H_6 activities were conducted. The following TFCMs with various selective layer materials as well as support used for membrane fabrication were investigated: PA, two types of PDMS, POMS, and microporous PAN. In order to find whether permeance deviations from values predicted by FVM originate from materials or whether they are related to the TFCM morphology of thick isotropic films of PA and PDMS, they were studied under the same conditions as TFCMs.

The results obtained during sorption and gas transport experiments with isotropic films of PA and PDMS showed no deviation from behavior, which has been numerously reported in the literature.

PAN microporous support showed, at the chosen experimental conditions, a gas flow mechanism that can be characterized as a Knudsen-type flow.

The investigation of all membranes with continuous defect-free layers of rubbery polymers, namely GL, PDMS, POMS, and PA, shows the presence of a permeance peak occurring at different condensable gas activities. The highest permeance at this peak was observed for the adhesive PDMS-based GL membrane, reaching 60 m^3(STP) m^{-2} h^{-1} bar^{-1} for CO_2 determined at 20 °C. The feed CO_2 activity at the peak permeance was the lowest among all the membranes under investigation.

The same peak behavior of the permeance curve in dependence of feed activity was observed for all other TFCMs, the peak position varying dependent on selective layer material, its thickness, and the temperature of the experiment.

It was found that for PA-based membranes, it is possible to observe significant changes in the permeance peak position in relation to the presence of the crystalline state of the poly(ethylene glycol) domains that are responsible for gas transport properties of the PA block copolymer.

The thermal effects on the permeate side of the membrane were investigated. It was found that an expansion of the penetrant causes a significant drop in the permeate temperature. Additionally, the temperature curve in dependence on feed gas activity is very complex and does not follow the dependence for the permeance, especially at high penetrant feed activity. A significant decrease in the permeate temperature was observed at a feed gas activity of $p/po \geq 1$, i.e., the penetrant on the feed side condensed to the liquid state. The following two conclusions should be drawn from this fact: (a) multiple experiments with various membranes showed that PDMS, POMS, and PA selective layers are stable under exposure to liquid CO_2 and C_2H_6; (b) sudden and significant temperature drops on the permeate side occur when the feed side pressure is equal to or exceeds the condensation pressure at the temperature of the feed side, meaning that the selective layer temperature is controlled by feed side conditions and information on temperature drops on the permeate interface of the selective layer is not transferred into the layer itself. This consideration can be valid only for the experimental conditions chosen for this study, where gas transport properties are studied as dead-end filtration and no penetrant movement occurs in the lateral direction to the membrane as in large-area membrane modules.

The reason for the permeance peak can be attributed to processes occurring within the rubbery selective layer of a TFCM during condensable penetrant transport through it: limited swelling accompanied by an increase in the free volume in accordance with the FVM; clustering of penetrant molecules, resulting in an increase in the effective kinetic diameter of a penetrant and simultaneously increased polymer swelling; free volume morphology, reaching the state when transporting through the rubbery swollen polymer matrix, according to the capillary condensation mechanism.

Further experimental and modeling efforts are necessary in order to investigate permeance peak phenomena for TFCMs with rubbery selective layers of varying thickness, glassy polymers, and polymers of high free volume. A comparison of properties should be conducted for TFCMs and integral asymmetric membranes. All this work is in progress and will be reported on shortly.

Supplementary Materials: The following supporting information can be downloaded at https://www.mdpi.com/article/10.3390/membranes14030066/s1. Figure S1: The scheme of the test cell build in the setup used for single gas experiments at high pressures. Figure S2: Scanning electron microscope (SEM) images of cross sections on TFCM membranes: (a) PDMS1280-PAN; (b) POMS6250-GL; (c,d) standard TFCM.

Author Contributions: The single gas experiments, the data analysis and curation, K.S., J.L. and S.S.; methodology, S.S. and J.P.; facility building and software, J.P.; writing—original draft preparation, K.S., S.S. and J.L.; writing—review and editing, S.S., J.L. and T.B.; supervision and project administration, T.B. All authors have read and agreed to the published version of the manuscript.

Funding: This research received no external funding.

Institutional Review Board Statement: Not applicable.

Data Availability Statement: The data presented in this study are available on request from the corresponding author.

Acknowledgments: The authors would like to thank Jan Wind and Juliana Clodt for purchasing the sample material and their valuable discussions. The authors also thank the Department of Instrumental Structure Analysis for the SEM-supported investigations.

Conflicts of Interest: The authors declare no conflicts of interest. K.S. is an employee of Camfil GmbH. The paper reflects the views of the scientists, not the company.

Appendix A

The density of solid thick films was determined at room temperature as follow:

$$\rho_S = \frac{A}{A-B}(\rho_0 - \rho_{air}) + \rho_{air}, \quad (A1)$$

where ρ_S is the density of sample (g cm^{-3}), A and B are the mass of the sample in air and in the auxiliary liquid, ρ_0 is the density of the auxiliary liquid (g cm^{-3}), and ρ_{air} is the air density at room conditions (0.0012 g cm^{-3}).

The density of the auxiliary liquid (FC-770) was determined using a sinker of known volume:

$$\rho_0 = \alpha\frac{A-B}{V} + \rho_{air}, \quad (A2)$$

where V is the volume of the sinker (cm^3).

The overall uncertainty of sorption measurements was estimated using the Gaussian law of propagation.

The combined standard uncertainty is the square root of the combined variance for the sorption experimental results is given by the following:

$$\sigma^2(f(x)) = \sum_{i=1}^{N} \left(\frac{\partial f(x)}{\partial x_i}\right)^2 \sigma^2(x_i), \tag{A3}$$

Here are the parameters involved in the uncertainty calculations of adsorption gas measurements: mass of polymer sample m_S obtained by weighting in vacuum (g), V_{SC} is the volume of the balance component holding the sample (cm^3), V_S is the volume of sample, and ρ^f is the sorptive gas density. Each measurement point was characterized by experimental standard deviation for minimum 100 series in observation in equilibrium [68].

References

1. Abetz, V.; Brinkmann, T.; Sözbilir, M. Fabrication and function of polymer membranes. *Chem. Teach. Int.* **2021**, *3*, 141–154. [CrossRef]
2. Brinkmann, T.; Lillepärg, J.; Notzke, H.; Pohlmann, J.; Shishatskiy, S.; Wind, J.; Wolff, T. Development of CO_2 Selective Poly(Ethylene Oxide)-Based Membranes: From Laboratory to Pilot Plant Scale. *Engineering* **2017**, *3*, 485–493. [CrossRef]
3. Dai, Z.; Deng, L. Membranes for CO_2 capture and separation: Progress in research and development for industrial applications. *Sep. Purif. Technol.* **2024**, *335*, 126022. [CrossRef]
4. Baker, R.W.; Low, B.T. Gas Separation Membrane Materials: A Perspective. *Macromolecules* **2014**, *47*, 6999–7013. [CrossRef]
5. Chen, X.; Liu, G.; Jin, W. Natural Gas Purification by Asymmetric Membranes: An Overview. *Green Energy Environ.* **2020**, *6*, 176–192. [CrossRef]
6. Shahid, S.; Nijmeijer, K. High pressure gas separation performance of mixed-matrix polymer membranes containing mesoporous Fe(BTC). *J. Membr. Sci.* **2014**, *459*, 33–44. [CrossRef]
7. Yuan, M.; Teichgraeber, H.; Wilcox, J.; Brandt, A.R. Design and operations optimization of membrane-based flexible carbon capture. *Int. J. Greenh. Gas Control.* **2019**, *84*, 154–163. [CrossRef]
8. Favre, E. Membrane Separation Processes and Post-Combustion Carbon Capture: State of the Art and Prospects. *Membranes* **2022**, *12*, 884. [CrossRef]
9. Kentish, S.E. 110th Anniversary: Process Developments in Carbon Dioxide Capture Using Membrane Technology. *Ind. Eng. Chem. Res.* **2019**, *58*, 12868–12875. [CrossRef]
10. Baker, R.R.W.; Cussler, E.L.E.; Eykamp, W.; Koros, W.W.J.; Riley, R.L.; Strathmann, H. *Membrane Separation Systems—A Research and Development Needs Assessment*; USDOE Office of Energy Research: Washington, DC, USA, 1990. [CrossRef]
11. Han, Y.; Ho, W.S.W. Recent advances in polymeric facilitated transport membranes for carbon dioxide separation and hydrogen purification. *J. Polym. Sci.* **2020**, *58*, 2435–2449. [CrossRef]
12. Lonsdale, H.K. The growth of membrane technology. *J. Membr. Sci.* **1982**, *10*, 81–181. [CrossRef]
13. Tong, Z.; Sekizkardes, A.K. Recent Developments in High-Performance Membranes for CO_2 Separation. *Membranes* **2021**, *11*, 156. [CrossRef]
14. Houben, M.; van Geijn, R.; van Essen, M.; Borneman, Z.; Nijmeijer, K. Supercritical CO_2 permeation in glassy polyimide membranes. *J. Membr. Sci.* **2021**, *620*, 118922. [CrossRef]
15. Tiwari, R.R.; Jin, J.; Freeman, B.D.; Paul, D.R. Physical aging, CO_2 sorption and plasticization in thin films of polymer with intrinsic microporosity (PIM-1). *J. Membr. Sci.* **2017**, *537*, 362–371. [CrossRef]
16. Alders, M. Membrane Gas Separation at Low Temperatures and High Activity Levels. Ph.D. Dissertation, Rheinisch-Westfälische Technische Hochschule Aachen, Aachen, Germany, 2020. [CrossRef]
17. Alders, M.; von Bargen, C.; König, A.; Wessling, M. Chilled Membranes—Efficient gas permeation at sub-ambient temperatures. *J. Membr. Sci.* **2019**, *576*, 171–181. [CrossRef]
18. Jordan, S.M.; Koros, W.J.; Fleming, G.K. The effects of CO_2 exposure on pure and mixed gas permeation behavior: Comparison of glassy polycarbonate and silicone rubber. *J. Membr. Sci.* **1987**, *30*, 191–212. [CrossRef]
19. Kim, S.; Lee, Y.M. High performance polymer membranes for CO_2 separation. *Curr. Opin. Chem. Eng.* **2013**, *2*, 238–244. [CrossRef]
20. Mukhtar, H.B.; Shariff, A.M.; Mannan, H.A. A Review: Fabrication of Enhanced Polymeric Blend Membrane with Amines for Removal of CO_2 from Natural Gas. *Int. J. Emerg. Technol. Adv. Eng.* **2013**, *3*, 529–551.
21. Penteado, A.; Esche, E.; Salerno, D.; Godini, H.R.; Wozny, G. Design and Assessment of a Membrane and Absorption Based Carbon Dioxide Removal Process for Oxidative Coupling of Methane. *Ind. Eng. Chem. Res.* **2016**, *55*, 7473–7483. [CrossRef]
22. Fang, S.-M.; Stern, S.A.; Frisch, H.L. A "free volume" model of permeation of gas and liquid mixtures through polymeric membranes. *Chem. Eng. Sci.* **1975**, *30*, 773–780. [CrossRef]
23. Vane, L.M. Effect of membrane performance variability with temperature and feed composition on pervaporation and vapor permeation system design for solvent drying. *J. Chem. Technol. Biotechnol.* **2022**, *97*, 2706–2719. [CrossRef]

24. Schuldt, K.; Pohlmann, J.; Shishatskiy, S.; Brinkmann, T. Applicability of PolyActive™ Thin Film Composite Membranes for CO_2 Separation from C2H4 Containing Multi-Component Gas Mixtures at Pressures up to 30 bar. *Membranes* **2018**, *8*, 27. [CrossRef]
25. Koch, K.; Górak, A. Pervaporation of binary and ternary mixtures of acetone, isopropyl alcohol and water using polymeric membranes: Experimental characterisation and modelling. *Chem. Eng. Sci.* **2014**, *115*, 95–114. [CrossRef]
26. Brinkmann, T.; Pohlmann, J.; Withalm, U.; Wind, J.; Wolff, T. Theoretical and Experimental Investigations of Flat Sheet Membrane Module Types for High Capacity Gas Separation Applications. *Chem. Ing. Tech.* **2013**, *85*, 1210–1220. [CrossRef]
27. Car, A.; Stropnik, C.; Yave, W.; Peinemann, K.-V. Tailor-made Polymeric Membranes based on Segmented Block Copolymers for CO_2 Separation. *Adv. Funct. Mater.* **2008**, *18*, 2815–2823. [CrossRef]
28. Ioannidi, A.; Vroulias, D.; Kallitsis, J.; Ioannides, T.; Deimede, V. Synthesis and characterization of poly(ethylene oxide) based copolymer membranes for efficient gas/vapor separation: Effect of PEO content and chain length. *J. Membr. Sci.* **2021**, *632*, 119353. [CrossRef]
29. Rahman, M.M. Multiblock copolymers containing polyether segments for separation of C4 hydrocarbons. *J. Membr. Sci.* **2023**, *666*, 121176. [CrossRef]
30. Favre, E.; Schaetzel, P.; Nguygen, Q.T.; Clément, R.; Néel, J. Sorption, diffusion and vapor permeation of various penetrants through dense poly(dimethylsiloxane) membranes: A transport analysis. *J. Membr. Sci.* **1994**, *92*, 169–184. [CrossRef]
31. Abeles, B.; Chen, L.F.; Johnson, J.W.; Drake, J.M. Capillary Condensation and Surface Flow in Microporous Vycor Glass. *Isr. J. Chem.* **1991**, *31*, 99–106. [CrossRef]
32. Lee, K.-H.; Hwang, S.-T. The transport of condensible vapors through a microporous vycor glass membrane. *J. Colloid Interface Sci.* **1986**, *110*, 544–555. [CrossRef]
33. Han, H.; Scofield, J.M.P.; Gurr, P.A.; Webley, P.A.; Qiao, G.G. Ultrathin membrane with robust and superior CO_2 permeance by precision control of multilayer structures. *Chem. Eng. J.* **2023**, *462*, 142087. [CrossRef]
34. Kárászová, M.; Zach, B.; Petrusová, Z.; Červenka, V.; Bobák, M.; Šyc, M.; Izák, P. Post-combustion carbon capture by membrane separation, Review. *Sep. Purif. Technol.* **2020**, *238*, 116448. [CrossRef]
35. Yave, W.; Car, A.; Wind, J.; Peinemann, K.-V. Nanometric thin film membranes manufactured on square meter scale: Ultra-thin films for CO_2 capture. *Nanotechnology* **2010**, *21*, 395301. [CrossRef] [PubMed]
36. Lillepärg, J.; Sperling, E.; Blanke, M.; Held, M.; Shishatskiy, S. Multicomponent Network Formation in Selective Layer of Composite Membrane for CO_2 Separation. *Membranes* **2021**, *11*, 174. [CrossRef] [PubMed]
37. Brennecke, F.; Clodt, J.; Pohlmann, J.; Abetz, C.; Brinkmann, T.; Abetz, V. Computational fluid dynamics simulation of the roll-to-roll coating process for the production of thin film composite membranes including validation. *J. Adv. Manuf. Process.* **2021**, *3*, e10076. [CrossRef]
38. Grünauer, J.; Filiz, V.; Shishatskiy, S.; Abetz, C.; Abetz, V. Scalable application of thin film coating techniques for supported liquid membranes for gas separation made from ionic liquids. *J. Membr. Sci.* **2016**, *518*, 178–191. [CrossRef]
39. Lillepärg, J.; Breitenkamp, S.; Shishatskiy, S.; Pohlmann, J.; Wind, J.; Scholles, C.; Brinkmann, T. Characteristics of Gas Permeation Behaviour in Multilayer Thin Film Composite Membranes for CO_2 Separation. *Membranes* **2019**, *9*, 22. [CrossRef]
40. Yave, W.; Car, A. *Polymeric Membranes for Post-Combustion Carbon Dioxide (CO_2) Capture*; Basile, A., Nunes, S.P., Eds.; Woodhead Publishing: Sawston, UK, 2011; pp. 160–183. [CrossRef]
41. Fuoco, A.; Monteleone, M.; Esposito, E.; Bruno, R.; Ferrando-Soria, J.; Pardo, E.; Armentano, D.; Jansen, J.C. Gas Transport in Mixed Matrix Membranes: Two Methods for Time Lag Determination. *Computation* **2020**, *8*, 28. [CrossRef]
42. De Angelis, M.G.; Merkel, T.C.; Bondar, V.I.; Freeman, B.D.; Doghieri, F.; Sarti, G.C. Hydrocarbon and fluorocarbon solubility and dilation in poly(dimethylsiloxane): Comparison of experimental data with predictions of the Sanchez–Lacombe equation of state. *J. Polym. Sci. Part B Polym. Phys.* **1999**, *37*, 3011–3026. [CrossRef]
43. Kamiya, Y.; Naito, Y.; Hirose, T.; Mizoguchi, K. Sorption and partial molar volume of gases in poly (dimethyl siloxane). *J. Polym. Sci. Part B Polym. Phys.* **1990**, *28*, 1297–1308. [CrossRef]
44. Lin, H.; Freeman, B.D. Gas solubility, diffusivity and permeability in poly(ethylene oxide). *J. Membr. Sci.* **2004**, *239*, 105–117. [CrossRef]
45. Merkel, T.C.; Bondar, V.I.; Nagai, K.; Freeman, B.D.; Pinnau, I. Gas sorption, diffusion, and permeation in poly(dimethylsiloxane). *J. Polym. Sci. Pol. Phys.* **2000**, *38*, 415–434. [CrossRef]
46. Sirard, S.M.; Green, P.F.; Johnston, K.P. Spectroscopic Ellipsometry Investigation of the Swelling of Poly(Dimethylsiloxane) Thin Films with High Pressure Carbon Dioxide. *J. Phys. Chem. B* **2001**, *105*, 766–772. [CrossRef]
47. Horn, N.R.; Paul, D.R. Carbon dioxide plasticization and conditioning effects in thick vs. thin glassy polymer films. *Polymer* **2011**, *52*, 1619–1627. [CrossRef]
48. Flichy, N.M.B.; Kazarian, S.G.; Lawrence, C.J.; Briscoe, B.J. An ATR−IR Study of Poly (Dimethylsiloxane) under High-Pressure Carbon Dioxide: Simultaneous Measurement of Sorption and Swelling. *J. Phys. Chem. B* **2002**, *106*, 754–759. [CrossRef]
49. Rahman, M.M.; Filiz, V.; Shishatskiy, S.; Abetz, C.; Georgopanos, P.; Khan, M.M.; Neumann, S.; Abetz, V. Influence of Poly(ethylene glycol) Segment Length on CO_2 Permeation and Stability of PolyActive Membranes and Their Nanocomposites with PEG POSS. *ACS Appl. Mater. Interfaces* **2014**, *7*, 12289–12298. [CrossRef] [PubMed]
50. Rahman, M.M.; Lillepärg, J.; Neumann, S.; Shishatskiy, S.; Abetz, V. A thermodynamic study of CO_2 sorption and thermal transition of PolyActive™ under elevated pressure. *Polymer* **2016**, *93*, 132–141. [CrossRef]

51. Blume, I.; Smit, E.; Wessling, M.; Smolders, C.A. Diffusion through rubbery and glassy polymer membranes. *Makromol. Chemie. Macromol. Symp.* **1991**, *45*, 237–257. [CrossRef]
52. Shamu, A.; Dunnewold, M.; Miedema, H.; Borneman, Z.; Nijmeijer, K. Permeation of supercritical CO_2 through dense polymeric membranes. *J. Supercrit. Fluids* **2019**, *144*, 63–70. [CrossRef]
53. Henis, J.M.S.; Tripodi, M.K. Composite hollow fiber membranes for gas separation: The resistance model approach. *J. Membr. Sci.* **1981**, *8*, 233–246. [CrossRef]
54. Knudsen, M. Die Gesetze der Molekularströmung und der inneren Reibungsströmung der Gase durch Röhren. *Ann. Der Phys.* **1909**, *333*, 75–130. [CrossRef]
55. Barrer, R.M.; Rideal, E.K. Activated diffusion in membranes. *Trans. Faraday Soc.* **1939**, *35*, 644–656. [CrossRef]
56. Jordan, S.M.; Koros, W.J. Permeability of pure and mixed gases in silicone rubber at elevated pressures. *J. Polym. Sci. Part B Polym. Phys.* **1990**, *28*, 795–809. [CrossRef]
57. Heidari, M.; Hosseini, S.S.; Nasrin, M.O.; Ghadimi, A. Synthesis and fabrication of adsorptive carbon nanoparticles (ACNs)/PDMS mixed matrix membranes for efficient CO_2/CH_4 and C_3H_8/CH_4 separation. *Sep. Purif. Technol.* **2019**, *209*, 503–515. [CrossRef]
58. Rahman, M.M.; Abetz, C.; Shishatskiy, S.; Martin, J.; Müller, A.J.; Abetz, V. CO_2 Selective PolyActive Membrane: Thermal Transitions and Gas Permeance as a Function of Thickness. *ACS Appl. Mater. Interfaces* **2018**, *10*, 26733–26744. [CrossRef] [PubMed]
59. Yu, C.; Xie, Q.; Bao, Y.; Shan, G.; Pan, P. Crystalline and Spherulitic Morphology of Polymers Crystallized in Confined Systems. *Crystals* **2017**, *7*, 147. [CrossRef]
60. Wijmans, J.G.; Baker, R.W. The solution-diffusion model: A review. *J. Membr. Sci.* **1995**, *107*, 1–21. [CrossRef]
61. Lundberg, J.L. Molecular clustering and segregation in sorption systems. *Pure Appl. Chem.* **1972**, *31*, 261–282. [CrossRef]
62. Nguyen, Q.T.; Favre, E.; Ping, Z.H.; Néel, J. Clustering of solvents in membranes and its influence on membrane transport properties. *J. Membr. Sci.* **1996**, *113*, 137–150. [CrossRef]
63. Sun, Y.-M.; Chen, J. Sorption/desorption properties of ethanol, toluene, and xylene in poly(dimethylsiloxane) membranes. *J. Appl. Polym. Sci.* **1994**, *51*, 1797–1804. [CrossRef]
64. Cardini, G.; Schettino, V.; Klein, M.L. Structure and dynamics of carbon dioxide clusters: A molecular dynamics study. *J. Chem. Phys.* **1989**, *90*, 4441–4449. [CrossRef]
65. Uchytil, P.; Loimer, T. Large mass flux differences for opposite flow directions of a condensable gas through an asymmetric porous membrane. *J. Membr. Sci.* **2014**, *470*, 451–457. [CrossRef]
66. Alpers, A.; Keil, B.; Lüdtke, O.; Ohlrogge, K. Organic Vapor Separation: Process Design with Regards to High-Flux Membranes and the Dependence on Real Gas Behavior at High Pressure Applications. *Ind. Eng. Chem. Res.* **1999**, *38*, 3754–3760. [CrossRef]
67. Da Conceicao, M.; Nemetz, L.; Rivero, J.; Hornbostel, K.; Lipscomb, G. Gas Separation Membrane Module Modeling: A Comprehensive Review. *Membranes* **2023**, *13*, 639. [CrossRef]
68. Keller, J.U.; Staudt, R. *Gas Adsorption Equilibria: Experimental Methods and Adsorptive Isotherms*; Springer Science & Business Media: New York, NY, USA, 2005; pp. 129–135.